THE WORLDS OF POSSIBILITY

The Worlds of Possibility

Modal Realism and the Semantics of Modal Logic

CHARLES S. CHIHARA

CLARENDON PRESS · OXFORD

This book has been printed digitally and produced in a standard specification in order to ensure its continuing availability

OXFORD
UNIVERSITY PRESS

Great Clarendon Street, Oxford OX2 6DP

Oxford University Press is a department of the University of Oxford.
It furthers the University's objective of excellence in research, scholarship, and education by publishing worldwide in

Oxford New York

Auckland Bangkok Buenos Aires Cape Town Chennai
Dar es Salaam Delhi Hong Kong Istanbul Karachi Kolkata
Kuala Lumpur Madrid Melbourne Mexico City Mumbai Nairobi
São Paulo Shanghai Taipei Tokyo Toronto

Oxford is a registered trade mark of Oxford University Press in the UK and in certain other countries

Published in the United States
by Oxford University Press Inc., New York

Reprinted 2004

ISBN 0-19-924655-6

Dedicated to
Ted, who showed us the way

This work was partially supported by funding from The National Endowment for the Humanities Fellowships for University Teachers (in the 1994–5 academic year), The President's Research Fellowships in the Humanities, University of California, and The Humanities Research Fellowships, University of California, Berkeley (both in 1996–7).

PREFACE

This is a book about Modal Realism—roughly, the ontological doctrine that our universe (the world we live in) is just one among many possible worlds. My book will supply a detailed exposition of Modal Realism, as well as a thorough examination and evaluation of the principal grounds that have been advanced in support of it. I shall also take up some of the outstanding objections that have been levelled at the view, and I shall assess various criticisms of these objections thrown up by the defenders of Modal Realism.

This book also investigates several possible worlds semantical systems, which have been devised to study the logic of the modal operators. One reason that my investigation into Modal Realism involves a study of possible worlds semantics is this: the fact that possible worlds semantics is such an effective instrument for evaluating and validating reasoning involving the modal operators is widely regarded as a strong argument for accepting some form of Modal Realism. In this work, the cogency of such an argument will be evaluated.

Also within the scope of these investigations are areas of philosophy of mathematics. Some arguments for Modal Realism are similar to those put forward for Mathematical Realism—the ontological doctrine that there are mathematical objects. This is suggested by David Lewis when he writes:

> [S]ystematic philosophy goes more easily if we may presuppose modal realism in our analysis. I take this to be a good reason to think that modal realism is true, just as the utility of set theory in mathematics is a good reason to believe that there are sets. (Lewis 1986: vii)

Utilitarian arguments of this sort will be investigated in some detail.

There are other reasons why topics in the philosophy of mathematics will be examined in this work. In a previous book (Chihara 1990), I developed a system of mathematics utilizing the notion of *constructibility quantifiers*—a notion that was explained in part by means of possible worlds semantics. This has led some scholars to infer that my system of mathematics presupposes the existence of possible worlds. One goal of this work is to assess the cogency of such an inference.

This book grew out of my "Perspectives in Philosophy" lectures given in March of 1991 at the University of Notre Dame. Some of the ideas in this work have found their way into lectures that I have delivered at the following: the Fifteenth International Wittgenstein Symposium held in Kirchberg am Wechsel; the University of California, Davis; the University of

California, Berkeley; the Centre for Philosophical Studies, King's College, London; the Graduate Center of the City University of New York; Keio University in Tokyo; the National Taiwan University; and the Philosophy Institute of National Chung-Chen University in Taiwan. I am grateful for the many helpful comments I received from those who attended these presentations.

In addition, I have used some of the material from this book in courses and seminars I have given at Berkeley. The suggestions, criticisms, and queries I received from those who took these courses (too many to list here) have been invaluable to me. In this connection, I should mention especially Jonathan Turkanis, Mathew Henkin, Richard Zach, Greg Ray, Scott Malone, William Goodwin, Kirk Ludwig, Peter Hanks, and Bendek Hansen.

The three philosophers whose work I discuss in most detail here are David Lewis, Alvin Plantinga, and Graeme Forbes. They were sent preliminary versions of relevant sections of this book for examination, and I was gratified to receive from them remarkably generous and helpful comments, despite the critical tone of much of my discussion. I should also like to express my thanks to Chris Menzel, Michael Jubien, and Greg Ray for their helpful responses to what I sent them. Thanks also go to Donald Gillies, with whom I have had many discussions relating to his possible worlds objection to my constructibility theory; to Vann McGee and Randolph Goldman, both of whom read the whole of versions of this work and made many valuable criticisms and suggestions; to the set theorist John Steel, with whom I carried on a useful dialogue on the epistemology and metaphysics of set theory in 1997; and to Martin Jones, who read and commented on the sections relating to Penelope Maddy's set theoretical realism.

In writing this book, I have been greatly aided by a number of very able Berkeley graduate students: Craig Bach, Lisa Hales, Jonathan Lebowitsch, and Greg Ray (all from the Group in Logic and the Methodology of Science), and Josh Dever (from the Philosophy Department). Each of them served wonderfully as my research assistant for at least a term; they have saved me from many errors and confusions. I have Lisa Hales to thank for the index and for improvements in the Fundamental Theorem.

I am also grateful to John Burgess, Hilary Putnam, and Brian Skyrms for their support and encouragement, and to Angela Blackburn for her excellent editorial assistance.

Finally, I would like to thank my wife, Carol, for aiding me in countless ways during the writing of this work, but especially for giving me the benefit of her expertise on all matters pertaining to biology and to the computer.

Berkeley, California
June 1997

C. S. C.

CONTENTS

NOTATIONAL CONVENTIONS

The notational conventions I use, in this work, are essentially those of my two previous books (Chihara 1973 and Chihara 1990). Thus, double quotation marks will be used for direct quotation and as scare quotes (i.e. as indicators that I am using the expression appearing between the quotation marks in a way that is accepted by some writers, but it is not something that I would ordinarily accept). Single quotation marks will be used to refer to linguistic items such as symbols, words, and sentences (i.e. to mention, as opposed to use, the linguistic item between the pair of quotation marks). For example, to refer to the name of the present President of the United States, I may enclose this name in a pair of single quotation marks in the following way:

'Bill Clinton' is the name of the President.

I will also sometimes use the convention of displaying the linguistic item or items in order to refer to them. Hence, to refer to the logical constants of the sentential calculus, I may simply display them in the following way:

$$- \vee \& \rightarrow \leftrightarrow$$

There are cases in which one could use either double or single quotes with very little difference in sense. For example, the sentence (which I display below, following the above convention):

Mates calls these items "English predicates"

differs little from the sentence:

Mates calls these items 'English predicates',

since the former utilizes double quotes to indicate that I am quoting the exact words Mates used and the latter mentions the words that Mates calls the items.

Greek letters will be used as meta-variables. The primitive symbols of the object-languages discussed in this work will frequently be used autonymously, that is, as referring to themselves. I shall also adopt the usual notational conventions in virtue of which the expression

$$(\theta \& \phi)$$

refers to the expression

$$(F^1a \mathbin{\&} G^2bc)$$

when θ has the value 'F^1a' and φ has the value 'G^2bc'.[1] In the later chapters of this work, I also employ script letters as special variables.

[1] See Mates 1972: ch. 2, sects. 3 and 6, for a more detailed description of the notational conventions I make use of in this connection.

INTRODUCTION

Imagine that a friend of yours tells you, in all sincerity, that he has made an important discovery: the Earth is not the only planet in existence that supports life—in fact, there are many planets very much like the Earth upon which intelligent life is to be found. He goes on to tell you that he has made this remarkable discovery, not on the basis of newly acquired astronomical data, but rather through philosophical reasoning. Would these not be amazing claims? Yet the distinguished Princeton philosopher David Lewis has made such claims,[1] and these claims are being taken quite seriously by many researchers in modal logic. According to Lewis, our "world"—the vast universe we live in—is just one of many such "possible worlds" that exist. Many of these possible worlds (or *possible universes*) contain planets just like those in our solar system, as well as real flesh and blood people, who live and love just as we do. These possible worlds form the basis of Lewis's analysis of the modal operators.

Lewis is by no means the first philosopher to believe in possible worlds. The extensive use which the great mathematician-philosopher Gottfried Leibniz made of possible worlds in his philosophical work still exerts a strong influence on metaphysical and theological research.[2] This can be seen in the writings of another well-known contemporary philosopher, Alvin Plantinga, who also analyses necessity and possibility in terms of possible worlds, and who also grapples with many thorny problems of Christian theology in terms of possible worlds.[3] However, for Plantinga, unlike Lewis, possible worlds are *abstract entities* of a kind that he calls "maximal possible states of affairs" and these abstract "worlds" are not thought to be literally made up of such things as people and planets. Lewis and Plantinga are just two of the many outstanding philosophers who, in varying degrees, believe in possible worlds of some kind or other. Among these "Modal Realists", Robert Stalnaker and Robert Adams should also be mentioned.[4]

How is it that these philosophers have come to believe in possible worlds? Although many different explanations can be given, it is clear that the striking success of a kind of semantic theory known as "possible

[1] In his Lewis 1986.

[2] For scholarly discussions of Leibniz's views on possible worlds, see Mates 1986, especially chs. 4, 6, and 7. See also Ishiguro 1990, especially ch. 9.

[3] See his Plantinga 1974.

[4] See Stalnaker 1984 and Adams 1979*b*.

worlds semantics", especially in the area of modal logic, has played a major role in making the postulation of possible worlds attractive. During the past twenty-five years or so, an enormous amount of philosophical theorizing has been carried out in terms of, or with the aid of, "possible worlds structures" of various kinds. Ever since the development of possible worlds semantics for modal, temporal, epistemic, and doxastic logics, it has been widely believed that possible worlds structures are an important, if not essential, tool of philosophical analysis.[5] For example, the well-known philosophers Jaakko and Merrill Hintikka have declared that possible worlds semantics "ought to be an absolutely vital part of every philosopher's repertoire".[6] And they explain their declaration with the following words:

> We consider it as self-evident as anything in philosophy that one cannot do justice to actual human experience without a conceptual system that includes *possibilia* . . . The need of considering unactualized *possibilia* means that our logic must be such that in it we can handle different and incompatible entities of the same logical type as facts: possible states of affairs and possible courses of events. And the model theory of these counterfactual entities is (in its basic aspects) precisely what is known as possible-worlds semantics.[7]

In a recent article, Christopher Menzel sets forth and analyses a kind of argument which seems to form a major part of the rationale underlying the acceptance of the ontology of possible worlds. According to this argument,

> The proponents of any semantic theory are committed to the existence of whatever entities are appealed to in its account of truth. Possible world semantics appeals to possible worlds [in its account of truth] . . . It follows that the proponents of the former are committed to the existence of the latter.[8]

Of course, philosophers who deny the existence of possible worlds may simply refrain from using possible worlds semantics in their account of the truth of modal sentences. But can such philosophers then legitimately go on to use possible worlds semantics in their analyses of the validity and invalidity of modal reasoning? Many philosophers doubt that they can.[9] (Their reasons will be examined in Chapter 3.) Furthermore, it would seem that the anti-Modal Realist must also forgo the explanations of modal reasoning that possible worlds semantics provides. Thus, J. Barwise and J. Perry have written: "A semantical theory is committed to the reality of the entities *it uses in its explanations* . . . The Montague grammarian, or

[5] Cf. Stalnaker 1984.
[6] Hintikka and Hintikka 1989: 73.
[7] Hintikka and Hintikka 1989: 73.
[8] Menzel 1990: 363.
[9] See, for example, Plantinga 1974: 125–6.

other possible worlds theorist, is committed to possible worlds . . ."[10] Jaegwon Kim makes a similar point when he writes: "If we take modal discourse realistically, and *if we choose to explain it in terms of possible worlds*, then I don't see an alternative to some form of *possible worlds realism*."[11] Philosophers who eschew including possible worlds in their ontology thus seem to be precluded from making use of possible worlds semantics in their account of the truth of modal propositions, in their assessment of the validity and invalidity of modal arguments, and in their explanations of modal principles and modal reasoning. For the Modal Realist, this is too high a price to pay for ontological economy. Thus, we find Raymond Bradley and Norman Swartz writing: "We feel driven to posit [possible worlds] because we seem unable to make ultimate sense of logic without them."[12]

Despite the power of the above considerations to move many philosophers to accept the reality of possible worlds, I have become convinced that, with a correct understanding of possible worlds semantics, a quite different conclusion should be drawn. In this study, I intend to explore the metaphysical bases, assumptions, and implications of applying possible worlds systems of modal logics. I intend to show, by means of a logical theorem obtained as a result of a detailed analysis of possible worlds semantics, that one can apply systems of modal quantificational logic to evaluating modal principles and to assessing modal arguments, without incurring a commitment to the existence of possible worlds.

This study will also be devoted to an examination of the Modal Realist's views and arguments, focusing in particular on the various Realist interpretations of the semantics of S5-type modal logics. But what is an "S5-type modal logic"? In Chapter 1, I shall present in some detail a standard version of S5 modal quantificational logic, complete with a system of derivation rules. At other parts of this work, however, I shall take up systems of modal logic that deviate markedly from the standard versions—so much so, in fact, that some logicians might well refuse to characterize such systems as S5 systems. However, since these systems have certain features that are characteristic of S5 systems, I classify them as "S5-type modal logics".

Why do I concentrate on S5-type modal logics? There are several reasons. First, S5 modal logics are widely held to formalize reasoning involving a notion of possibility that is absolutely central to much philosophical reasoning. Second, S5 modal logics have a kind of simplicity that is especially attractive. Third, a number of major figures in the area of

[10] Barwise and Perry 1985: 116, italics mine.
[11] Kim 1986: 612.
[12] Bradley and Swartz 1979: 64.

Modal Realism have produced or greatly influenced important work on S5-type modal logics.

I shall not attempt to discuss each and every important philosopher or logician who has written on the metaphysics of S5 modal logics. The literature on this topic is much too vast to allow such complete coverage in the sort of work I have in mind. Instead, I have decided to focus most of my investigations on only major figures in the area, who have written extensively on the target topic (at least one important book and many widely read papers) and who have developed an interesting position on the *semantics of modal logic*. In addition, I shall focus on authors who have advanced clearly articulated ontological theses and carefully presented arguments regarding these theses. The three philosophers I have selected for special study, David Lewis, Alvin Plantinga, and Graeme Forbes, in my opinion satisfy these conditions.

This is not to say that I shall only consider the writings of this small group of authors—far from it. But the views of these three will be given the most detailed examination. Thus, one chapter of this work will consist of an investigation in depth of the principal arguments put forward by the Modal Realists Lewis and Plantinga in support of their ontological views; it will provide a discussion of the claim that the kinds of uses of possible worlds semantics discussed above require an implicit acceptance of possible worlds. Another chapter will examine a sustained attempt by Forbes to develop an anti-Modal Realist theory of possible worlds semantics. An analysis of Forbes's account will serve to introduce and clarify my own anti-Modal Realistic account of possible worlds semantics, by emphasizing the striking differences in the two accounts (I also intend to investigate the views of Modal Fictionalists, for a similar reason).

This project may appear to be a substantial departure from my previous work in the philosophy of mathematics. However, it can be seen to be a natural development from this work. In Chihara 1990, I examined various arguments put forward by Mathematical Realists for belief in the existence of mathematical objects. These arguments are similar to the Modal Realist's argument in several ways: they all imply that denying the existence of mathematical objects forces one to pay a very high theoretical price. One would have to deny, they argue, that the theorems of classical mathematics are true. Another respect in which the Mathematical Realist's arguments are similar to the Modal Realist's will be discussed at some length in Chapter 3.

My earlier work is also related to the present one because of my response to the arguments of the Mathematical Realist: there, I develop a system of mathematics in which no existence assertions are made. In place of the existential assertions of classical mathematics, constructibility assertions (assertions to the effect that such-and-such is constructible) are to be found in this system. Since constructibility assertions are modal in character, I found it useful, for heuristic and didactic purposes, to make use of possible worlds semantics in explaining and clarifying the primitive terms of the system.

Donald Gillies, reviewing my book for the *British Journal for the Philosophy of Science*, agrees with most of what I do in the book, but he worries that I may have eliminated the need to postulate mathematical objects in mathematics only at the expense of presupposing, in some unintended way, possible worlds. Since possible worlds are far more repugnant to common sense than are the abstract objects of mathematics, Gillies finds this worry to be a serious one (Gillies 1992: 269–70). A similar (but more extreme) doubt is expressed by Jan Wolenski, in a review of my book in *History and Philosophy of Logic*, when he writes: "Though Chihara does not need to appeal to 'classical' Platonic objects, his ontology is committed to possible worlds" (Wolenski 1992: 234). Such doubts and worries have prompted me to investigate the use of possible worlds models in our modal reasoning. Thus, this present work can be seen to be, in part, a response to these doubts and worries. It needs to be emphasized, however, that almost everything I propose in this work, regarding the semantics of S5-type modal logics, is completely independent of my anti-Realist views about the ontology of mathematics. I should think that a Mathematical Realist could, in principle, accept practically all of my positive analyses of possible worlds semantics.

Finally, I would like to offer a suggestion about how to read this work. There will no doubt be some readers who will be able to start this work from the beginning and to work their way, chapter by chapter, through all the technical details right to the end. For many readers, however, I believe it would be best to skip the more complicated proofs on first reading. I especially recommend that the proof of the Fundamental Theorem (in Chapter 7) be skipped on first reading, so that the overall structure of my position not be lost in a thicket of details. The lengthy proofs can be studied later.

1

Possible Worlds Semantics

Much of what follows will be concerned with the philosophical theories and presuppositions of applying possible worlds semantics in order to draw conclusions about matters involving what are called 'conceptual' or 'broadly logical' possibilities and necessities. Every purely logical truth is necessary in this sense, but the set of conceptually necessary truths will include much more: mathematical truths and "analytic truths" are also held to be necessary in this broadly logical sense.[1] As a rough guide, Graeme Forbes suggests that 'it is possible that P' be taken to mean: "There are ways things might have gone, no matter how improbable they may be, as a result of which it would have come about that P" (Forbes 1985: 2). There are many different systems of modal logic, but the type of system that is generally believed to correctly formalize the logical features of this broadly logical sense of necessity is S5.[2]

It will be useful to have a specific version of an S5 quantificational modal logic at hand, in terms of which my philosophical investigations can be carried out. This system is being presented here not as "the right and true" system of S5 modal logic, but rather as a sort of *standard* S5 system from which other S5-type systems I shall be discussing will be seen to diverge.[3] Later, in Chapter 7, I shall defend a system that will be very close to this standard system.

I shall present this formal system in Section 1 of this chapter, and then I shall undertake a preliminary investigation of the significance of various semantical features of its language. For purposes of simplicity and perspicuity, this initial modal language will not include sentential letters or

[1] See Plantinga 1974: ch. 1, sect. 1, and Forbes 1985: ch. 1, sect. 1, for more examples and discussion.

[2] Thus, Kit Fine writes: "S5 provides the correct logic for necessity in the broadly logical sense", in Fine 1978: 151. For an interesting discussion of the development of S5 modal logic, see Kneale 1962: ch. 9, sect. 4.

[3] The system to be presented is similar to the one given by Kit Fine in Fine 1978, which was based upon Kripke 1971. Fine's decision to base his modal semantics on Kripke's arose from his belief that "the Kripke semantics is correct, both in the sense of being coherent and in the stronger sense of providing the most fundamental means of formalizing modal discourse" (Fine 1978: 127). The semantics of my system is also very close to Forbes's S5 semantics given in Forbes 1985: ch. 2, sect. 2, which is another version of Fine's system.

the identity relation. Later, I will discuss various ways in which identity can be incorporated into the language; and, in Chapter 6, a sentential modal language will be given. In Section 2 of this chapter, I shall consider some objections, raised by Alvin Plantinga and others sympathetic to his approach, to the sort of modal logic developed in Section 1.

1. An S5 Modal Quantificational Logic

The language of the system of modal logic I present here will be called '**M**'. In presenting the rules of this language, object-language expressions and symbols will be regarded frequently as autonymous (that is, as denoting themselves). Greek letters will be used as meta-variables.

The Syntax of **M**

Primitive Symbols

1. *Variables.* Variables are italicized lower-case letters u through z, with or without numerical subscripts (the numerical subscripts being arabic numerals of positive integers).

2. *Constants.* There are two types of constants: non-logical constants and logical constants. There are two types of non-logical constants: individual constants and predicates.
   (a) *Individual constants* are italicized lower-case letters a through t, with or without numerical subscripts.
   (b) *Predicates* are italicized upper-case letters A through Z, with numerical superscripts and with or without numerical subscripts. (The superscript gives the *degree of the predicate*).
   (c) There is, however, one exception to this classification of predicates as non-logical constants: E^1 is a special constant, and is classified as a logical constant.
   (d) The *logical constants* are:

$$- \vee \; \& \rightarrow \leftrightarrow (\,) \exists \Box \Diamond E^1$$

3. *Individual Symbols.* Any variable or individual constant is an individual symbol; and nothing else is an individual symbol. In other words, all and only italicized lower-case letters, with or without subscripts, are individual symbols.

Formulas

1. *Atomic Formulas.* If ϕ is a predicate of degree n (or is, as I shall sometimes say, an n-ary predicate), and $\alpha_1, \alpha_2, \alpha_3, \ldots, \alpha_n$ are individual symbols, then

$$\phi\alpha_1\alpha_2\alpha_3 \ldots \alpha_n$$

is an atomic formula.

2. *Formulas.*
(a) All atomic formulas are formulas.
(b) If ϕ and ψ are formulas and α is a variable, then the following are also formulas:

$$-\phi$$
$$(\phi \,\&\, \psi)$$
$$(\phi \vee \psi)$$
$$(\phi \rightarrow \psi)$$
$$(\phi \leftrightarrow \psi)$$
$$(\alpha)\,\phi$$
$$(\exists\alpha)\phi$$
$$\Box\,\phi$$
$$\Diamond\,\phi$$

(c) Nothing is a formula that is not a formula according to (a) and (b).

3. *Additional terminology.*
(a) I assume the usual definitions of *bound* and *free occurrences of variables* in a formula. For those who would like an intuitive explanation of this distinction, the following is given in Mates 1972: "an occurrence of a variable α in a formula ϕ is *bound* if it is within an occurrence in ϕ of a formula of the form $(\alpha)\psi$ or $(\exists\alpha)\psi$; otherwise it is a *free* occurrence" (p. 45).
(b) A *sentence* is a formula in which no variable occurs free.

Here are a few examples of formulas that are not atomic formulas:

$$\Diamond F^1 a,\ (\exists x)\,\Box H^2 yx\,,\ \Box(F^1 a \rightarrow \Diamond G^2 ba\,)$$

Of these formulas, the first and third are sentences, but not the second.

Notational Conventions

I shall frequently omit superscripts from predicates and the outermost parentheses from formulas when there is little danger of confusion.

The Semantics of **M**

S5 Interpretations

An S5 interpretation is an ordered sextuple <**W**, **D**, **w***, **d**, **r**, **e**>, where:

W is a non-empty set;
D is a non-empty set;
w* is a member of **W**;

d is a function with domain **W** and a range that is a subset of the power set of **D** (that is, for every w that is a member of **W**, **d**(w) is a subset of **D**); furthermore, **D** = ⋃**d**(w), for all $w \in$ **W**;

r is a function that has as its domain the set of individual constants and that has a range that is a subset of **D**; that is, **r** assigns to each individual constant a unique referent from **D**;

e is a function that has as its domain the Cartesian product of the set of predicates with **W**: For any predicate ϕ of degree n and any member w of **W**, **e**(ϕ, w) is a subset of the nth Cartesian product of **D**.[4] However, the logical constant E^1 is special: **e**(E^1, w) must be **d**(w).

Comments

According to what Alvin Plantinga calls "the canonical conception" of possible worlds semantics (to be discussed in the following section),[5] one should think of **W** as the set of possible worlds; of **D** as the set of possible "individuals" or things; of **w*** as the actual world; of **d**(w) as the set of "individuals" that exist in world w; of **r**(α) as the referent of the individual constant α; and of **e**(ϕ, w) as the extension of the predicate ϕ in world w. Notice that, since **D** is the union of the **d**(w), every possible individual must be in at least one possible world. The logical constant E^1 can be regarded as expressing existence; that is, the things that are in the extension of E^1 at world w are just the things that exist in w. Since **M** is an S5 modal language, I have omitted the *accessibility relation* between worlds.[6] It should be noted that **d**(w) may be the empty set and that an individual constant may be assigned a referent that is not a member of **d**(**w***). As a result, an individual constant may "name" a merely possible object, that is, a possible object that does not actually exist.

4 The Cartesian product of sets α and β is the set of all ordered pairs $<x, y>$ such that $x \in \alpha$ and $y \in \beta$. The nth Cartesian product of **D** is the set of all ordered n-tuples $<x_1, x_2, x_3, \ldots, x_n>$ such that $x_i \in$ **D**, $i = 1, 2, 3, \ldots, n$. The first Cartesian product of **D** can be taken to be the set of elements of **D**, i.e. **D** itself. $\mathbf{D}^2 = \mathbf{D} \times \mathbf{D}$, the second Cartesian product of **D**.

5 Plantinga 1979: 254.

6 The accessibility relation between worlds is used to express the idea of a world being possible relative to a world. To say that world u is possible relative to world v (or that u is accessible from v) is to say that every proposition true in u is possible in v (or equivalently: for any proposition ϕ, if $\Box\phi$ is true in v, then ϕ is true in u). In a system with an accessibility relation, specifying an interpretation for the modal language would involve specifying a relation **A** on **W** which satisfied certain conditions. A minimum condition is reflexivity, since every world is possible relative to itself. What other conditions one would require would depend upon the specific kind of possibility the modal system was intended to capture. In the case of S5 modal logic, devised to capture the notion of conceptual possibility, it is generally agreed that the accessibility relation should be an equivalence relation. It turns out, however, that setting up the more complicated semantics for the system with an accessibility relation that is an equivalence relation yields the same set of validities as a system whose semantics is formulated as is that of **M**'s, without an accessibility relation. See Forbes 1985: ch. 1, sect. 3, for details.

The approach being taken here is the *free logical* one.[7] In free logic, one does not assume, as one does in classical logic, that the range of the quantifiers is non-empty or that each individual constant stands for something that exists in the range of the quantifiers. Thus, from

$$F^1a$$

one will not be able to infer, in this system,

$$(\exists x)F^1x.$$

Why does one adopt a free logic instead of the more familiar logic of standard quantification theory? Here is one problem the free logical approach avoids: suppose that one used standard classical quantificational laws and one accepted the law of necessitation. Then, the following derivation could be constructed:

$\varnothing$	(1)	$(x)(\exists y)x = y$	*Classical law*
$\varnothing$	(2)	$(\exists y)a = y$	(1)*US*
$\varnothing$	(3)	$\Box(\exists y)a = y$	(2) *N*
$\varnothing$	(4)	$(x)\Box(\exists y)x = y$	(3) *UG*

As a result, there would be a theorem of such a modal logic implying that, for every object x that exists in the actual world, there would be something in every world that is identical to x. That is, we would have modal logic implying that every object that actually exists necessarily exists. The free logical approach we take avoids this conclusion by blocking the *US* and *UG* steps.[8]

Allowing $\mathbf{d}(w)$ to be empty prevents the sentence

$$\Box(\exists x)E^1x$$

from being a valid sentence of this system, which, to me, is intuitively satisfactory. It seems to me that it is a contingent fact that anything exists at all.[9]

Requiring **D** to be non-empty guarantees that

$$\Diamond(\exists x)E^1x$$

[7] For an elementary discussion of the philosophical implications and applications of free logic, see Lambert 1972: Pt. 5. For a more advanced treatment, see Lambert 1991. See especially Garson 1991.

[8] See Deutsch 1990 for a more detailed discussion of the free logical approach to the above problem.

[9] Lewis, however, maintains that it is necessary there is something. He is driven to this position by his realistic doctrine that possible worlds are maximal mereological sums of spatiotemporally interrelated things. He admits that his own position on this matter is counter intuitive, but his strategy is to accept the "unwelcome consequences" of his realistic doctrines and to claim that these consequences are not as bad as one might think (Lewis 1986: 73–4). I shall discuss this feature of Lewis's Modal Realism in more detail in Chapter 3.

is a valid sentence of this system. Some have questioned taking this last displayed sentence to be a truth of S5 logic; but for me, this sentence is a reasonable validity. I know that something exists. So I know that possibly something exists. I conclude that necessarily, possibly something exists. The system I am presenting here simply takes 'possibly something exists' as broadly logically necessary.

Notice that, as I have characterized the function **e**, the things that belong to the extension of a unary predicate in world w need not exist in w. Also the things that are related by a binary predicate in world w need not exist in w. This has seemed unacceptable to some philosophers (who are classified as "Serious Actualists"). How can a thing have a property if it does not exist? How can a thing be related to something if neither exists? I shall discuss Serious Actualism in more detail in the last section of this chapter and in a later chapter (Chapter 7), but here is one way in which one might interpret the system so as to make some sense of such possibilities. One usually thinks of the actual world as having a long history, i.e. as being extended over time. According to the suggested conception, each possible world is just an instantaneous snapshot of what one ordinarily regards as a world. What then is the "actual world"? Let it be the actual world as it is . . . *now*. According to this conception of actuality, "worlds" that existed at past moments are "gone" and, hence, not actual. Future "worlds" are also not actual: they are not yet. Only the present "world" actually exists. I am in existence, so I am one of the things that exists in **w***. Let the individual constant a denote me. My father is no longer living: he does not exist in **w***. Yet we do want to say that my father is one of my ancestors. To express this thought in this system, we can let the individual constant b denote my father, and we can let A^2 denote the relation: x is an ancestor of y. Then it is not hard to see why one might want to allow 'A^2ba' to be true at **w*** under this interpretation, even though **r**(b) is not an element of **d(w*)**.

Notational Conventions

Suppose that interpretation **I** = <**W**, **D**, **w***, **d**, **r**, **e**>, and let $w \in$ **W**. Then, if α is an individual constant, I shall use the notation '**I**(α)' to denote that entity that **I** assigns to α. More specifically, **I**(α) will stand for **r**(α). Similarly, if ϕ is a predicate, then **I**(ϕ, w) will be that set of n-tuples that is **e**(ϕ, w).

Truth at a World under an Interpretation. Let **I** = <**W**, **D**, **w***, **d**, **r**, **e**> be an interpretation, and let $w \in$ **W**.

(a) If ϕ is a predicate of degree n, and the α_i are individual constants, then the atomic sentence

$$\phi\alpha_1\alpha_2\alpha_3 \ldots \alpha_n$$

is true at w under **I** if, and only if,

$$<\mathbf{r}(\alpha_1), \mathbf{r}(\alpha_2), \mathbf{r}(\alpha_3), \ldots, \mathbf{r}(\alpha_n)> \in \mathbf{I}(\phi, w).$$

(b) If ϕ is the sentence $-\sigma$, then ϕ is true at w under **I** if, and only if, σ is not true at w under **I**.

(c) If ϕ is the sentence $(\sigma \,\&\, \tau)$, then ϕ is true at w under **I** if, and only if, both σ and τ are true at w under **I**.

(d) If ϕ is the sentence $(\sigma \vee \tau)$, then ϕ is true at w under **I** if, and only if, either σ is true at w under **I**, or τ is true at w under **I**, or both.

(e) If ϕ is the sentence $(\sigma \rightarrow \tau)$, then ϕ is true at w under **I** if, and only if, either σ is not true at w under **I**, or τ is true at w under **I**, or both.

(f) If ϕ is the sentence $(\sigma \leftrightarrow \tau)$, then ϕ is true at w under **I** if, and only if, either both σ and τ are true at w under **I** or both σ and τ are not true at w under **I**.

(g) If ϕ is the sentence $\Box\sigma$, then ϕ is true at w under **I** if, and only if, for every $w^{\#}$ that is a member of **W**, σ is true at $w^{\#}$ under **I**.

(h) If ϕ is the sentence $\Diamond\sigma$, then ϕ is true at w under **I** if, and only if, for some $w^{\#}$ that is a member of **W**, σ is true at $w^{\#}$ under **I**.

Notice that w plays no role in the modal truth conditions: if either $\Box\phi$ or $\Diamond\phi$ is true at some world w, then it will be true at every world.

The conditions for the quantificational cases will be given after a few preliminary definitions are set forth.

β-variants. Suppose that $\mathbf{I}^{\#} = <\mathbf{W}^{\#}, \mathbf{D}^{\#}, \mathbf{w}^{\#}, \mathbf{d}^{\#}, \mathbf{r}^{\#}, \mathbf{e}^{\#}>$ is an interpretation, and suppose that β is an individual constant. If $\mathbf{I}^{\#}$ differs from **I** at most in what is assigned to β, then $\mathbf{I}^{\#}$ is a β-variant of **I**.

A Notational Definition. If α is a variable, β is an individual symbol, and ϕ is a formula, then $\phi\alpha/\beta$ is the result of replacing all free occurrences of α in ϕ with β.

The conditions for the quantifier cases can now be given as follows:

(i) If ϕ is the sentence $(\alpha)\psi$ and β is the first individual constant not occurring in ψ, then ϕ is true at w under **I** if, and only if, $\psi\alpha/\beta$ is true at w under $\mathbf{I}^{\#}$, for every β-variant $\mathbf{I}^{\#}$ of **I** that assigns to β an element of $\mathbf{d}(w)$.

(j) If ϕ is the sentence $(\exists\alpha)\psi$ and β is the first individual constant not occurring in ψ, then ϕ is true at w under **I** if, and only if, $\psi\alpha/\beta$ is true at w under $\mathbf{I}^{\#}$, for some β-variant $\mathbf{I}^{\#}$ of **I** that assigns to β an element of $\mathbf{d}(w)$.

Comment

The clause 'every β-variant $\mathbf{I}^{\#}$ of **I** that assigns to β an element of $\mathbf{d}(w)$' shows that the quantifiers are being interpreted to range over the "inner w domain", that is the domain of the things that exist in w, as opposed to the "outer domain" **D**. One could develop the system in such a way that the existential statement is to be true at w if and only if there is something in **D** (not necessarily in $\mathbf{d}(w)$) that satisfies ψ. In that case, the quantifiers would be regarded as "Possibilist" quantifiers, as opposed to the "Actualist" quantifiers of the present system. One could, of course, also develop a system in which both Actualist and Possibilist quantifiers would be used. However, one can use '$\Box(x)$' in **M** to obtain the effect of a Possibilist universal quantifier. For example, to express the idea that every possible object possibly exists, one could use '$\Box(x)\Diamond E^1x$'—a sentence that is easily derivable from the empty set of premisses in the system to be given below.[10]

Truth Under an S5 Interpretation. A sentence ϕ is true under an S5 interpretation **I** if, and only if, ϕ is true at **w*** under **I**.

Structures and Models

Many mathematical logicians use the term 'structure' to refer to the sort of thing I have called an 'interpretation'.[11] From a mathematical realist's point of view, an S5 interpretation is a mathematical object—an ordered sextuple, the members of which are also mathematical objects. At times, I too will want to call the sextuple a 'structure'. When I wish to emphasize or call attention to the fact that what we are talking about can be regarded as a mathematical object, I shall use the term 'structure' (in this case, 'S5 structure') to refer to the sextuple. Sentences of **M** will be said to be true (or false) *in a structure*.

If a sentence ϕ is true in a structure (or under an interpretation) **I**, then **I** will be said to be a *model of* ϕ. Similarly, if all the members of a set of sentences Γ are true in a structure **I**, then **I** will be said to be a model of Γ. Since any model of a set of sentences is a structure and any structure is a model of some set of sentences, one will find writers using the term 'model' to refer to what I have called a 'structure'.[12] In many contexts, it will not make much difference whether one uses the terms 'interpretation', 'structure', or 'model'. Indeed, at times I use these terms interchangeably. However, in the later parts of this work, it will be important to remember that, as I use these terms, a model is a model of a sentence or set of sentences, and a structure is an interpretation regarded as a mathematical object.

As I have defined 'S5 interpretation', in order to specify an S5 interpre-

[10] For a discussion of the philosophical significance of the difference between Actualist and Possibilist quantifiers, see Forbes 1989: ch. 2.

[11] See, for example, Enderton 1972: sect. 2.2. Thus, to speak somewhat loosely, a structure, for Enderton, is a mapping from a set of linguistic entities into a set theoretical structure. It should also be mentioned that logicians also use the term 'structure' to refer to just the set theoretical structure. Roughly, a structure, in this latter sense, is the range of the mapping. Generally, this ambiguity does not lead to significant confusions.

[12] See, for example, Forbes 1985: ch. 2, sect. 2. The term 'model' is also used to refer to "structures" in the other sense described in the previous footnote. Thus, the term 'model theory' is frequently used to refer to the theory of "structures" in this latter sense.

tation one must assign a denotation for each and every individual constant of **M** and one must assign an extension (at every world) to each and every predicate of **M**. It turns out that, for many purposes, assigning denotations and extensions at worlds to some small finite set of nonlogical constants will be sufficient. For example, to show that some set of sentences has a model, it will suffice to specify the worlds and the set of things that exist in each world, in addition to making an assignment of a denotation to each individual constant occurring in a member of that set of sentences and an extension at every world to each predicate occurring in a member of that set. There would be no need to make additional assignments.

The Derivational System **ML**

The derivational system presented here is a natural deduction system.[13] A *derivation* is a finite sequence of consecutive numbered lines, each consisting of a sentence together with a set of numbers (called the *premiss-numbers* of the line), the sequence being constructed according to the rules of inference given below. If a numbered line consists of a number *n*, a sentence ϕ, and a set of premiss-numbers Γ, then ϕ is said to be *on* line n. If *m* is a premiss-number of line *n*, and if sentence ϕ is on line *m*, then ϕ is called a *premiss* of line *n*.

Rules of Inference

P (Introduction of Premisses). Any sentence may be entered on a line, with the line number taken as the only premiss-number.

T (Tautological Inference). Any sentence may be entered on a line if it is a tautological consequence[14] of a set of sentences that appear on earlier

[13] For a version of this natural deduction system for first-order logic, see Mates 1972: ch. 7.

[14] Those readers unfamiliar with the notion of *tautological consequence* can find a detailed exposition of this notion in Mates 1972: ch. 6. The following unrigorous explanation of the notion will give the reader the main ideas. Roughly speaking, an assignment of truth values to all the sentences of **M** is a *normal assignment* (or Boolean valuation) iff the assignment to the molecular sentences is "consistent with" the definition of 'true under an interpretation'. The idea is that the atomic, general, and initially modal sentences (sentences that begin with a modal operator) may be assigned truth values arbitrarily by a normal assignment **N**. However, every molecular sentence must be assigned by **N** the truth value that it would have under any interpretation **I** which was such that those atomic, general, and initially modal sentences occurring in the molecular sentence were true under **I** iff they were assigned true by **N**. Now any sentence ϕ is a tautological consequence of a set Γ of sentences iff ϕ is assigned true by every normal assignment that assigns true to all the members of Γ. A tautology is a sentence of **M** that is assigned true by every normal assignment. Regarding the above rule *T*, it should be noted that there is an effective procedure for determining if a sentence ϕ is a tautological consequence of Γ. See Mates 1972: ch. 6 for an indication of how such a procedure could be specified.

lines; as premiss-numbers of the new line take all premiss-numbers of those earlier lines. In particular, any tautology may be entered on a line; as premiss-numbers of this line, take the empty set.

C (Conditionalization). The sentence $(\phi \rightarrow \psi)$ may be entered on a line if ψ appears on an earlier line; as premiss-numbers of the new line, take those of the earlier line, with the exception (if desired) of any that is the line number of a line on which ϕ appears.

E (Existential Quantification). Suppose that the sentence σ is like τ except for containing an occurrence of the sentence $(\exists\alpha)\psi$ where τ contains an occurrence of $-(\alpha)-\psi$. Then σ may be entered on a line if τ appears on an earlier line; or τ may be entered on a line if σ appears on an earlier line. As premiss-numbers of the new line, take those of the earlier line.[15]

US (Universal Specification). The sentence $(E^1\beta \rightarrow \psi\alpha/\beta)$ may be entered on a line if $(\alpha)\psi$ appears on an earlier line; as premiss-numbers of the new line, take those of the earlier line.

UG (Universal Generalization). The sentence $(\alpha)\psi$ may be entered on a line if $(E^1\beta \rightarrow \psi\alpha/\beta)$ appears on an earlier line and $\Box(\alpha)\psi$ may be entered on a line if $\Box(E^1\beta \rightarrow \psi\alpha/\beta)$ appears on an earlier line—so long as β does not occur in ψ or in any premiss of that earlier line. As premiss-numbers of the new line, take those of the earlier line.

A (Axioms). Any of the sentences

$$\Box(\phi \rightarrow \psi) \rightarrow (\Box\phi \rightarrow \Box\psi)$$
$$\Box\phi \rightarrow \phi$$
$$\Diamond\phi \rightarrow \Box\Diamond\phi$$
$$\Diamond E^1\beta$$

may be entered on a line (β being any individual constant). As premiss-numbers of the new line, take the empty set.

N (Necessitation). The sentence $\Box\phi$ may be entered on a line if ϕ appears on an earlier line with the empty set of premisses. As premiss-numbers of the new line, take the empty set.

PI (Possibility Interchange). Suppose that the sentence σ is like τ except for containing an occurrence of the sentence $\Diamond\psi$ where τ contains an occurrence of $-\Box-\psi$. Then σ may be entered on a line if τ appears on an earlier line; or τ may be entered on a line if σ appears on an earlier line. As premiss-numbers of the new line, take those of the earlier line.

Comments

No attempt has been made to obtain the weakest set of derivation rules that would "do the job". For example, rule *E* could be weakened without

[15] It should be noted that this rule is more powerful than the rule called 'Existential Quantification' in the system of Mates 1972.

loss of deductive power.[16] Ease of use was a more weighty consideration than parsimony in logical strength.

Axioms. This rule could easily be stated in a way that does not bring to mind the axiomatic approach of the Hilbert school. I chose to give it in this way mostly to save space and to keep the number of rules small.

The first axiom, known as 'Modal Distribution' (since it tells us that box distributes over the conditional), is one of the three standard axioms of the S5 sentential calculus, and can be shown to be valid in a straightforward way. The second of these three axioms can be regarded as expressing the idea that a necessary statement is true. It is clearly valid. The first two axioms are to be found in most systems of modal logic. The third axiom—the "characteristic formula of S5 modal logic"—can be shown to be valid almost directly from clauses (g) and (h) of the definition of 'truth at a world under an interpretation'. For if $\Diamond\phi$ is true at **w***, then it must be true at every world. That the fourth axiom is true at **w*** under any interpretation **I** can be seen from the fact that each individual constant β must denote some element of **D**, and hence, there must be some w which is such that the element of **D** denoted by β is an element of **d**(w).

Other Rules. UG has two forms: the non-modal and the modal form. The two forms correspond to restricted and unrestricted versions of the universal generalization rule. The non-modal version allows one to infer a generalization over some restricted domain, i.e. a **d**(w), whereas the modal version allows one to infer a generalization over the whole of **D**. Thus, *US* and the non-modal *UG* are restricted free logical versions of the Universal Specification and Universal Generalization rules of first-order logic. Given the free logical approach taken above, it is easy to see why these restrictions are necessary. The quantifiers range over **d**(w); but the individual constants can refer to individuals that are not members of **d**(w). Thus, if all the things that exist in w have the property F, it does not follow that b has the property F, since b may not exist in w, but it does follow that if b exists in w, it has the property F.

The soundness of rule *E* follows from the fact that $(\exists\alpha)\phi$ is equivalent to $-(\alpha)-\phi$. Similarly, the soundness of rule *PI* reflects the equivalence of $\Diamond\phi$ with $-\Box-\phi$.

Terminology

A Derivation of ϕ *from* Γ. If ϕ is on the last line of a derivation and all the premisses of that last line are members of Γ, then the derivation is called a derivation of ϕ from Γ. To express the fact that it is possible to construct a derivation of ϕ from Γ, one may write: $\Gamma \vdash \phi$. To express that fact, one may also say: ϕ is derivable from Γ.

[16] See Mates 1972: 113 for a statement of a weaker rule that would yield a system with the same deductive power.

Theorem of Modal Logic. A sentence ϕ is a theorem of modal logic if, and only if, ϕ is derivable from the empty set of premisses. To express the fact that ϕ is a theorem of modal logic, one may write: $\vdash \phi$.

Consequence. A sentence ϕ is a consequence of a set of sentences Γ if, and only if, ϕ is true under every interpretation under which all the members of Γ are true. To express the fact that ϕ is a consequence of Γ, one may write: $\Gamma \vDash \phi$.

Valid. A sentence ϕ is valid if, and only if, ϕ is true under every interpretation. To express the fact that ϕ is valid, I may sometimes write: $\vDash \phi$.

Satisfiable. A set of sentences Γ is satisfiable if, and only if, there is an interpretation **I** such that every member of Γ is true under **I**.

The following schematic derivation illustrates how any instance of what is called 'Brouwer's Axiom' can be shown to be a theorem of modal logic:

$\varnothing$	(1)	$\Box{-\phi} \rightarrow -\phi$	*A*
$\varnothing$	(2)	$\phi \rightarrow -\Box{-\phi}$	(1)*T*
$\varnothing$	(3)	$\phi \rightarrow \Diamond\phi$	*PI*
$\varnothing$	(4)	$\Diamond\phi \rightarrow \Box\Diamond\phi$	*A*
$\varnothing$	(5)	$\phi \rightarrow \Box\Diamond\phi$	(1)(2) *T*

Sound. A system of inference rules is sound if, and only if, for any sentence ϕ and any set of sentences Γ, if $\Gamma \vdash \phi$, then $\Gamma \vDash \phi$.

Complete. A system of inference rules is complete if, and only if, for any sentence ϕ and any set of sentences Γ, if $\Gamma \vDash \phi$, then $\Gamma \vdash \phi$.

Theorems, Short-Cut Rules, and Metatheorems

It was noted earlier, in discussing the semantics of **M**, that the sentences '$\Box(x)\Diamond E^1x$' and '$\Diamond(\exists x)E^1x$' are valid. The former can easily be shown to be valid by constructing a derivation from the empty set of premisses as follows:

$\varnothing$	(1)	$\Diamond E^1b$	*A*
$\varnothing$	(2)	$E^1b \rightarrow \Diamond E^1b$	*C*
$\varnothing$	(3)	$(x)\Diamond E^1x$	*UG*
$\varnothing$	(4)	$\Box(x)\Diamond E^1x$	*N*

However, to prove that the latter sentence is a theorem of modal logic is somewhat more complicated. Since I shall make use of the fact that this sentence is valid in the next chapter, it may be instructive to show just how this sentence can be derived from the empty set of premisses.

$\varnothing$	(1)	$\Diamond E^1b$	*A*
2	(2)	$(x)-E^1x$	*P*
2	(3)	$E^1b \rightarrow -E^1b$	(2)*US*

2	(4)	$-E^1b$	(3)T
$\varnothing$	(5)	$(x)-E^1x \rightarrow -E^1b$	(2)(4)C
$\varnothing$	(6)	$E^1b \rightarrow -(x)-E^1x$	(5)T
$\varnothing$	(7)	$\Box(E^1b \rightarrow -(x)-E^1x)$	(6)N
$\varnothing$	(8)	$\Diamond E^1b \rightarrow \Diamond-(x)-E^1x$	*Theorem of S5 sentential logic*
$\varnothing$	(9)	$\Diamond-(x)-E^1x$	(1)(8)T
$\varnothing$	(10)	$\Diamond(\exists x)E^1x$	(9)E

I have skipped several steps in going from (7) to (8). Those familiar with S5 modal sentential logic will see that I have made use of the well-known theorem

$$\Box(\phi \rightarrow \psi) \rightarrow (\Diamond\phi \rightarrow \Diamond\psi).$$

I give below a schematic derivation showing that all sentences of this form are indeed theorems of modal logic:

$\varnothing$	(1)	$(\phi \rightarrow \psi) \rightarrow (-\psi \rightarrow -\phi)$	T
$\varnothing$	(2)	$\Box((\phi \rightarrow \psi) \rightarrow (-\psi \rightarrow -\phi))$	(1)N
$\varnothing$	(3)	$\Box((\phi \rightarrow \psi) \rightarrow (-\psi \rightarrow -\phi)) \rightarrow$ $(\Box(\phi \rightarrow \psi) \rightarrow \Box(-\psi \rightarrow -\phi))$	A
$\varnothing$	(4)	$\Box(\phi \rightarrow \psi) \rightarrow \Box(-\psi \rightarrow -\phi)$	(2)(3)T
$\varnothing$	(5)	$\Box(-\psi \rightarrow -\phi) \rightarrow (\Box-\psi \rightarrow \Box-\phi)$	A
$\varnothing$	(6)	$(\Box-\psi \rightarrow \Box-\phi) \rightarrow (-\Box-\phi \rightarrow -\Box-\psi)$	T
$\varnothing$	(7)	$\Box(\phi \rightarrow \psi) \rightarrow (-\Box-\phi \rightarrow -\Box-\psi)$	(4)(5)(6)T
$\varnothing$	(8)	$\Box(\phi \rightarrow \psi) \rightarrow (\Diamond\phi \rightarrow -\Box-\psi)$	(7)PI
$\varnothing$	(9)	$\Box(\phi \rightarrow \psi) \rightarrow (\Diamond\phi \rightarrow \Diamond\psi)$	(8)PI

The task of constructing derivations in this system will be rendered considerably easier and more natural by the addition of various short-cut rules. These are rules that do not add any genuine power to the theorem proving capabilities of the system—anything provable with the addition of these rules can be proved without the rules—but they enable us to skip steps in proofs. The following four rules are especially useful.[17]

EG (Existential Generalization). The sentence $(\exists\alpha)\phi$ may be entered on a line if $(E^1\beta \,\&\, \phi\alpha/\beta)$ appears on an earlier line; as premiss-numbers of the new line, take those of the earlier line.

ES (Existential Specification). Suppose either that $(\exists\alpha)\psi$ appears on line i of a derivation and that $(E^1\beta \,\&\, \psi\alpha/\beta)$ appears (as a premiss) on a later line j, or that $\Diamond(\exists\alpha)\psi$ appears on line i of a derivation and that $\Diamond(E^1\beta \,\&\, \psi\alpha/\beta)$ appears (as a premiss) on a later line j. And suppose that ϕ appears on a still later line k; and suppose further that the constant β occurs neither in ϕ, ψ, nor in any premiss of line k other than $\psi\alpha/\beta$; then ϕ may be

17 For a justification of classical versions of *EG* and *ES*, see Mates 1972: ch. 7, sect. 3. The reader can easily construct justifications of the free-logical versions of these rules by making suitable modifications of Mates's justifications.

entered on a new line. As premiss-numbers of the new line, take all those of the lines i and k, except the number k.

Q (Quantifier Exchange). Suppose that the sentence σ is like τ except for containing an occurrence of the sentence $-(\exists\alpha)\psi$ where τ contains an occurrence of $(\alpha)-\psi$, or containing an occurrence of the sentence $(\exists\alpha)-\psi$ where τ contains an occurrence of $-(\alpha)\psi$, or containing an occurrence of the sentence $-(\exists\alpha)-\psi$ where τ contains an occurrence of $(\alpha)\psi$. Then σ may be entered on a line if τ appears on an earlier line; or τ may be entered on a line if σ appears on an earlier line. As premiss-numbers of the new line, take those of the earlier line.

RAA (Reductio Ad Absurdum). ϕ may be entered on a line if ψ and $-\psi$ appear on earlier lines; as premiss-numbers of the new line take all those of the aforementioned earlier lines, with the exception (if desired) of any line number of a line on which $-\phi$ appears.

The Modal Reduction Property

One striking feature of S5 modal logic that distinguishes it from other standard systems is the modal reduction property. If a sentence θ consists of a sentence ϕ, preceded by a string of modal operators, for example

$$\Diamond\Diamond\Box\Diamond\Diamond\Box\Diamond\Box\phi,$$

then θ is equivalent to the sentence consisting of ϕ preceded by just the last operator in the string, so that, in this case, θ is equivalent to

$$\Box\phi.$$

This feature of the system can be established in standard (but somewhat involved) ways, so I shall not go through the required derivations here.[18] In so far as we are attempting to develop a formal system of modal logic for the "broadly logical" notion of necessity, it does seem appropriate that S5 have this feature.[19] It would seem that if a proposition is possible (in this broadly logical sense) in the actual world, then the proposition should be possible in every possible world.[20] Thus,

$$\Diamond\phi \rightarrow \Box\Diamond\phi$$

should be a valid principle of the modal logic—which is what is needed to obtain the reduction property.

[18] The reader can find a good discussion of this feature in Chellis 1980; see especially sect. 4.4.

[19] See Kneale 1962: ch. 9, sect. 4 for a philosophical defence of the reduction property.

[20] Plantinga provides a more detailed argument along these lines in Plantinga 1974: ch. 4, sect. 6. But see Philip Quinn for critical comments about Plantinga's reasoning (Quinn 1982). Quinn argues that Plantinga does not provide a "decisive metaphysical argument" for the truth of the principle under discussion (p. 451). But even if we grant Quinn this point, that does not mean that we cannot attribute *some* intuitive force to arguments such as the one I gave above.

The above derivational system **ML** can be shown to be sound in the usual straightforward way, by means of direct semantical analyses of the individual rules. Proving the completeness of the system, however, is a more complicated matter. The establishment of two short-cut rules can simplify the completeness proof. These rules will be formulated, using the abbreviation '$\phi \Rightarrow \theta$' for strict implication, '$\Box(\phi \rightarrow \theta)$', as follows:

GUS (Generalized Universal Specification). Given that sentence $\theta = (E^1\beta \rightarrow \psi\alpha/\beta)$ and sentence $\tau = (\alpha)\psi$, or that sentence $\theta = \Box(E^1\beta \rightarrow \psi\alpha/\beta)$ and sentence $\tau = \Box(\alpha)\psi$, then the sentence $\phi_1 \Rightarrow (\phi_2 \Rightarrow \ldots (\phi_n \Rightarrow \theta)) \ldots)$ may be entered on a line if $\phi_1 \Rightarrow (\phi_2 \Rightarrow \ldots (\phi_n \Rightarrow \tau)) \ldots)$ appears on an earlier line; as premiss-numbers of the new line, take those of the earlier line.

GUG (Generalized Universal Generalization). Given that sentence $\theta = (E^1\beta \rightarrow \psi\alpha/\beta)$ and sentence $\tau = (\alpha)\psi$, or that sentence $\theta = \Box(E^1\beta \rightarrow \psi\alpha/\beta)$ and sentence $\tau = \Box(\alpha)\psi$, then the sentence $\phi_1 \Rightarrow (\phi_2 \Rightarrow \ldots (\phi_n \Rightarrow \tau)) \ldots)$ may be entered on a line if $\phi_1 \Rightarrow (\phi_2 \Rightarrow \ldots (\phi_n \Rightarrow \theta)) \ldots)$ appears on an earlier line, so long as β does not occur in ψ or in any premiss of that earlier line. As premiss numbers of the new line, take those of the earlier line.

When the sentences θ and τ in the above rules of derivation are restricted to the non-modal versions, the two quantifier rules, augmented by the derivation rules of some appropriate modal sentential calculus (such as S4), yields the system GS of (Garson 1991), the completeness of which is proven. To show in a reasonably compact fashion that the above two short-cut rules can be established for **ML** requires a significant amount of proof theoretical development—one needs, for example, to prove a *replacement metatheorem* that allows one to replace a formula, occurring in a sentence appearing on an earlier line, with a proved equivalent formula. I shall not undertake such a development here. However, the following sketch will indicate how *GUG* (with non-modal θ and τ) can be justified (*GUS* is relatively easy).

I start with the supposition that '$\Box(\phi \rightarrow (E^1\beta \rightarrow \psi\alpha/\beta))$' appears on a line with 'ζ' as its premiss. Thus, we have:

{1}	(1)	ζ	
{1}	(2)	$\Box(\phi \rightarrow (E^1\beta \rightarrow \psi\alpha/\beta))$	

We can then show that '$\Box(\phi \rightarrow (\alpha)\psi)$' can be derived from line (1), using the derivation rules of **ML**, as follows:

{3}	(3)	$-\Box(\phi \rightarrow (\alpha)\psi)$	*P* (for *RAA*)
{3}	(4)	$\Diamond-(\phi \rightarrow (\alpha)\psi)$	*NI*
{3}	(5)	$\Diamond(\phi \,\&\, -(\alpha)\psi)$	*Replacement*
{3}	(6)	$\Diamond(\phi \,\&\, (\exists\alpha)-\psi)$	*Q*
{3}	(7)	$\Diamond(\exists\alpha)(\phi \,\&\, -\psi)$	*Replacement*
{8}	(8)	$\Diamond(E^1\beta \,\&\, \phi \,\&\, -\psi\alpha/\beta)$	*P* (for *ES*)

{1}	(9)	$-\Diamond-(\phi \rightarrow (E^1\beta \rightarrow \psi\alpha/\beta))$	*PI*
{1}	(10)	$-\Diamond(\phi \,\&\, -(E^1\beta \rightarrow \psi\alpha/\beta))$	*Replacement*
{1}	(11)	$-\Diamond(E^1\beta \,\&\, \phi \,\&\, -\psi\alpha/\beta)$	*Replacement*
{1, 8}	(12)	$\zeta \,\&\, -\zeta$	*T*
{1, 3}	(13)	$\zeta \,\&\, -\zeta$	*ES*
{1}	(14)	$\Box(\phi \rightarrow (\alpha)\psi)$	*RAA*

Once one has established the above two short-cut rules, a proof of the completeness of **ML** can be given that is basically similar in structure to Leon Henkin's famous proof of the completeness of first-order logic.[21] We thus have:

> For every sentence ϕ and every set of sentences Γ, $\Gamma \vdash \phi$ if, and only if, $\Gamma \vDash \phi$

and

> For every sentence ϕ, ϕ is a theorem of modal logic if, and only if, ϕ is valid.

Various Compactness results follow almost immediately. For example, we have:

> For every set of sentences Γ, if every finite subset of Γ is satisfiable, then Γ is satisfiable.

2. Objections Raised by Plantinga and his Supporters

In this section, I shall seek to give the reader a deeper understanding of the system developed in Section 1 by discussing some objections that have been raised to this sort of system. It will be seen that these objections arise from far-reaching metaphysical views.

Alvin Plantinga and some of his supporters have developed a rival view of modal quantificational logic that I shall be discussing in several parts of this work. A discussion of this rival view is especially relevant to the aims of this work, since Plantinga's metaphysical views lead to the development of an S5-type modal logic that differs both formally and philosophically from the more standard versions of S5 modal logic that I shall be advocating. Furthermore, my investigations will illustrate both how a philosopher's metaphysical beliefs can influence his or her choice of the semantics for modal logic, and also how modal logic can be used in the ar-

[21] See Mates 1972: ch. 8 for a clear version of Henkin's theorem. Jonathan Lebowitsch has given such a Henkin style completeness proof for an S4 version of the Serious Actualistic modal system developed in Chapter 7 below. See Fine 1978 for a related system and proof of completeness of that system.

ticulation, comparison, evaluation, and systematization of a philosopher's modal principles.

Pure and Applied Semantics

In Plantinga 1974, Plantinga ponders the significance of the sort of possible worlds semantics I have given above, asking: "[H]ow, exactly, are we to take such a semantics for modal logic? What is the point of offering such a system? Why do we need one and what is it designed to do for us?"[22] In attempting to answer these questions, Plantinga draws a distinction between what he calls "pure" and "applied" semantics. What I have presented above is pure semantics. An interpretation of **M** is a purely set theoretical construction and, as such, need have no obvious connection with modality. One can regard the study of such interpretations as a purely mathematical activity. Pure semantics, according to Plantinga, commits itself only to a fragment of set theory.[23] Furthermore, this pure semantics fails to assign any meaning to the modal operators or tell us what such sentences as

[1] $(\exists x)\Box x$ is a person

mean (Plantinga 1974: 127). For this reason, Plantinga argues, there is need for an "intended or associated applied semantics" according to which the set **W** of the structure is not just any set but is the set of all possible worlds, **a*** is not just any element of **W** but is the actual world, etc. In other words, it is thought that one of the possible worlds structures talked about in pure semantics must be the *intended applied semantics*. It is this intended applied semantics that gives meaning to the modal operators and allows us to use the logic to draw concrete modal inferences involving such sentences as [1] above. Thus, in order to obtain genuine insights into the nature of necessity, Plantinga suggests, we must turn to applied semantics. Needless to say, it is through the use of this intended applied semantics that the commitment to possible worlds is supposedly made.

The Canonical Conception

Let us reconsider the language **M** discussed in Section 1, and imagine the intended applied semantics of this language. The set **W** of the intended applied semantics is not just any non-empty set: It is, supposedly, the set of all possible worlds. And **D** is supposed to be the set of all possible individuals; **d**(w) is the set of all possible individuals that exist in possible world w; **w*** is the actual world, etc. As was noted earlier, the view of possible worlds and possible worlds semantics that arises from this kind of intended interpretation Plantinga calls "the canonical conception".[24]

[22] Plantinga 1974: 125. [23] Plantinga 1974: 127. [24] Plantinga 1979: 254.

Can this Talk of Possible Worlds be Dispensable?

It might be thought that we can regard the talk of possible worlds and possible objects as mere imagery, useful heuristically perhaps, but dispensable in the end. Plantinga doubts that such a position is truly defensible. However, before I explain Plantinga's doubts, it is necessary that I explain a distinction he makes use of between *de re* and *de dicto* modal sentences and assertions.

As Plantinga explains the distinction, a modal assertion is *de dicto* if it predicates a modal property of another *dictum* or proposition (Plantinga 1974: 9). For example, the assertion

It is necessary that all men are rational

predicates necessity of the proposition that all men are rational. On the other hand, a modal assertion is *de re* that predicates a modal property of a thing that is not a dictum or proposition. For example, the assertion

9 has the necessary property of being odd

is *de re*.

To give a syntactic criterion for distinguishing the *de re* sentences of **M** from the *de dicto* sentences of that language, we can say that a modal sentence ϕ is *de re* iff some formula θ that occurs in ϕ consists of a modal operator followed by a formula containing either a variable that is free in θ or an individual constant. By classifying every occurrence of an individual constant as a free occurrence of the constant, we can say that a modal sentence ϕ is *de re* iff some formula that occurs in ϕ consists of a modal operator followed by a formula containing at least one free occurrence of an individual symbol. A modal sentence of **M** that is not *de re* is *de dicto*. Thus, the sentence '$\Box(x)(F^1x \rightarrow G^1x)$' is *de dicto*: The modal operator can be regarded as applying to the dictum '$(x)(F^1x \rightarrow G^1x)$'. On the other hand, the sentence '$(\exists x)\Box F^1x$' is *de re*, since the formula '$\Box F^1x$' consists of a modal operator followed by a formula in which there is an occurrence of the variable 'x' that is free in '$\Box F^1x$'.

Let us now consider Plantinga's doubts that one can regard all talk of possible worlds as mere dispensable imagery. He gives an example from the writings of St Thomas in which it is claimed that the sentence

Whatever is seen to be sitting is necessarily sitting

is true, if it is regarded as a *de dicto* modal sentence, but false, if it is taken to be a *de re* modal sentence. In terms of the possible worlds semantics we have been studying, we can understand this claim as asserting that the above sentence can be understood as the *de dicto*

In every possible world, whatever is seen to be sitting is sitting

or as the *de re*

> Whatever is seen (in the actual world) to be sitting is sitting in every possible world,

and that the former is true, whereas the latter is false. Plantinga is sceptical that one can understand this explanation if one adopts the view, described above, that all our talk of possible worlds here is mere dispensable imagery.

Plantinga gives another example to illustrate and bolster his case: The statement 'Johan is essentially a person but contingently a philosopher' can be explained to be the assertion that *Johan has both properties, and has the former, but not the latter, in every possible world in which he exists.* Plantinga expresses doubt that an anti-realist can legitimately make use of possible worlds semantics to give such an explanation. His view is that, if we do not avail ourselves of possible worlds, "it requires a well-trained eye to see just what our explanation accomplishes."[25] Plantinga concludes: "[I]f we look at a semantics of this kind as a sober and literal account of modality—one whose talk of possible [worlds] and all the rest is to be taken seriously—then what is crucial is the applied, not the pure, semantics."[26]

Plantinga's Actualism

It can be seen that if we accept the canonical conception, we are driven towards an acceptance of possible worlds. For Plantinga, this is no serious problem, since (in some sense) he believes in possible worlds. However, there is a serious problem for Plantinga's position to which this conception gives rise. Plantinga is an "Actualist". *An Actualist does not believe in the existence of merely possible objects*—things that do not exist (in the actual world) but that do exist in other possible worlds. More precisely, Plantinga defines 'Actualism' to be the view that "there neither are nor could have been things that do not exist" (Plantinga 1985*a*: 314). Of course, Plantinga is not denying that there could have been things that do not, in fact, exist. Perhaps it would be less misleading to characterize an Actualist as someone who holds both that there is no thing that does not (actually) exist and also that there could not have been something that did not (actually) exist.[27]

[25] Plantinga 1979: 126.

[26] Plantinga 1974: 128.

[27] The opposing view is called 'Possibilism'. One problem with the characterization of the opposing views I have given here is that the dispute involves the very meaning of the word 'actual'. For the Actualist, a thing is actual (actually exists) iff it exists; the Possibilist believes that some things exist which are not actual. So the question arises: what does the word 'actual' mean? See Jubien 1996 for a discussion of this question and an attempt to specify a sense of the term which will not beg the question for either side.

The Ontological Flaw in the Canonical Conception

Now, unless one is willing to require the intended interpretation to be such that the set **d(w*)** = **D**, the applied semantics seems to require a belief in merely possible individuals. But requiring **d(w*)** to be **D** seems to be unacceptable because *it yields the counterintuitive result that no possible world could contain individuals that do not exist in the actual world*, which implies that there could not have been any individuals that do not actually exist. To avoid this absurdity, we seem to be driven to accept the existence of merely possible objects. Plantinga concludes that "the canonical conception of possible worlds exacts a substantial ontological toll".[28]

Later, Plantinga described his reasoning this way. The canonical conception presupposes that there are or could have been objects that do not in fact exist: "[I]t presupposes that in addition to all the things that exist, there are some more—"possible objects"—that do not exist but could have." Such a presupposition, it is argued, is "resoundingly false".[29]

Plantinga's response to this reasoning is to reject the canonical conception. He believes that one can achieve the basic insights obtained by means of the canonical conception without accepting the various presuppositions and implications of this conception. To achieve these aims, Plantinga develops a competing conception of possible worlds and possible worlds semantics, which gives rise to a different set of validities and a very *different sort of intended interpretation of the formalism.*[30]

Plantinga's Possible Worlds

The basic (primitive) notion of Plantinga's ontology is *state of affairs*. A state of affairs is an abstract object of some sort: it is incapable of activity or of any causal relationships. Plantinga does not define 'state of affairs' but he does gives us some examples: *Socrates being shorter than Plato* is one such. (Evidently, one refers to states of affairs in English by means of gerund clauses.) Some states of affairs *obtain* or are *actual*. Since Socrates was in fact shorter than Plato, the state of affairs *Socrates being shorter than Plato* did obtain (i.e. was actual). Clearly, the state of affairs *Socrates being able to play the violin* did not obtain (was not actual). However, according to Plantinga, that state of affairs exists. Hence, we need to distinguish the existence of a state of affairs from its obtaining or being actual. States of affairs are like propositions: they have *complements*. Presumably, the complement of the state of affairs *Socrates being shorter than Plato* would be *Socrates not being shorter than Plato*.

28 Plantinga 1974: 257.

29 Plantinga 1985: 91. On the other hand, Anil Gupta claims that we can name merely possible objects; and he believes that merely possible objects can bear relations to actual objects. See Gupta 1980: 20 n. I shall discuss this view of Gupta's in more detail later.

30 This is done explicitly in Plantinga 1979. Much of what I say in what follows is based upon this article.

Among the states of affairs, there are some that are *possible*. If a state of affairs could have obtained, then it is a possible state of affairs. In what follows, let *S* and *S** be states of affairs. *S precludes S** if and only if it is not possible that both *S* and *S** obtain. It follows, of course, that if *S* precludes *S**, then *S** precludes *S*. To say that *S* precludes *S** is analogous to saying that *S* and *S** are incompatible states of affairs. *S includes S** if and only if it is not possible that both *S* obtain and *S** not obtain. Thus, one might say that *S* includes *S** if and only if the obtaining of *S* entails the obtaining of *S**. *S properly includes S** if and only if *S* includes *S** and *S** does not include *S*. *S* is a *maximal state of affairs* if and only if, for every state of affairs *S**, *S* includes either *S** or the complement of *S** (a maximal state of affairs is like a *complete* deductive theory formalized in the predicate calculus). A *possible world,* then, is a maximal possible state of affairs.[31]

Possible Objects and Essences

Having thus defined 'possible world', Plantinga has a way of interpreting (or understanding) the **W** of an interpretation **I** of the language **M**. But to have a reasonable way of understanding the possible worlds semantics of **M**, we need more. After all, **D** is supposed to be the set of all possible individuals. How can Plantinga give an Actualist interpretation to **D**? The short answer to this question is: via essences. But this answer requires some explanation.

Plantinga maintains that, in addition to individuals, states of affairs, and sets, there exist *properties.* For example, there is the property of *being married to Xantippe*. Properties are possessed by individuals. Thus, Socrates possesses (or has) the above property. One can also say that the property is *exemplified* by Socrates. According to Plantinga, there are possible worlds in which Socrates possesses the property of *not being married to Xantippe*. What does it mean to say that Socrates has some property *F* in a world *w*? It is to say, according to Plantinga, that Socrates would have had the property *F* had *w* been actual. Now Socrates possesses the property of *being married to Xantippe* in the actual world **w***. Corresponding to this fact, Plantinga thinks that there is a "world-indexed property":

[31] There is one modification that Plantinga has made recently: strictly speaking, a possible world is taken to be "a temporally invariant state of affairs that is maximal with respect to temporally invariant states of affairs" (Plantinga 1979: 91). Plantinga made this revision in response to John Pollock's suggestion that some states of affairs, such as *Nixon's running for President*, may obtain at some times and not at others; whereas some states of affairs, such as *Nixon's running for President in 1970*, obtain at all times if they obtain at all. Pollock had suggested that only "non-transient states of affairs enter into the construction of possible worlds" (Pollock 1985: 122). In what follows, the distinction between transient and non-transient states of affairs will play no role, so I shall not qualify my references to states of affairs and maximal possible states of affairs in terms of this distinction.

being married to Xantippe in **w***. In some possible world w in which Socrates exists, Socrates may lack the property of *being married to Xantippe*, but even so, Socrates will still possess, in w, the property of *being married to Xantippe in* **w***. The latter is an *essential property* of Socrates. Since no actual individual other than Socrates has this property, it is impossible that there should have been any individual who had this world-indexed property. One might reason to this conclusion as follows. Suppose that, in some world w, there is some individual x which is such that x has the property of *being married to Xantippe in* **w***. Then x would be married to Xantippe in the actual world. But there is only one person married to Xantippe in the actual world, and that person is Socrates. So x would have to be Socrates.

The world-indexed property *being married to Xantippe in* **w*** is a property that has the following characteristics: (1) It is possessed by Socrates in every possible world in which he exists; (2) There is no possible world in which some individual other than Socrates possesses this property. Such a property is called an "essence of Socrates".[32] More generally,

> An *essence* is a property F which is such that there is some world w in which there is an individual x which is such that: (a) x has F *essentially*; and (b) there is no world $w^{\#}$ in which some individual distinct from x has F. If there is such a world in which there is such an individual x, F is an essence of x.

What does it mean to say that an individual x has a property essentially? It means that x has the property in every possible world in which x exists.

For every possible world w, every individual x that exists in w possesses an essence: x possesses the property of being identical to x, and this property is clearly an essence of x. However, world-indexed properties allow Plantinga to specify essences that are not trivial essences of the sort illustrated by this last example.

Plantinga's Applied Semantics

According to Plantinga's possible worlds semantical views, the intended interpretation of the modal language will assign to **W** the set of possible worlds, just as it does according to the canonical conception. (Of course, a possible world in Plantinga's system is very different from what is generally regarded as a possible world.) Essences are the key elements in Plantinga's revision of the canonical conception of possible worlds. The principal idea of the revision comes to this: take **d**(w) to be not the set of possible individuals that exist in world w, but rather the set of essences exemplified in w. Then **D** will not be the set of possible individuals, as it is according to the canonical conception; instead it will be the set of all

[32] Plantinga 1974: 72.

essences exemplified in at least one world. Essences are held to be necessary beings: they exist in every possible world. So *Plantinga's conception does not require that there be possible individuals that do not exist; it only presupposes the existence of essences, some of which are not exemplified.* (Since Plantinga's semantical system does not assume the existence, in any world, of any merely possible beings, it is classified as an Actualist system.) The reference function **r** will then assign to each individual constant an essence; and the extension function **e** will assign to each ordered pair consisting of an *n*-ary predicate ϕ and world w the set of n-tuples of essences coexemplified with ϕ in w. The element $\mathbf{w}^*$ will again be the actual world.[33]

To indicate how truth conditions for modal sentences are given by Plantinga, I need to give one more definition. Property F is *coexemplified* with property G in world w if, and only if, some individual would have had both properties had w been actual. Then the sentence

$$Fa$$

is true at world w iff the essence denoted by 'a' is coexemplified with the property denoted by 'F' in w. Thus,

$$\Diamond(\exists x)Fx$$

is true if, and only if, there is a possible world w which is such that there is some essence G that is coexemplified with F in w. The sentence

$$(\exists x)\Diamond Fx$$

is true if, and only if, there is an essence G which is such that it is exemplified in the actual world and, in some possible world w, G is coexemplified with F. From these definitions, one can extract a good idea of how Plantinga understands possible worlds semantics.

Some Formal Differences From ***M***

Consider now the sentences known as *the Barcan Formula*:

$$\text{[BF]} \quad (x)\Box Fx \rightarrow \Box(x)Fx,$$

and *the Converse Barcan Formula*:

$$\text{[CBF]} \quad \Box(x)Fx \rightarrow (x)\Box Fx.$$

Are these valid sentences of **M**? If so, they would seem to show that *de re* and *de dicto* necessities can be interchanged freely. However, [BF] is clearly not valid. Suppose, for example, that only one being exists and that this being has the property of being necessarily benevolent. Would it follow

33 This way of giving the applied semantics of **M** was obtained by extrapolation from what is done in Jager 1982—a work soundly endorsed by Plantinga.

that, necessarily, everything that exists is benevolent? Surely not. The following diagram shows in more detail that [BF] is not valid.[34]

	$\mathbf{w}^*$	v
d	{1}	{1, 2}
F	{1}	{1}

This diagram captures the essential elements of an interpretation **I** = <**W**, **D**, **w***, **d**, **r**, **e**> under which [BF] is false. It tells us that the components of the interpretation are such that $\mathbf{W} = \{\mathbf{w}^*, v\}$, $\mathbf{D} = \{1, 2\}$, $\mathbf{d}(\mathbf{w}^*) = \{1\}$, $\mathbf{d}(v) = \{1, 2\}$, $\mathbf{e}(F, \mathbf{w}^*) = \{1\}$, and $\mathbf{e}(F, v) = \{1\}$.

Most currently employed S5 quantificational systems of modal logic reject [BF]. [CBF], however, is more controversial. [CBF] is not a valid sentence of **M**, as can be seen from the following diagram:[35]

	$\mathbf{w}^*$	v
d	{1, 2}	{1}
F	{1, 2}	{1}

'$\Box(x)Fx$' is true under this interpretation since everything is *F* in all the worlds; however '$(x)\Box Fx$' is not true, because 2 is not *F* in *v* and hence is not necessarily *F*. It is clear that we can obtain a counterexample to [CBF] in the system under discussion because a possible object 2 is deemed to lack a property *F* in a world *v*, even though that object does not exist in that world. In other words, in **M**, for 2 to possess $\Box F$, 2 must have *F* in every world, even those in which 2 does not exist. But does it make sense for an object to possess a property in a world in which it does not exist?

Imagine that, in the above diagram, 1 stands for Plato and 2 stands for Socrates; and suppose that *F* is the predicate 'is a living human being'. Then, the diagram gives us an interpretation according to which Socrates is not a living human being in world *v*. It seems reasonable that Socrates is not a living human being in *v* since Socrates does not exist in *v*.

Still, one might wonder how Socrates could have any property in *v* (even the property of not being a living human being) if he does not exist in *v*. Arthur Prior maintained that to say an individual *s* exists is just to say that there are facts about *s*. So, according to Prior's doctrine, in any world *v*, if *s* does not exist in *v*, then there cannot be any such fact as is expressed by '*s* is *F* in *v*'.[36] Plantinga agrees with Prior to this extent: he is willing to affirm that "an object has no *properties* in a world in which it does not exist".[37] Plantinga would thus agree that if '*F*' expresses a property, there

[34] This diagram is essentially the one given in Forbes 1985: 36.

[35] This diagram is given in Forbes 1985: 36, but it is essentially the interpretation given by Kripke in Kripke 1971: 67–8.

[36] Prior's doctrine regarding the connection between existence and facts is discussed in some detail by Christopher Menzel (Menzel 1991). See also Deutsch 1990.

[37] Plantinga 1974: 60, italics mine.

cannot be any such fact as is expressed by '*s* is *F* in *v*', if *s* does not exist in *v*. Such an ontological view is what I classified earlier as "Serious Actualism", and this view will be discussed in much more detail in Chapter 7.

Jager's System of Quantificational Modal Logic

Various systems of quantificational modal logic, based on the Actualistic doctrines of Plantinga, have been developed.[38] In particular, Thomas Jager's system of quantificational modal logic was designed to conform to Plantinga's Serious Actualistic point of view. I shall discuss the details of this system in Chapter 3 (see Section 10). For now, it will be sufficient for my purposes to note that the truth conditions of this system are such that:

> '$\Box Fa$' is true at world w under interpretation $\mathbf{I} = \langle \mathbf{W}, \mathbf{D}, \mathbf{w}^*, \mathbf{d}, \mathbf{r}, \mathbf{e} \rangle$ iff $\mathbf{r}(a) \in \mathbf{d}(w)$ and, for every $u \in \mathbf{W}$, if $\mathbf{r}(a) \in \mathbf{d}(u)$, then $\mathbf{r}(a) \in \mathbf{e}(F, u)$.[39]

Thus, '□' has been given the "weak" sense of necessity, according to which '$\Box F$' stands for the property of having *F essentially* (as contrasted with the property of being necessarily *F*, which is how the formula can be understood in the system **M**). In possible worlds terms, if '□' has the "weak" sense, then '$\Box Fa$' is true iff *a* is *F* at every possible world in which *a* exists; whereas, if '□' has the "strong" sense, then '$\Box Fa$' is true iff '*Fa*' is true at every possible world. It follows, of course, that the above diagram used to show that [CBF] is not a theorem of **M** cannot be used to show that the formula [CBF] is not a theorem of Jager's system.

As an illustration of these concepts, 'Socrates is essentially rational' ('$\Box Fa$' where '□' has the weak sense of necessity) is true iff Socrates has the property of rationality at every world in which he exists; whereas 'Socrates is necessarily rational' ('$\Box Fa$' where '□' has the strong sense of necessity) is true iff 'Socrates is rational' is true at every world.

Actualistic Objections to [CBF]

Let us investigate the diagram used to show that [CBF] is not a valid sentence of **M**. Interpret '*b*' to stand for Socrates and 'F^1' to be the predicate 'is a living human being'. Then 'F^1b' is to be understood to express the sentence 'Socrates is a living human being'—a false sentence according to the diagram. Now some philosophers may wonder how the sentence '$-F^1b$' can be true at world *v* if Socrates does not exist in that world. As Plantinga analyses the situation, to claim that '$-F^1b$' is true at world *v* is just to claim that '$-F^1b$' would have been true, had *v* been actual. But had *v* been actual, Socrates would not have existed. And had Socrates not existed, then the

38 An S5 quantificational modal logic based on this doctrine is the system **Q** set forth in Menzel 1991. For a discussion of Plantinga's form of Serious Actualism, see Pollock 1985: sect. 2 and Plantinga 1985: sect. B.

39 Jager 1982: 340.

sentence 'Socrates is not a living human being' would have been neither true nor false, according to the Frege–Strawson position. This line of reasoning suggests that '$-F^1b$' should be taken to be neither true nor false at *v*. Perhaps, **M** should have "truth value gaps". (Of course, there would not be this sort of problem for the Plantinga–Jager system since '*b*' denotes the essence of Socrates in that system, and this essence would have existed, even if *v* had been actual).

I admit that the above line of reasoning has some plausibility to it. But this plausibility derives, to a significant extent, from adopting Plantinga's way of understanding the modal language. As I view the language in question, the sentence '$-F^1b$' being true at world *v* does not presuppose or imply that we could truly assert '$-F^1b$' in *v* had *v* been actual, as may appear to be the case given Plantinga's gloss; rather, we should take '$-F^1b$' to be a sort of description of *v* from the point of view of someone outside the world. We should ask: is *v* such that '$-F^1b$' is a true statement of what is the case in that world? Here, the answer seems to be clear. In *v*, Socrates does not exist, so he is not a living human being.

Let us investigate how some Actualists have come to the conclusion that [CBF] is a valid principle of modal logic. Konyndyk's argument for the acceptability of [CBF] is the following:

> [I]f we want □ as attached to open sentences to express possessing a property essentially, then it can be seen that the Converse Barcan Formula should be a theorem of our system. For the CBF simply expresses what is a necessary truth: If it is true in every possible world that everything is *A* , then everything has *A* in every world in which it exists.[40]

Plantinga has similarly defended the validity of [CBF] from the point of view of Actualism as follows:

> [L]et *x* be an object—any actually existing object, that is—and *W* any world in which *x* exists. *Everything is F* is necessarily true and hence true in *W*. *x*, accordingly, has the property of being *F* in *W*. So *x* has that property in every world in which it exists—that is, *x* has it essentially.[41]

Jager's system of quantified modal logic, which is regarded by Plantinga as an adequate formal development of his version of applied semantics, excludes [BF] but includes [CBF] among the set of its valid sentences.[42]

An examination of the above arguments in favour of [CBF] makes it abundantly clear that the box operator in [CBF] has been given the "weak" sense of necessity, so that '□*Fa*' expresses not 'necessarily *Fa*' but rather '*a* is essentially *F*'.[43] Thus, what the sentence [CBF] expresses in

40 Konyndyk 1986: 102. 41 Plantinga 1974: 59.

42 See Jager 1982. Cf. Konyndyk 1986: sect. 4.14, where a system of natural deduction rules for Jager's system is given.

43 It should be noted that some authors use the expression 'weak necessity' in a way that

such systems as **M** is not what it is taken to express in Jager's system. In **M**, [CBF] simply does not say that if it is true in every possible world that everything is A, then everything has A in every world in which it exists. What, in **M**, does assert what Konyndyk suggests [CBF] says, is the sentence

$$\Box(x)Ax \rightarrow (x)\Box(Ex \rightarrow Ax).$$

And that sentence is valid, as can be seen from the following derivation:

$\varnothing$	(1)	$(x)Ax \rightarrow (Ea \rightarrow Aa)$	*US and C*
$\varnothing$	(2)	$\Box((x)Ax \rightarrow (Ea \rightarrow Aa))$	*Necessitation*
$\varnothing$	(3)	$\Box(x)Ax \rightarrow \Box(Ea \rightarrow Aa)$	*Modal Distribution*
4	(4)	$\Box(x)Ax$	*P*
4	(5)	$\Box(Ea \rightarrow Aa)$	(3)(4) *T*
4	(6)	$Ea \rightarrow \Box(Ea \rightarrow Aa)$	(5)*C*
4	(7)	$(x)\Box(Ex \rightarrow Ax)$	(6) *UG*
$\varnothing$	(8)	$\Box(x)Ax \rightarrow (x)\Box(Ex \rightarrow Ax)$	(4)(7)*C*

In short, the apparent disagreement regarding the validity of [CBF] can be found in the different ways in which '□' is being understood. And once it is seen that the valid principle involving weak necessity can be proven in **M**, the objection seems to lose its force.

Konyndyk's Argument for Interpreting Box as Meaning Essentially

Konyndyk has given an argument for treating '□' as meaning 'essentially' (as opposed to meaning 'necessarily'). Since reading '□' as meaning 'essentially' comes to giving it the weak sense of necessity, this argument functions as a kind of argument giving the box operator the weak sense. It is thus worthy of our consideration.

Here is how the argument is given:

> [W]e have been reading (interpreting) □ as meaning "necessarily" or "it is necessarily true that." But this reading will not do when we apply the □ to predicates (or to symbolic representations of predicates). Reading $(x)\Box\mathscr{A}x$ as "for anything *x*, *x has* $\mathscr{A}$ is a necessary truth," is unsatisfactory. For one thing, the expression '*x* has $\mathscr{A}$' does not express a truth (or a falsehood), let alone a necessary one. So how can an operator like □, as we have been reading it, be sensibly attached to it? The answer seems to be that it cannot. (Konyndyk 1986: 83)

Konyndyk goes on to say that he favours reading '□' as meaning 'essentially'.

Let us investigate the above reasoning. We begin by taking '□' to mean 'it is necessarily true that' (we can take that to mean 'it is necessarily the

is quite different from the way I have been using the term. For example, Konyndyk says that a proposition is weakly necessary if it is true in every world in which it (the proposition) exists (Konyndyk 1986: 63).

case that'). Then it is objected that when we apply the □ to predicates, we get something unsatisfactory. In particular, we end up reading '$(x)\Box\mathcal{A}x$' as 'for anything x, *x has* $\mathcal{A}$ is a necessary truth'—something that is clearly unsatisfactory.

But if '□' is applied to predicates, why must we read '$(x)\Box\mathcal{A}x$' as 'for anything x, *x has* $\mathcal{A}$ is a necessary truth'? Why can we not read '$(x)\Box\mathcal{A}x$' as, for example, 'for anything x, it is necessarily the case that *x has* $\mathcal{A}$'? Konyndyk has not given us any substantial reason for accepting the above crucial step in his reasoning. Indeed, his reasoning seems to involve an unwarranted shift. It begins by taking '□' as meaning 'it is necessarily the case that' (the operator sense); then after it is asked whether the box can be applied to predicates, it takes the box to mean 'is a necessary truth' (the predicate sense). No wonder we get something unsatisfactory. But the unsatisfactory result seems to be due to Konyndyk's unwarranted shift. I conclude that, if we take Konyndyk's reasoning above as an objection to using the strong sense of necessity in quantificational modal logic, then it is unconvincing.

Why Give Box the Strong Sense?

Still, the reader might wonder if I have any reason for preferring not to use '□' as weak necessity. Suppose that we took '□' to have the weak sense. Then clearly the following sentence would be valid:

$$\Box(Fa \to (\exists x)Fx).$$

But '$\Box Fa \to \Box(\exists x)Fx$' would not. Hence, a fundamental law of standard S5 modal logic, namely, modal distribution, would not be valid. True, a restricted version would hold.[44] But I prefer having the full unrestricted version of the law, since such a system would be compatible with standard systems of the modal sentential calculus.

The Jager Objection

Consider now an objection that Jager raises to Kripke's system, which might be thought to be applicable to **M**. Jager considers the possibility that one could revise Kripke's system so that it incorporated Plantinga's version of applied semantics. But even if we construed the possible worlds semantics of Kripke's system in the way advocated by Plantinga, the truth conditions of the system are such that '□' has the strong sense of necessity, so that the formula '$\Box F$' must be understood as standing for the property being F in every world. (As noted above, the truth conditions of **M** coincide with Kripke's in this regard.) Now the formula

$$(\exists x)\Box Fx \to \Box(\exists x)Fx$$

[44] See Jager 1982: 343.

will still be invalid in such a system. But it ought to be valid, says Jager, since if there is something that has *F* in every possible world, then in every possible world, this thing must exist and have *F*.[45] For this reason, Jager concludes that such systems possess "an irremovable defect".[46]

But why does Jager believe that, if there exists some object that has *F* in every world, then, in every world, this object must exist and have *F*? The reason is to be found in his Serious Actualistic perspective. From that point of view, if an object has a property in a world, then that object must exist in that world. Certainly, given the basic assumptions of the Serious Actualist, Jager's principle must be accepted.

But can we not accommodate this perspective without changing the definition of 'true under an interpretation' or making some drastic change in the semantics of the system? Logical systems such as **M** have been devised to allow a variety of metaphysical views. There is nothing preventing a Serious Actualist using **M** from declaring that, in this system, the n-place primitive predicates are to express genuine "properties" that can only be satisfied in a world by things that exist in that world. The Serious Actualist can then assert, as a principle of her system, all sentences of the form:

$$\Box(x_1)\Box(x_2)\ldots\Box(x_n)\Box(\theta_n x_1 x_2 \ldots x_n \rightarrow (Ex_1 \,\&\, Ex_2 \,\&\ldots\&\, Ex_n)),$$

where θ_n is an *n*-ary predicate of **M** ($n = 1, 2, 3, \ldots$).

Following Kit Fine, we could call the above a "Falsehood Principle".[47] Since the considerations that support various Falsehood Principles are philosophical and open to dispute,[48] there are advantages, *at this part of the work* at least, in not guaranteeing the truth of such principles by the way we set up our semantical rules. (In Chapter 7, I shall set forth a variation on **M** that conforms to both Actualism and Serious Actualism.)

The Plantinga–Konyndyk Ontological Objection

There does remain, however, the most formidable of the objections that these Actualists have raised to the sort of system I presented in Section 1 of this chapter. This is the sort of objection, discussed earlier, which was suggested by Plantinga in his critique of the "canonical conception". Konyndyk puts forward a version of this objection by claiming that such systems of modal logic as the one I gave earlier cannot express the truths

> There could have been an object distinct from every actual object

and

> There could have been more objects than there are,

[45] Jager 1982: 342. [46] Jager 1982: 342.

[47] Fine 1981: sect. 3. The name of the principle is derived from the implication of the principle that each primitive predicate is false of the non-existents in each world.

[48] See Pollock 1985: sect. 2.

without having both to presuppose non-existent objects and also to ascribe properties to objects in worlds in which they do not exist.[49] In other words, expressing the above truths with sentences of the sort of system I gave earlier seems to require violating both Actualism and Serious Actualism.

It may seem that one can easily express such truths as the above if one makes a small addition to the vocabulary of the modal language. For example, we can expand the vocabulary of **M** to include an actuality operator '$\mathcal{A}$'. Thus, if σ is a sentence of the new expanded language $\mathbf{M}^{\#}$, then $\mathcal{A}\sigma$ is also sentence of $\mathbf{M}^{\#}$. The definition of 'truth at a world under an interpretation' can then be revised to include another clause that reads:

(h′) If ϕ is the sentence $\mathcal{A}\sigma$, then ϕ is true at w under **I** iff σ is true at **w*** under **I**.

The first of the above problematic sentences can then be expressed in $\mathbf{M}^{\#}$ as:

$$\Diamond(\exists x)\text{–}\mathcal{A}E^{1}x$$

It is clear, however, that although we can indeed express the above truth in our new language, we will not have avoided the conflict with Actualism. For the above formal sentence is true under some interpretation **I** = <**W**, **D**, **w***, **d**, **r**, **e**> iff, for some member of **W**, say w, there is an element of **d**(w) which is not an element of **d**(**w***). Thus, according to the canonical conception, for the formal sentence to be true, there would have to be some possible object that did not actually exist.

This is a deep objection to which any proponent of possible worlds semantics should respond. My own response will be given only much later in this work (in Chapter 8), after I have provided a great deal of logical development.

This last objection is closely related to another doctrine that these Actualists have adopted. This is the view that the Plantinga–Jager Actualistic system is *superior on ontological grounds* to such systems as the one put forward in Part 1. In Chapter 3, I shall examine in some detail the ontological underpinnings of the Plantinga–Jager system.

[49] See Konyndyk 1986: 117.

2

Transworld Identity

In this chapter, I take up various attempts to state *the problem of transworld identity*. I then explore various responses to such statements. The responses to be considered are: Graeme Forbes's propagule criterion of transworld identity, Alvin Plantinga's state of affairs semantical analysis of modality, and David Lewis's counterpart-theoretic analysis of modality. A sketch of a version of counterpart-theoretic possible worlds semantics for a modal quantificational logic will be given to provide the reader with a basis for comparing the counterpart-theoretic possible worlds semantics of modal logic with the more standard system discussed in the previous chapter—this to further our investigation into the semantics of modal logic.

The problem of transworld identity is relevant to the ontological doctrine known as Modal Realism. As we shall see in this chapter, both the full-blown version of Modal Realism advocated by Lewis and the ersatz version promulgated by Plantinga allow attractive responses ("solutions") to the problem. I suspect that Lewis was led to his counterpart-theoretic version of Modal Realism by his researches into this problem. Indeed, one might argue that the fact that Modal Realism allows such nice responses to the problem is itself a strong argument for adopting some version of Modal Realism. However, to see the force of such an argument, one needs to appreciate just how difficult it is to meet the challenge posed by the problem in the straightforward way the anti-realist Forbes does.

1. The Problem of Transworld Identity

Let us reconsider the semantics of **M**. According to the semantical definitions of Chapter 1, given an interpretation $\mathbf{I} = \langle \mathbf{W}, \mathbf{D}, \mathbf{w}^*, \mathbf{d}, \mathbf{r}, \mathbf{e} \rangle$, with w and $w^{\#}$ elements of **W** and $w \neq w^{\#}$, something in $\mathbf{d}(w)$ could also be in $\mathbf{d}(w^{\#})$. That is, according to the canonical conception, a possible thing could exist in distinct possible worlds. This is thought by some philosophers to give rise to "the problem of transworld identity". The problem seems to be that of giving sense to (or making sense of) *the same thing* *being in two distinct possible worlds*. The classic statement of the problem is generally thought to be that of David Kaplan's in Kaplan 1979. Kaplan supposes that we are faced with the problem of determining whether 'x is bald' is true of Bob Dylan in different possible worlds. But this seems to give rise to the problem of *identifying* Bob Dylan in some other world. How do we do that? Kaplan writes:

> I'll even let you peep in at this other world through my Jules Verne-o-scope. Carefully examine each individual, check his fingerprints, etc. The problem is: which one, if any, is Bob Dylan? (Kaplan 1979: 93)

When he wrote those words, Kaplan considered this problem to be "the central problem of philosophical interest in the development of intensional logic" (Kaplan 1979: 94).

What would it be for Bill Clinton to be in another possible world? After all, Bill Clinton need not have, there, the properties and features he possesses in the actual world. He need not be married, or over six feet tall, or President of the United States. He need not be living in the same country he does in the actual world; he could even be living on a different planet in a different solar system. He might even be living in a completely different time period from the one in which he actually lives. In that world, he need not be anything like he in fact is. Perhaps he might not even be male. Indeed, some have thought that he need not even be a human being.[1] But in that case, what would it mean to say that that strange thing in the other possible world *is* Bill Clinton? What is the *content* of that transworld identity statement? As Michael Loux sees it, "since we have no criteria of identity and difference across worlds, talk of identity in this context can only be incoherent" (Loux 1979: 39).

The problem seems to arise most critically in the case of *de re* modal sentences. In order to determine if some *de re* sentence, say '$\Box F^1 a$', is true under an interpretation, it would seem that we would need to know that, for every world w, $\mathbf{r}(a)$ is in the extension of F^1 in w. And how could we know that, if we have no criteria for deciding, for any object x in w, whether or not x is $\mathbf{r}(a)$?

Jaakko Hintikka seems to be posing what is essentially the same problem by noting that his own account of modal notions "presupposes that we can make what might be called cross-identifications, that is to say, identifications across the boundaries of possible worlds". For Hintikka, this requires having a "method of *individuation*" that can be applied across possible worlds. Since quantification into modal contexts often makes sense, he concludes that "we obviously must have as a part of our normal

[1] See Lewis 1986.

conceptual structure such method [sic] of individuation" (Hintikka 1969: 169). He does not worry about just what such a method must be, since he does not believe that it is the job of the logician to characterize such things.

Quine puts his own twist on this problem to attack quantified modal logic. I shall give Quine's objection in terms of a *tense logic* that is taken from Forbes (Forbes 1985: ch. 1, sect. 3). The syntax of the language of this version of tense logic will have the temporal operators **F**, **P**, and **A**, instead of the modal operators, but otherwise it will be identical to **M.** An interpretation of this language will be an ordered septuple:

$$\langle \mathbf{T}, \mathbf{D}, \mathbf{t}^*, \mathbf{d}, \mathbf{r}, \mathbf{e}, \mathbf{P} \rangle,$$

where the first six elements are specified as they were in the case of modal logic. However, according to the canonical conception, **T** will be regarded as the set of moments of time instead of the set of worlds; and **t*** will be taken to be the present moment instead of the actual world. The seventh item is specified to be a binary relation among the elements of **D** that is a linear ordering of **D** (that is, **P** must satisfy the trichotomy, anti-reflexive, and transitive laws). **P** is called the 'precedent relation' and, according to the canonical conception, it tells us what moments of time precede what moments of time. The truth conditions (at a time under an interpretation **I**) for the temporal operators are given as follows:

For any sentence ϕ,

> $\mathbf{F}\phi$ is true at t under **I** iff there is some $t^{\#} \in \mathbf{T}$ which is such that $\mathbf{P}(t, t^{\#})$ and ϕ is true at $t^{\#}$ under **I**;
>
> $\mathbf{P}\phi$ is true at t under **I** iff there is some $t^{\#} \in \mathbf{T}$ which is such that $\mathbf{P}(t^{\#}, t)$ and ϕ is true at $t^{\#}$ under **I**;
>
> $\mathbf{A}\phi$ is true at t under **I** iff every $t \in \mathbf{T}$ is such that ϕ is true at t under **I**.

F is called the 'future tense operator', **P** is called the 'past tense operator', and **A** is called the 'all-times operator'.

Corresponding to the problem of transworld identity, there should be, one might think, a problem of transtemporal identity. Take some sentence 'W^1a' which we interpret as saying that Bill Clinton is in Texas. Suppose this sentence is true at **t***. Now consider the *de re* sentence '$\mathbf{F}W^1a$'. To determine if this is true at **t***, we need to be able to identify, at future moments, momentary objects as Bill Clinton (i.e. Bill Clinton at moment *m*). We can do this, Quine thinks, because such momentary objects "are declared to be stages of the same body by considerations of continuity of displacement, continuity of deformation, continuity of chemical change".[2] One cannot understand transworld identity in these terms, however, because (he writes) "you can change anything to anything by easy stages through some connecting series of possible worlds". The point is amplified by the following:

> The devastating difference is that the series of momentary cross-sections of our real world is uniquely imposed on us, for better or for worse, whereas all manner of paths of continuous gradation from one possible world to another are free for the thinking up.[3]

One idea that lies behind Quine's reasoning can be detailed along the following lines. Take the momentary objects that exist at some moment *m*, and call the mereological sum of these momentary objects 'the momentary world at *m*'. Then, the momentary worlds can be seen to be linearly ordered by the precedent relation; so we have a situation like that of a movie film in which one can have continuity of displacement, continuity of distortion, and continuity of chemical change. But in the case of modal logic, there is no ordering of the possible worlds that will do for transworld identity what the precedent relation does for transtemporal identity. It is as if the movie film was cut into individual frames, so that any frame could be placed next to any other frame.

2. Forbes's Solution

The Quinean Challenge

Forbes regards Quine's argument as amounting to a sort of "challenge" to practitioners of quantified modal logic to provide *criteria of transworld identity* that are as good as those Quine puts forward for transmoment identity (i.e. continuity of displacement, continuity of deformation, and continuity of chemical change). Forbes believes that he can meet this challenge by showing that there are essential properties, possessed at least by certain sorts of entities, that furnish us with non-trivial conditions of transworld identity of these entities. Forbes's response to Quine's challenge can be seen as yielding a kind of solution to the problem of specifying the content of transworld identity statements: if one has a view of meaning or content that some Wittgensteinians and Dummettians have advocated, according to which the meaning of statements or sentences is linked to *criteria*,[4] criteria of transworld identity would, on this

[2] Quine 1976: 860.
[3] Quine 1976: 861.
[4] See Canfield 1986 for a sampling of what some advocates and critics of this view have argued.

view, furnish us with at least one dimension of the content of transworld identity statements.

The Principle of Grounding

The basis upon which Forbes attempts to build his criteria of transworld identity for biological organisms is his *Principle of Grounding.*

> [I]t is part of the content of our concept of identity, whether transworld or transtemporal, that there are no ungrounded facts about such identities.[5]

What is being asserted by this principle? What, in particular, is meant by a 'grounded fact about identity'? What is it for a fact about identity to be grounded or ungrounded? As a first approximation, we can answer that the principle rules out the existence of bare identities and differences: for each instance of an identity or failure of identity, *there must be some fact (distinct from the mere fact of the identity or difference) in virtue of which that instance obtains.*[6]

Some idea of what is being asserted can be gleaned from the considerations that Forbes musters in support of the principle. Imagine a game that is played on a chequered board that is no different from that upon which chess is played; that is played with the same pieces that are used in playing chess; that is played according to the same rules according to which chess is played; and so on; but that is not chess! That would be, Forbes would contend, an ungrounded difference—an ungrounded difference that simply could not obtain. And such a difference would be ruled out by the Principle of Grounding.

Forbes asks us to consider the supposition that there is a steel tower in the place the Eiffel Tower is actually located. According to this supposition, this steel tower is not the Eiffel Tower, yet it is made of the same material out of which the Eiffel Tower is made, it has the same design and designer, even the same history. The only respect in which this tower is supposed to differ from the Eiffel Tower is in being a different thing. Forbes concludes: "The extent to which such a difference seems unintelligible is some measure of the plausibility of the view that transworld identities must be grounded."[7]

Another example that Forbes gives in support of his principle is that of the splitting of an amoeba. We do not believe, he claims, that the amoeba that splits is identical to one of the amoebae that results from the split, for we cannot cite any feature in virtue of which the splitting amoeba is identical to one of the resulting amoebae and not to the other. Examples of this sort are supposed to convince us that there cannot be a bare transworld identity: there must be something in virtue of which the things in the different worlds are identical.

[5] Forbes 1985: 128. [6] Forbes 1985: 130. [7] Forbes 1985: 128.

Two Principles of Set Identity

The Principle of Grounding is used to justify two principles regarding set identity. The first, which Forbes calls 'Membership Rigidity' or '(MR)', states:

For any possible set Σ and any possible object α, if there is a world in which $\alpha \in \Sigma$, then in every world W, if Σ exists in W, α is a member of Σ in W.

In short, a set will always have the same members in any world in which it exists. The second principle, called 'Crossworld Extensionality' or '(CE)', states:

For any possible sets Σ and Θ, and for any possible worlds W and W*, if Σ has the same members in W that Θ has in W*, then $\Sigma = \Theta$.

With these two principles, Forbes believes that he could refute Quine's claim that anything in one world can be changed to anything in another world by easy stages. In the case of sets, at least, we would have a criterion of transworld identity, namely, having the same members. Forbes allows that, for the case of sets of urelements (non-sets), we would need to apply a different criterion of transworld identity to determine when these urelements are identical across worlds, but, he claims, "however that should go, the question about the transworld identity of the original sets . . . has been completely resolved" (Forbes 1985: 123).

One wonders whether one could give the sort of "refutation" that Forbes gives by arguing that we have a criterion of transworld identity of natural numbers: natural number α in world W is identical to natural number β in world W* iff either α has the same predecessor in W that β has in W* or else neither α nor β have predecessors in their respective worlds. The problem is that our notion of predecessor seems to be purely structural. We have the idea of a totality of objects, the natural numbers, a distinguished object, zero, and an ordering relation on these objects, the predecessor relation, which satisfies certain fundamental laws (say, the five Peano axioms). But this grasp of these ideas is not sufficient to determine whether or not the natural numbers are, say, the von Neumann finite ordinals. Thus, it is questionable that the above "criterion of transworld identity" of natural numbers gives us a way of picking out the natural number two in another possible world. If the problem is to develop a "way of telling" whether an arbitrary object is or is not the number two, then the above is not a genuine criterion: It does not even allow us to determine whether or not two is the set $\{\varnothing,\{\varnothing\}\}$.[8] Similarly, many, if not most,

[8] I have learned from Forbes (personal communication) that he would prefer to describe the problem as that of giving an account of what the transworld identity of natural numbers consists in, rather than the problem of developing a way of telling whether an object in a world is or is not identical to some natural number. Thus, put in Forbesian terms,

philosophers of mathematics would question whether our grasp of the membership relation extends beyond the structural.[9] If not, then Forbes's criterion of transworld identity of sets does not yield a genuine criterion, which would tell us whether an arbitrary object in another world is or is not identical to some particular set.

The Necessity of Origin

In any case, I doubt that Quine would find Forbes's refutation very damaging to his position. I doubt that he had in mind abstract entities when he argued that anything can be changed to anything by easy stages. Much more impressive would be a refutation that produced criteria of transworld identity of physical objects of some sort. Realizing this, Forbes argues in support of the following analogue of (MR), which he calls '[K]' (in honour of Kripke, the person Forbes credits with first putting forward this principle):[10]

$$\Box(x)\Box(y)\Box(\text{Prop}(x,y) \rightarrow \Box(E^1 y \rightarrow \text{Prop}(x,y))),$$

where 'Prop(x,y)' abbreviates 'x is a propagule of y'.[11] The term 'propagule' is coined by Forbes to refer to the antecedent entity from which a biological organism develops: he tells us that an oak tree's propagule is its acorn and a human's propagule is its zygote (the fertilized egg from which the human developed). Thus, principle [K] implies that, in any possible world w in which Bill Clinton exists, he must have developed from that very zygote from which he developed in the actual world. Similarly, [K] implies that, if an oak tree actually developed from some acorn x, then in every possible world in which that oak tree exists, it will have developed from x. Principle [K] has come to be known as "the Necessity of Origin".[12]

Consider now the reasoning by which Forbes arrives at his principle. Suppose that, in the actual world w_1, there is an oak tree (which I shall call

the above point can be stated: the above principle about natural numbers does not give an acceptable account of what the transworld identity of natural numbers consists in.

9 As we shall see in the next chapter, there are other grounds for questioning the depth of our grasp of the membership relation.

10 Forbes's justification for [K] is clearly similar to one given by Kripke for his analogous principle. For a penetrating analysis of Kripke's attempted justification, see Salmon 1981: ch. 7.

11 In a later work (Forbes 1986), Forbes allows that there are some plausible counterexamples to this principle (of the "science fiction" variety). See Forbes 1986: 7. How Forbes attempts to revise [K] so as to immunize his essentialism from such counterexamples need not concern us here, since his revisions do not affect the logic of his justification of [K].

12 If one distinguishes essentiality from necessity, as was done in the previous chapter, the doctrine should be called "the Essentiality of Origin" K (as Forbes has noted in Forbes 1986).

FIGURE 1

'Ollie'), which grew up from some acorn **a** that was planted near a building. Now suppose that, in world w_2, a different acorn, **b**, which is identical in genetic make-up to **a**, is planted in exactly the spot that **a** was planted in the actual world; and suppose that **b** develops into an oak tree that is, in every apparent respect, except origin, identical to Ollie: the tree has a trunk, branches, and leaves which are identical in shape, colour, texture, functioning, etc. to the trunk, branches, and leaves, respectively, of Ollie's. The only difference between Ollie and this tree in w_2 seems to be the difference in origin. (See Fig. 1.)

Then question (1) arises: can the mere difference in origin be sufficient for us to conclude that the tree that developed from **b** in w_2 is *not* identical to Ollie? A philosopher sceptical of the Necessity of Origin (henceforth to be called 'the sceptic') will deny this. She will be willing to assert:

> [S] This tree in w_2 that developed from **b** is identical to Ollie, even though Ollie came from acorn **a** in w_1.

Forbes believes that the sceptic will be forced into the position of having to defend an ungrounded difference.

To see this, Forbes asks us to consider a world, w_3, which is exactly like w_2 except for the fact that, in w_3, acorn **a** is also planted some distance from the spot acorn **b** is planted, **b** being planted in w_3 in exactly the spot it was planted in w_2, and **a** then develops into an oak tree which is also very much like the actual Ollie. (See Fig. 2.)

Then question (2) arises: is the tree that develops from **a** in w_3 identical to Ollie? Suppose that an affirmative answer is given. Then the sceptic will be forced to maintain that the tree that came from **b** in w_3 is not identical to the tree that came from **b** in w_2, since by [S] that tree in w_2 is identical to Ollie. But in that case, Forbes argues, we would have a difference that is not

FIGURE 2

grounded in any *intrinsic* differentiating features—features that are causally relevant to the identity of the tree—for the tree that came from **b** in w_3 is, in all relevant respects, identical to the tree that came from **b** in w_2, the only difference being the existence in w_3 of the tree that came from **a** (and surely that *extrinsic* fact cannot ground this difference). This, Forbes believes, is an absurdity that cannot be maintained: "How can what goes on concerning *another* acorn affect the identity of *this* tree?" (Forbes 1985: 141).

Suppose, then, that the sceptic offers a negative answer to question (2). We are to suppose this time that the tree that develops from **a** in w_3 is not Ollie. In that case, the sceptic will be asked to consider another world, say w_4, which is exactly like w_3 except for the fact that there is no acorn **b** (and hence there is no tree that has sprung from acorn **b**). This world is almost identical to the actual world: it seems to differ from the actual world only in so far as the acorn **a** has been planted some distance from the spot that this acorn has been planted in the actual world. (See Fig. 2.) Now the sceptic will be asked question (3): is the tree that springs from **a** in w_4 identical to Ollie or not? If an affirmative answer is given, then the sceptic is faced with the implication that this **a**-tree in w_4 is different from the **a**-tree in w_3 that seems to be in all relevant respects identical to it—the only difference being the existence in w_3 of the other tree that came from **b**, which is an *extrinsic* feature.[13] She would again be faced with a difference that is not intrinsically grounded. On the other hand, if she opts for denying that the tree that springs from **a** in w_4 is Ollie, then she runs into an equally un-

[13] Forbes 1985: 141. There, he argues that our concept of identity "does not permit its application to be extrinsically grounded".

promising implication.[14] Why should it be that the tree that has sprung from **a** in w_4 differs from the tree that springs from **a** in the actual world, since the only significant difference between the two is the place in which acorn **a** is planted in the two worlds? Again, this option forces the sceptic to accept a difference that is only extrinsically grounded.

To see how implausible it would be to deny that the **a**-tree in w_4 is Ollie, let us imagine this time a world w_5 in which the acorn **a** is planted in the exact spot it is planted in the actual world, but that after a little seedling has developed from the acorn, it is dug up and replanted at the exact location the acorn is planted in world w_4. We can suppose that the tree that grows up from this little seedling is in all apparent respects identical to the tree that develops from **a** in w_4. There seems to be no non-extrinsic ground for denying that this tree is Ollie. Since it is clearly possible that Ollie could have been transplanted when it was a little seedling to the spot in question, there must be such a possible world as w_5. But then the previous position would require denying that this tree in w_5 is identical to the tree in w_4. And the only difference between the two trees is where the acorn was first planted. Surely that small difference is not sufficient to ground the non-identity of the two trees. Thus, Forbes concludes, all the possibilities lead to unacceptable consequences when [S] is asserted. By reasoning of this sort, he attempts to convince his readers that [K] cannot reasonably be rejected.[15]

The Principle of Propagule-and-sort Indiscernibility

Even with [K], however, Forbes still does not have an answer to Quine's challenge: he has not yet provided us with a criterion of transworld identity. At most, he has given us an essential property that each biological organism possesses. To fill this gap, he puts forward another principle, which he calls 'propagule-and-sort indiscernibility':

> [PSI] For all worlds u and v, if x is an organism in world u with exactly the propagules $z_1, z_2, \ldots, z_k$, and y is an organism at world v with exactly the propagules $t_1, t_2, \ldots, t_j$ then x is the same organism as y iff (i) $\{z_n\} = \{t_m\}$; and (ii) the sort of x at u is the same as the sort of y at v.

[14] It should be noted that my version of this acorn argument differs slightly from Forbes's. In his version, the world w_4 is specified to be one in which the actual tree is the one that springs from **a**.

[15] Forbes's justification for his principle [K] may seem to have its source in Kripke's argument for his principle that if a material object has its origin from a "certain hunk of matter", then it could not have had its origin in "any other matter" (Kripke 1980: 114 n. 56). However, I have learned from Forbes (letter of 31 May 1995) that his acorn argument was developed quite independently of Kripke's. Still, the reader may find it enlightening to attempt to uncover the respects in which Forbes's argument mirrors Kripke's. For an excellent discussion of Kripke's argument, see Salmon 1981.

[PSI] gives necessary and sufficient conditions for a biological organism in one world to be the same entity as a biological organism in another world, so long as they are of the same sort: whether they are the same or different depends simply on whether they develop from the same propagules.

But what does it mean to say that *x* and *y* are *of the same sort*? And why is this condition on transworld identity dependent upon their being of the same sort? The latter question is answered by Forbes by appeal to what David Wiggins calls 'sortal concepts'.[16] These are concepts that determine the persistence conditions of a kind of thing: they mark the limits of change that a thing can undergo without going out of existence. These limitations, it is held, "can be regarded as mapping out the contingent properties of a single individual; these limitations make the sort to which a thing belongs, or the kind of substance it is, essential to it".[17] Thus, to say that *x* and *y* are of the same sort is to say that both *x* and *y* fall under the same sortal concept. However, Forbes admits that his principle is "vague" in so far as it is nowhere specified how the sort of an entity is to be determined. He brushes off the thought that, because of this vagueness in his principle, he has not come up with a satisfactory answer to Quine, claiming that we have here only a "boundary problem" which our concepts are simply not sufficiently sharply defined to settle.[18]

3. Doubts about Forbes's Solution

The Principle of Grounding and the Justification of [K]

Since Forbes's essentialistic solution to the problem is based upon the Principle of Grounding, philosophers assessing this solution will certainly analyse his use and justification of the principle. It is striking that the Principle of Grounding undergoes some subtle changes in Forbes's book. When the principle is first stated, there is no mention of intrinsic features. The principle is simply that there are no ungrounded facts about identity or differences. It was this principle that the example of the Eiffel Tower was supposed to illuminate and illustrate. However, when the principle was brought in to justify [K], it was strengthened to assert that no identity or difference could be merely extrinsically grounded. Later, in discussing Adams's example of the globes (which I discuss below), Forbes seems to require that assertions of identity or difference have even more than just intrinsic grounds.

But even granting Forbes his use of his various principles of ground-

[16] Wiggins 1980: ch. 3. See also Wiggins 1967: Pt. 2.

[17] Forbes 1985: 147.

[18] Forbes 1985: 148.

ing, is his acorn–tree argument in support of [K] really decisive? What can the sceptic say in reply to this complicated argument? Note, first of all, that this argument is supposed to generalize to all organisms that develop from propagules. So consider a world *u* in which a man develops from a propagule, **b**, that is different from the propagule, **a**, from which Bill Clinton develops in the actual world. Now **b** differs from **a** only in so far as the egg, **e**, that was fertilized by a sperm to form **b** differs from the egg, **e***, that was fertilized by this same sperm in the actual world to form **a**. However, **e** is genetically identical to **e***, so that, in *u*, the person that develops from **b** is genetically indistinguishable from the actual Bill Clinton. Furthermore, this person in *u* has the same parents as does the actual Bill Clinton; and he goes on to live a life that is indistinguishable from the life that Bill Clinton leads in the actual world. The sceptic declares that this person is Bill Clinton.

Now imagine Forbes attempting to apply his acorn–tree argument against this sceptic. We have, it would seem, the situation depicted in Figure 1; so the argument would proceed by describing the sort of worlds depicted in Figure 2. This would involve describing a world, say *v*, that is very similar to w_3: *v* is to be exactly like *u* except for the fact that, in *v*, a person develops from propagule **a**. But how could there be such a world? One and the same sperm cannot fuse with both **e** and **e***—at least, not if the biological laws of *v* are to be identical to those of the actual world. And if the biological laws of *v* differ from those of *u*, then *v* will not be exactly like *u* except for the additional fact about **a** described above. So one could not argue in the way Forbes did that the sceptic would be forced to accept a difference that is not intrinsically grounded.

Forbes has responded to the above argument by suggesting that "the very same sperm molecule" might be reconstructed and brought together to form the very same sperm that fused with **e**; in which case, this sperm could be "reused" to fertilize **e***.[19] There are several problems with this response. For one thing, one gets into the problem of establishing transworld criteria identity and difference of molecules, atoms, electrons, etc., since we would need some way of determining when some molecules are "the very same sperm molecules" as those used to fertilize **e**. There is an even more serious problem with this suggestion. If sperm could be reconstructed and "reused" in the above fashion, then surely propagules could also be so reconstructed and reused. Then we could have a case in which two organisms existing side by side could come from the very same propagule. There could even be two people who develop from the same zygote—something that should be impossible according to [PSI].

[19] Personal communication, 31 May 1995. This kind of case was raised by Tom McKay in McKay 1986. See Forbes's reply in Forbes 1994.

There is another way in which the sceptic could attack Forbes's reasoning. Forbes is an anti-Modal Realist.[20] As such, he does not believe in the existence of possible worlds. Presumably, then, his talk about possible worlds is not to be taken literally. Let us, then, try to reconstruct Forbes's reasoning in terms of counterfactuals instead of possible worlds. Thus, instead of considering a possible world *w* in which acorn **b** is planted in place of **a**, giving rise to a tree that looks exactly like Ollie, we should consider *what would have been the case had acorn* **b** *been planted instead of* **a**, *thereby giving rise to a tree that looks exactly like Ollie.* What difference does it make whether we frame the question of the identity of the **b**-tree in terms of such counterfactual situations or in terms of possible worlds? Framed in terms of possible worlds, the question of whether the **b**-tree is or is not Ollie seems to require a 'Yes' or 'No' answer, as does the question of whether the man in the straw hat at June's garden party yesterday was Tom. The question of whether this tree in world *w* is identical or not to the **a**-tree in w_2 seems to be simply a factual question of identity. However, when the kind of questions of identity Forbes poses to the sceptic are framed in terms of counterfactual situations, it is by no means clear that there ought to be straightforward 'Yes' or 'No' answers. Suppose that the sceptic takes the position that, as the above counterfactual situation is described, whether the tree that develops from acorn **b** would have been Ollie or not is simply not determined. It could be argued, with some plausibility, that the concept of *same organism* or of *same person* is a "this-worldly" concept, adequate for all practical purposes, given the general framework of regularities and natural laws within which the concept generally functions in the actual world, but simply not sharp enough to be definitely applicable or definitely not applicable for a whole range of cases when this general framework is removed, as it is when we try to apply the concept in other worlds. Thus, the sceptic could maintain then that, in the counterfactual situation described above, there is simply no fact of the matter as to whether or not the tree that develops from **b** would be Ollie. Of course, if one describes a counterfactual situation in which a tree develops from **b** and then stipulates that this tree is Ollie, it would not be surprising that this identity turned out to be ungrounded: one is simply stipulating that this identity holds in this situation. In short, arriving at ungrounded identities of this sort is not obviously a *reductio ad absurdum*, as the above reasoning presupposes. So there is still room for the sceptic to manœuvre without being forced to accept [K].[21]

[20] This aspect of Forbes's thought will be explored in Chapter 4.

[21] Forbes has suggested to me (in his letter of 31 May 1995) that he believes he could get a *reductio* of the sceptic's position by getting the sceptic to admit that, if **b** had been planted instead of **a**, then the resulting tree *could* have been Ollie. Unfortunately, the details of such a *reductio* were not communicated to me; so I remain sceptical.

Doubts About [PSI]

I find it interesting that Forbes does not even attempt to provide any justification for [PSI], as he did for [K]. Perhaps he believes that since [K] has been justified, it is reasonable to suppose that the Necessity of Origin will furnish us with both necessary *and sufficient* conditions for transworld identity. But reflection will make it evident to most readers that justification for [PSI] is sorely needed. Imagine a world *w* in which the propagule that had actually developed into Bill Clinton develops very differently because the laws of physics in *w* are different. Suppose that, in *w*, the person who develops from this propagule is nothing like the actual Bill Clinton either in personality or in physical characteristics. (We can imagine that this person looks and acts exactly like Saddam Hussein.) Must this person be Bill Clinton? What reasons do we have for thinking so? Surely we should not accept such a conclusion just on the basis of Forbes's belief in [PSI].

Has Forbes given us a Criterion of Transworld Identity?

But even if we subdue our doubts about [PSI], we will still want to know if Forbes has really given us a truly acceptable criterion of transworld identity (so that transworld identity is no more problematic than transtemporal identity). One response to Forbes's principle that comes to mind almost immediately is:

> The principle [PSI] only pushes the problem back one step; for now we need a criterion of transworld identity of propagules. How is it to be determined, one might well ask, whether any particular propagule in one world is or is not identical to a propagule in another world? Lacking such a criterion, we are in no better position to determine if any particular organism in one world is or is not identical to an organism in another world.

Branching Worlds

Forbes has anticipated this sort of objection, and he has given a sort of reply in terms of possible worlds that are identical to each other up to some time, at which point they branch from each other; that is, these are worlds that are regarded as being one and the same world up to some time, whereupon there is a deviation in the course of events in the respective worlds. His idea is that, for a pair of such worlds, *u* and *v*, the principles [K] and [PSI] "will enable all facts about transworld identity between *u* and *v* for objects which come into existence after the branch point, to be completely fixed by the content of the initial segment which *u* and *v* have in common".[22] Presumably, Forbes believes that the initial segment that is

[22] Forbes 1985: 151.

common to both worlds will provide a ground upon which the two principles can be applied to yield a criterion of transworld identity.

Forbes introduces his concept of branching worlds in the course of discussing Robert Adams's argument (in Adams 1979*a*) that there are "thisnesses" that are not reducible to "suchnesses". Let me explain this terminology. A "thisness", Adams tells us, is the sort of property traditionally called a 'haecceity'.[23] More specifically, a thisness is the property of being identical with an *individual*, that is, it is a property expressible by 'is identical to α', where α refers to a particular, such as a person, a physical object, or an event (Adams 1979*a*: 6). "Suchnesses", on the other hand, are supposed to be "purely qualitative properties", in a sense to be explained. The idea is that a *purely qualitative property* should not involve individuals, and hence should not be a property of the sort that involves being related to a particular (such as *being married to Hillary Clinton*). Adams suggests that one might try to capture the idea by claiming that a suchness is a property which would be expressed in a "sufficiently rich" language, without the aid of such referential devices as proper names, proper adjectives (e.g. 'Leibnizian'), indexical expressions, and referential uses of definite descriptions. Adams does allow as being purely qualitative the property of *being a homeowner*, which in some sense does involve being related to a particular thing—at least implicitly—but it supposedly does not involve being related to a particular explicitly in the way that the property of *owning the house at 111 Telegraph Avenue, Berkeley* does. As Adams classifies these properties, the latter is not a suchness, but the former is. Note that Plantinga's examples of world-indexed essences (such as *being a homeowner in the actual world*) would not be suchnesses, since the property in question involves the referential use of the definite description 'the actual world'.

Adams's thesis that there are thisnesses that are not reducible to suchnesses is not a trivial one. Leibniz, for example, supposedly held that all haecceities are conjunctions of simpler, logically independent suchnesses (Adams 1979*a*: 21). To justify his thesis, Adams describes a world *w* in which there are two qualitatively indiscernible iron globes, called 'Castor' and 'Pollux', that have always existed and that always will exist. That is all there is to this world. That these globes are everlasting is a contingent fact, however, so there must be a world in which one goes out of existence at some point. Let such a world be *u*. Similarly, there must be world *v* in

[23] Adams attributes the invention of this term 'haecceity' to Duns Scotus (see Adams 1979*a*: 6 for references). The reader should be warned that contemporary philosophers use the term 'haecceity' in several closely related but different ways. For example, Alan McMichael takes haecceities to be "properties which necessarily characterize single individuals but which are not reducible to general properties" (McMichael 1983*b*: 59). Thus, as McMichael uses the term, Adams's thesis that there are haecceities that are not reducible to suchnesses is trivial, if it is allowed that there are haecceities.

which the other goes out of existence at the same point. Then, Adams argues, there is nothing qualitative by means of which the eternal globe of *u* can be distinguished from the eternal globe of *v*.

Such examples are taken by Forbes to threaten his Principle of Grounding, because we appear to have an ungrounded transworld non-identity: he seems to think that there is no intrinsic ground for the non-identity of the eternal globe of *u* and the eternal globe of *v*. In order to save his Principle of Grounding from this sort of counter-example, Forbes introduces a relation between worlds which he calls 'branching'.

Before examining his branching conception, we should investigate the question of why Forbes feels a need to respond to Adams's example. Consider again the Eiffel Tower example that Forbes used to support his Principle of Grounding. We are supposed to have, in some possible world, a tower that is exactly like the Eiffel Tower in all respects: it is located in the exact spot the Eiffel Tower is located in the actual world, is even made of the identical metal that the Eiffel Tower is made of, etc. Notice that, in this example, the two towers are supposed to be made of the same metal—not the same kind of metal, mind you, but the *identical metal*. This is given as an example of an ungrounded non-identity. The absurdity of this supposition is supposed to support the Principle of Grounding. But notice that much of the force of this example depends upon the sameness of the metal of the two towers: different material, evidently, would have grounded the difference in towers.

Returning to the Adams example, notice that the Castor and the Pollux of world *w* are clearly made of different hunks of iron: after all, in *w*, they are not in the same place. Now this difference of material is an *intrinsic* difference between the Castor of *u* and the Pollux of *v*. Hence, one might think that this example poses no threat to Forbes's Principle of Grounding, since the difference in metal can ground the non-identity of the two globes. However, Forbes does not seem to think that this difference is a sufficient ground for the non-identity, and his reasoning is evidently tied to the fact that these hunks of iron are, by stipulation, qualitatively indistinguishable.[24] Thus, we have some reason for thinking that Forbes understands his principle to require more than what was originally indicated by his discussion of such examples as the one involving the Eiffel Tower. The Principle of Grounding seems to require that all identities or non-identities be, ultimately, in some respect or other, *qualitatively grounded*.

In any case, here is how Forbes responds to the Adams example.

[24] I have learned from Forbes (letter of 31 May 1995) that he regards the Adams example as posing a threat to his Principle of Grounding because, although one can ground the non-identity of the globes in the non-identity of the material, one has no way of grounding the non-identity of the material.

Arguing that the two worlds *u* and *v* branch from *w*, he attempts to ground the identities in terms of the branching relation. Since there is nothing qualitative in the worlds *u* and *v* by means of which one can ground the transworld identities in question, Forbes postulates a transworld link (branching) between each of the worlds and *w*, and he then claims that this link provides the ground that is needed.

I find the idea of branching worlds puzzling. Is not this talk of worlds branching from one another just metaphorical? How can two distinct worlds be one and the same up to some time? How can distinct worlds literally overlap? Forbes suggests that we can think of a world as a course of events. Perhaps we can, but the mere fact that we can think of a world this way does not imply that the world just is a course of events.

What does Forbes mean by 'possible world'? To put the question a bit differently, what is a possible world supposed to be for Forbes? If by 'possible world', Forbes means essentially what Lewis means by the term, then it would make no sense to say that two distinct worlds are literally identical up to some time and then branch.[25] On the other hand, if Forbes has in mind what Plantinga means by the term, then, as will become apparent, the problem of transworld identity should not even get off the ground. So, it would seem, Forbes must have in mind some third conception of possible world. But what is that? He never makes it clear.

Let us now consider in more detail Adams's argument for his thesis that there are thisnesses that are not *metaphysically equivalent* to any suchnesses.[26] Consider again world *w*. This is the world in which both Castor and Pollux exist eternally. Now each is a contingent being. So it must be possible for Pollux not to have existed at all. Thus, there must be a world in which just Castor exists. Let that world be u^*. Similarly, it must be possible for Castor not to have existed; and there must be a world in which just Pollux exists. Let that world be v^*. But these cannot be the same world, since the possibility that Pollux not have existed is not the same as the possibility that Castor not have existed. We cannot let the fact that we cannot tell which world is which dictate the metaphysical conclusion that $u^* = v^*$. Thus, the Castor of $u^* \neq$ the Pollux of v^*, even though the two worlds are qualitatively indistinguishable. Since the thisness of Castor ≠ the thisness of Pollux, we can conclude that *these two thisnesses cannot be metaphysically equivalent to suchnesses.* This, then, is how Adams hoped to establish his thesis that there are thisnesses that are not metaphysically equivalent to suchnesses.

Forbes sees this argument as posing an additional threat to his own

[25] See Lewis 1986: sect. 4.2.

[26] For any propositions ϕ and θ, ϕ is metaphysically equivalent to θ just in case ϕ is true at *w* iff θ is true at *w*, for every metaphysically possible world *w*.

Principle of Grounding. His reasoning seems to be this. With the worlds u^* and v^*, we seem to have a case of transworld non-identity (Castor in $u^* \neq$ Pollux in v^*) which is not grounded in any intrinsic features of Castor and Pollux at the relevant worlds u^* and v^*. Since neither u^* nor v^* shares an initial segment with w, it would seem that the branching device cannot be employed to ground the identity of the Castor of u^* with the Castor of w.

To obviate this potential example of an ungrounded transworld identity, Forbes fashions a new twist to his conception of branching worlds: he claims that the existence for all time of Castor in w is "a separable course of events" and, since u^* simply "consists in" this course of events, we can count u^* as a world branching from w.[27] Thus, Forbes implies, the identity of the Castor of u^* with the Castor of w is grounded by the branching of u^* from w.

I find this response to be unsatisfactory for several reasons. First, there are reasons for questioning Forbes's appeal to *separable courses of events* in order to count u^* as a world branching from w.[28] Second, Forbes has not provided the sort of intrinsic grounding that his Principle of Grounding requires. Recall that Forbes first argued that transworld identity statements must be *intrinsically grounded*, in the sense that, "whether or not the identity relation holds across worlds between x and y must be settled by intrinsic features of x and y at the relevant worlds".[29] But the kind of grounding he appeals to in the above reply to Adams is not one in which the non-identity is settled by the intrinsic features of Castor in u^* with those of Pollux in v^* (*one cannot determine if the Castor of u^* is identical to the Pollux of v^* just by examining the two candidate globes in their respective worlds*). What we have here is further evidence that something about the Principle of Grounding is so obscure that even Forbes loses track of what its content is.

But there are other serious problems with Forbes's reply. With his mysterious doctrine of branching, Forbes has moved to the position of maintaining that the identity between the Castor of w and the Castor of u^* is grounded in the fact that the course of events that u^* consists in is identical to the separable course of events that the Castor of w participates in. *But what grounds the latter identity*? What grounds the identity of the course of events that u^* consists in with the separable course of events that the Castor of w participates in? Obviously, one cannot ground that identity in the identity of the Castor in u^* with the Castor in w, since that would involve us in a rather tight circularity: that would be to ground the

[27] Forbes 1985: 151.
[28] See Mills 1991.
[29] Forbes 1985: 141.

identity of transworld individuals in the identity of transworld courses of events, while at the same time grounding the identity of the transworld courses of events in the identity of the transworld individuals.[30]

Doubts about Forbes's Criteria

Let us pause here to see where we are in our assessment of Forbes's attempt to forge a criterion of transworld identity that is at least as good as those we have for transtemporal identity. I began by questioning the acceptability of various principles upon which Forbes justifies his criteria of transworld identity. Then, I raised several objections to Forbes's Principle of Grounding, upon which some of his principles were based. I shall now investigate the very criteria of transworld identity that Forbes believes will answer Quine.

Let us start with Forbes's propagule criterion for the transworld identity of biological organisms. Forbes claims that every biological organism x "comes from" some organic antecedent, and that, if that antecedent is a "single thing", it would be the propagule of x.[31] We have been given a very vague definition or characterization of 'propagule', but despite this vagueness, we can judge that, if a propagule, itself, has a propagule, the criterion could be reapplied to the propagule of the propagule to give a partial answer to the above critic. This will then push the question back another step. Of course, if the point at which the worlds in question branch is sufficiently close to the time at which x comes into existence, then there may be no sensible question of what in both worlds is the propagule of the propagule of x. But what if the branch point is nowhere near the time at which x begins to exist? What if the branch point is a thousand years earlier than that? One might well begin to wonder if this process of reapplying [PSI] can ever come to a satisfactory conclusion.

What if a propagule comes from two things, as is the case of a human's propagule? The human's propagule, Forbes tells us, is her zygote, which is the fusion of the sperm and egg from which the human develops. So, to apply [PSI] to determine the transworld identity of a person, we will need a criterion of transworld identity for the sperm and egg.

Now a human sperm develops from a particular stem cell (known as a spermatogonium) found in the testes. Thus, given Forbes's vague characterization of 'propagule', it would seem that the spermatogonium is the propagule of a sperm that develops from it. Even if a particular sperm could be linked, in the actual world, to its particular spermatogonium, we would still be faced with the question of the transworld identity of these

30 See Yablo 1988: 333–4 for more on this sort of objection.
31 Forbes 1985: 133.

stem cells. This question, however, raises problems of another order entirely, since spermatogonia are continuously dividing, making two new cells, one of which is a stem cell, while the other, called a spermatocyte, goes on to develop mitotically into two secondary spermatocytes. Then, each of the two secondary spermatocytes continues through mitosis to form two spermatids, each of which becomes a sperm cell.[32] Thus, if a particular sperm *s* has developed in the actual world from a particular line of stem cells that originated from the spermatogonium *a*, then even if we could identify cell *a* in another world *w*, how are we to identify which secondary spermatocyte that comes off this line is the one that is the propagule of *s*? After all, there are *two* secondary spermatocytes that give rise to the four sperm, only one of which can be *x*. What in world *w* determines which of these secondary spermatocytes is the relevant one? Clearly, propagule identity will not be sufficient, since the two secondary spermatocytes have the same propagule. Besides, as was noted above, each secondary spermatocyte gives rise to two sperm. So even if we could identify somehow the crucial secondary spermatocyte, we would still be faced with the question of which of the two sperm is *s*. Obviously, one cannot use propagule identity to identify *s*.

This problem is related to another, perhaps more formidable problem. Whatever vagueness may infect the principle [PSI], it seems clear that the two sperm that develop from each of the secondary spermatocytes fall under the same sortal concept. Thus, applying [PSI] to the case of two sperm, *s* and *t*, that developed from a spermatagonium *e*, it would seem that since *s* and *t* have the same propagule, they would have to be one and the same thing! This absurdity cannot be avoided by insisting that sperm must be genetically identical to be "of the same sort"; for a pair of sperm that are genetically identical can, on rare occasions, form from the same secondary spermatocyte.

Forbes might try to avoid this absurdity by revising [PSI] so that the worlds *u* and *v* referred to in the principle are required to be different worlds. Will such a revision save the principle from the above sort of absurdity? Consider a pair of worlds *r* and *t* which are such that *s* and s^* come from the secondary spermatocyte *y* in *r* and that *y* exists in *t* and gives rise to the two sperm *z* and z^* in that world. Since *s* has the same propagule in *r* that *z* has in *t*, it follows by [PSI] that $s = z$; but by parity of reasoning, we also have $s = z^*$. Thus, we obtain the absurdity that $z = z^*$.

A similar problem seems to arise in the case of identical twins, since each twin has the same propagule as the other. Being genetically identical to each other, the twins are evidently "of the same sort". Applying [PSI] to these humans, we again seem to be able to conclude that the twins are one.

[32] Balinsky 1981: 49. See also Walbot and Holder 1987: 17.

Again, we cannot obviate the absurdity by restricting the principle to individuals from different worlds (as was made clear above).

The only way out of this absurdity, so far as I can see, is for Forbes to revise his account of what is to count as a propagule; so that, when identical twin humans develop from a zygote, the zygote is not a propagule of either twin. Suppose, for example, that the twins develop separately from the two cells that are produced by the first mitotic division of the zygote. Forbes could stipulate that, in such cases, each of these two resulting daughter cells are the propagules of the resulting twins. In that case, let *w* be a world in which identical twins develop in the above manner from the zygote that is the propagule of Bill Clinton in the actual world. This zygote in *w* would not be the propagule of either twin, so [PSI] would not require that the twins be numerically identical to each other. One remarkable consequence of such a revision, given Forbes's propagule principles, seems to be that it would be metaphysically impossible for Bill Clinton to be an identical twin (at least, by a natural process). This may seem very counterintuitive, but Forbes insists that our intuitions regarding such a possibility are just wrong.[33]

Consider a set of identical twins, say John and Ed O'Brien. Let us suppose that they developed from daughter-cell propagules in the manner described above. Given that their ontogenic development was actually as I have supposed, then is it metaphysically possible that they could have developed in another way, say, by developing from clumps of cells forming after the fifth mitotic division? Evidently not, if this revised version of Forbes's doctrine of transworld identity is to be believed; since such clumps of cells could not be identical to the daughter cells formed in the actual world at the second mitotic division.

But there are consequences of the revised view that are even more counterintuitive. Consider a possible world *w* in which the zygote, from which the O'Brien twins actually developed, implants in a different place in the uterus of the mother. That there is such a possible world can hardly be denied, since the place at which a zygote implants is definitely not an essential property of the zygote. Since there is no reason that this zygote should be identical to the actual one in every respect, we can imagine that its genetic makeup, in *w*, differs in some relatively small ways from how it is in the actual world. Furthermore, the mother herself could be slightly different in *w*, in ways that affect the rate of mitosis of the developing zygote. Now suppose that, in *w*, this zygote mitotically divides, resulting in daughter cells that separate and implant, thus starting a biological process

[33] In Forbes 1980, Forbes argues that an untwinned person could not have been an identical twin. It can be seen that he would also maintain that an identical twin could not have been an untwinned person.

which ends in the birth of identical twins that look very much like the O'Brien twins. Which (if any) of these two daughter cells is identical to the propagule (in the actual world) of Ed? Which of John? Is there any way of deciding? It would seem that there is not. Any such identity claim would appear to be an ungrounded one. Should we insist, therefore, that these twins are not the O'Brien twins? But what (intrinsic) grounds would we have for asserting that? Would we not have another ungrounded non-identity statement? Either way, we seem to have a counterintuitive position.

Finally, it should be noted that, even if Forbes's criterion of transworld identity of biological organisms were to be accepted, despite all its difficulties, this would still only supply us with the transworld identity of a minute portion of the things being quantified over in modal logic. What about the vast majority of things that are supposed to be in the range of the bound variables and for which we have been given no such criterion?

All in all, taking into account all the many difficulties that plague Forbes's account of transworld identity, the above path of attempting to answer Quine's challenge does not seem to be very promising.

4. Plantinga's Solution

Plantinga's approach to the problem of transworld identity is very different from that of Forbes. He makes no attempt to meet Quine's challenge. Rather, he sets out to show that the sorts of questions about transworld identity that give rise to the puzzles are either senseless or easily answered.[34]

Once Plantinga has developed his version of applied semantics of modal logic, he can make short shrift of the problem of transworld identity. How does Plantinga deal with this problem? In a sense, for

[34] Plantinga's response to the problem may bring to mind Kripke's highly suggestive comments in Kripke 1980 about the problem (see especially pp. 49–53). Some commentators have taken Kripke to be maintaining in these passages that the "problem" is only a pseudo-problem. This is suggested, in the preface to the paperback edition, by his description of the "spurious problems of 'transworld identity'", and by what he says about the dice example, when he suggests that he is doubtful that there is a genuine problem of transworld identity and that the demand for some criterion of transstate identity is "confused" (p. 17). It is noteworthy that he writes: "There are no special grounds to suppose that possible worlds must be given qualitatively, or that there need be any genuine problem of 'transworld identification'" (p. 19). Still, Kripke does not provide enough details for me to state with any confidence just what his "resolution" of the problem comes to. In this vein, Tony Roy has complained that "Kripke says surprisingly little about worlds, and what he does say is difficult to fit into a cohesive whole" (Roy 1993: 357). Also, there is no consensus regarding Kripke's claim that possible worlds are stipulated rather than discovered—see Hazen 1996: 165–6; and compare Salmon 1996, in which it is claimed that Kripke's contention "has been seriously misunderstood" (p. 203).

Plantinga, there are no possible worlds. There are, of course, things in Plantinga's ontology that he calls "possible worlds", but these things, we have seen, are abstract entities that are more like propositions than the concrete worlds that these propositions describe (and that would have existed if these propositions had been true). Hence, asking for the content of a statement of transworld identity is a bit like asking for the content of a statement that the Socrates of the proposition that *Socrates was married* is identical to the Socrates of the proposition that *Socrates took hemlock.* One wonders just what it is that is being asked.

We can characterize Plantinga's "solution" to the problem, roughly, in these terms. The problem of transworld identity depends upon a certain conception of possible worlds. We imagine a world distinct from the actual world in which some concrete thing, say a person, exists; and then we ask how can it be determined that the person is or is not identical to Bill Clinton. But this conception is simply mistaken. There is only one concrete world in which concrete things like real flesh-and-blood people exist. All the other "worlds" are mere abstract entities (states of affairs) and there is no such thing as looking into these worlds and finding some real person who may or may not be Bill Clinton.

Taking "possible worlds" to be Plantinga's maximal possible states of affairs, it would seem that one can no longer even describe a context in which it is sensible to ask: what is the content of the identity statement that *that* person in *that* possible world is identical to the Socrates who lived in the actual world? Thus, Plantinga writes:

> There is no such thing as "looking into" another possible world to see what is going on there. There is no such thing as inspecting the inhabitants of another possible world with a view to deciding which, if any, is Socrates. A possible world is a possible state of affairs.[35]

5. Lewis's Solution

Consider now how David Lewis deals with the problem of transworld identity. According to Lewis's counterpart theory of modality, there is (in some sense) no such relation as transworld identity: no thing exists in more than one possible world; each possible object is "world-bound".[36] So there is no problem of making sense of one thing in one world being

[35] Plantinga 1974: 96.

[36] In Lewis 1986: 69, Lewis qualifies his doctrine of world-bound possible things by allowing that, assuming that universals exist, they exist in many possible worlds. Lewis uses the term 'universals' for those things "that are wholly present as non-spatiotemporal parts

identical to a thing in a different world. According to Lewis, to say that George Bush could have decided to end the Gulf War a week later than he did, is not to say that in some possible world, George Bush decides to end the Gulf War a week later than he did in the actual world; rather, it is to say that, in some possible world, there is some person who is not identical to George Bush, but who is, in some appropriate way, very much like George Bush—so much so that he is a *counterpart* of George Bush—and this counterpart decides to end the Gulf War a week later than Bush did in the actual world.[37]

I shall leave for the following chapter a more detailed discussion of the metaphysical aspects of Lewis's counterpart theory. Here, I shall concentrate on logical matters. Lewis did not present his ideas on counterpart theory by developing a counterpart-theoretic possible worlds semantics for modal logic. Instead, he developed (in Lewis 1979*a*) his counterpart theory of modal reasoning within the framework of classical first-order logic. I shall present the details of this theory in a later chapter. Here, I shall continue my investigation of the possible worlds semantics of modal logic by sketching how Lewis's ideas on counterparts can be embodied in the semantics of a modal quantificational language.

It is easy to see how the main ideas of such a semantic theory can be abstracted from Lewis 1979*a*, since he tried to show how his counterpart theory and quantified modal logic are so "intertranslatable" that "every sentence of quantified modal logic has the same meaning as a sentence of counterpart theory".[38] I shall present, here, the principal features of a Lewis type counterpart-theoretic semantics of modal logic, based primarily on Forbes's account.[39] The syntax of this system will be identical to that of **M**.

Counterpart Modal Semantics

A counterpart-theoretic interpretation of the language will be an ordered septuple <**W**, **D**, **w***, **d**, **r**, **e**, **c**>. The clauses explaining the first six elements of this septuple differ from those of the corresponding elements of the septuple of Chapter 1 only in the following respect:

> **d** must be such that, for any w and $w^{\#}$ that are elements of **W**, if

in each of the things that instantiate some perfectly natural property" (Lewis 1986: 67). However, he remains neutral about whether there are universals.

[37] Evidently, Leibniz advanced the view that each individual possible thing exist in one and only one world but has counterparts in many worlds. See Konyndyk 1986: 115.

[38] Lewis 1979*a*: 113–14. Lewis makes clear, however, that not every sentence of his counterpart theory is the translation of a sentence of quantified modal logic.

[39] See Hazen 1979 and Forbes 1985: ch. 3, sect. 4.

$w \neq w^{\#}$, then **d**(w) and **d**($w^{\#}$) are disjoint. (Thus, no possible thing can exist in more than one possible world.)

As for **c**, we can regard it as the *counterpart* function that assigns to each possible thing x and each possible world w, the set of things that are counterparts of x in w. I thus add the clause:

> **c** is a function whose domain is the Cartesian product of **D** with **W** and whose range is a subset of the power set of **D**, subject to the following two constraints:
>
> (a) **c**(x, w) must be a subset of **d**(w),
>
> and
>
> (b) if $x \in$ **d**(w), then **c**(x, w) must be $\{x\}$.

[*Hence,* **c** *must be a function that takes as arguments ordered pairs whose first element x is a possible object and whose second object w is a possible world, and the value of the function for argument* <*x, w*> *must be a set of possible objects that exist in w (the "counterparts of x at w"), where each possible object that exists in w will have only one counterpart in w, namely itself.*][40]

I shall not give fully the definition of 'true at a world under an interpretation'. The truth conditions given in counterpart-theoretic modal semantics differ from those given in standard modal semantics only in the case of *de re* modal sentences. For my purposes, it will be sufficient to give the definition of 'truth at a world under an interpretation' for the case of a sentence that consists of a modal operator followed by an atomic sentence (this will give the essential idea of how the counterparts work).

To explain how this definition works, let us first see how one might have defined 'truth at a world under an interpretation' had our view of counterparts been simpler. Suppose that, for every object α and every world w, we could be assured that α had one and only one counterpart in w. We could then specify an interpretation in such a way that, instead of the **c** function of the above septuple, we would have a **g** function that takes elements of **D** × **W** to **D**. In other words, **g**(x, w) would be some element of **D**. We could then say that **g**(x, w) is the counterpart of x in w. Thus, if α is

[40] I have not defined the counterpart function in the way given in Forbes 1985; instead, I have attempted to capture Lewis's original views on counterparts. Thus, condition (a) was derived from Lewis's statement that "something has for *counterparts* at a given world those things existing there that resemble it closely enough in important respects"; whereas condition (b) corresponds to the condition expressed by the words, "anything is its own unique counterpart at its own world" (Lewis 1979*a*: 126). Later, I shall discuss Forbes's way of defining counterparts.

an individual constant, then $\mathbf{g}(\mathbf{r}(\alpha), w^{\#})$ would be the counterpart of the referent of α in $w^{\#}$. Then, we could say that the sentence

$$\Diamond\phi\alpha_1\alpha_2 \ldots \alpha_n$$

is true at w under **I** iff there is a $w^{\#} \in \mathbf{W}$ such that

$$<\mathbf{g}(\mathbf{r}(\alpha_1), w^{\#}), \mathbf{g}(\mathbf{r}(\alpha_2), w^{\#}), \ldots, \mathbf{g}(\mathbf{r}(\alpha_n), w^{\#}) > \in \mathbf{e}(\phi, w^{\#}).$$

Unfortunately, we cannot assume that every object has one and only one counterpart in every world. An object may have no counterparts in a world w, or it may have many such counterparts. In what follows, let ϕ be an n-ary predicate and let $\alpha_1, \alpha_2, \ldots, \alpha_n$ be individual constants (not necessarily distinct).

The sentence

$$\Diamond\phi\alpha_1\alpha_2 \ldots \alpha_n$$

is true at w under **I** iff there is a $w^{\#} \in \mathbf{W}$ and there is some y_1 that is an element of $\mathbf{c}(\mathbf{r}(\alpha_1), w^{\#})$, and some y_2 that is an element of $\mathbf{c}(\mathbf{r}(\alpha_2), w^{\#}), \ldots,$ and some y_n that is an element of $\mathbf{c}(\mathbf{r}(\alpha_n), w^{\#})$, such that $<y_1, y_2, \ldots, y_n> \in \mathbf{e}(\phi, w^{\#})$

The sentence

$$\Box\phi\alpha_1\alpha_2 \ldots \alpha_n$$

is true at w under **I** iff for every $w^{\#} \in \mathbf{W}$ and every y_1 that is a member of $\mathbf{c}(\mathbf{r}(\alpha_1), w^{\#})$, and every y_2 that is a member of $\mathbf{c}(\mathbf{r}(\alpha_2), w^{\#}), \ldots,$ and every y_n that is a member of $\mathbf{c}(\mathbf{r}(\alpha_n), w^{\#})$, $<y_1, y_2, \ldots, y_n> \in \mathbf{e}(\phi, w^{\#})$.

Truth under an interpretation **I** is then defined as truth at $\mathbf{w}^*$ under **I**.

Comment. The basic idea underlying the above truth definition for *de re* sentences can be gleaned from the following:

> '$\Diamond Fa$' is true under **I** iff there is a world u and there is some object x such that x is a counterpart of the referent of 'a' at u and $x \in \mathbf{I}(F, u)$. '$\Box Fa$' is true under **I** iff, for every world u, and for every object x, if x is a counterpart of the referent of 'a' at u, then $x \in \mathbf{I}(F, u)$.

A special feature of the above definition can be brought out by considering the conditions under which the sentence '$\Diamond Raa$' would be true. Given an interpretation $\mathbf{I} = <\mathbf{W}, \mathbf{D}, \mathbf{w}^*, \mathbf{d}, \mathbf{r}, \mathbf{e}, \mathbf{c}>$, the sentence would be true under this interpretation if, for some $w^{\#} \in \mathbf{W}$, there is a counterpart $c \in \mathbf{c}(\mathbf{r}(a), w^{\#})$, and there is a counterpart $d \in \mathbf{c}(\mathbf{r}(a), w^{\#})$—*where d may be a different object from c*—such that $<c, d> \in \mathbf{e}(R, w^{\#})$. In other words, in evaluating the truth of '$\Diamond Raa$' under **I**, we regard the two token occurrences of 'a' as "independent" so that a counterpart of the referent of the first occurrence may be different from a counterpart of the referent of the second occurrence. This unusual feature of the definition is taken from Forbes 1985: 63. I shall point out later how this feature of the definition yields some rather counterintuitive results and how these results can be avoided by adopting a definition of 'truth under an interpretation' that is closer in spirit to the views of Lewis 1979*a*. I shall attempt to explain later why Forbes adopted this unusual feature.

How truth values are determined can be illustrated by means of the following example. Let **J** be the interpretation given by Figure 3.

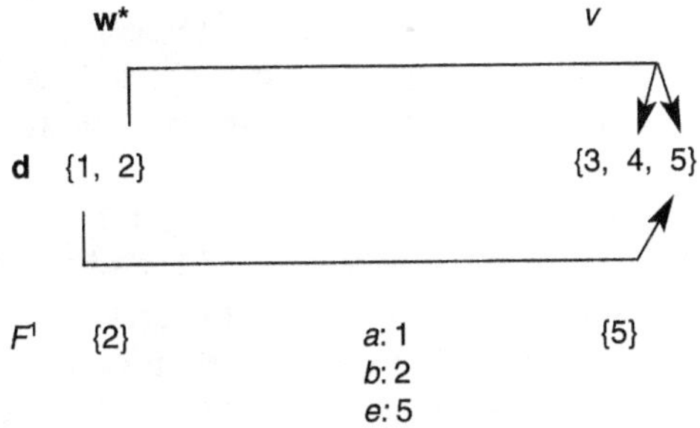

(The arrows indicate what the counterparts are.)

FIGURE 3

Then '$\Box F^1 b$' is false under **J** because not every counterpart of **r**(b) in v, in particular 4, is an element of **e**(F^1, v). On the other hand, '$\Diamond Fa$' is true under **J** because there is a world, namely v, and a counterpart of **r**(a) at v, namely 5, which is such that it is an element of **e**(F^1, v).

Some Apparent Anomalies

Now consider the sentence '$\Box F^1 e$'. Is this true under **J**? Yes. At world v, every counterpart of **r**(e) at v, namely 5, is an element of **e**(F^1, v); and at world **w***, it is trivially the case that every counterpart of **r**(e) at **w*** is an element of **e**(F^1, **w***), since there are no counterparts of **r**(e) at **w***. Surprisingly, although '$\Box F^1 e$' is true under **J**, '$F^1 e$' is not. Hence, not every sentence of the form '$\Box\phi \rightarrow \phi$' is valid. This is a very counterintuitive feature of this system. It shows that, in this counterpart-theoretic system, one cannot infer ϕ from $\Box\phi$. Yet, this inference is one of the very fundamental ones that one tries to capture in systems of modal logic, as is evidenced by the fact that the sentence '$\Box\phi \rightarrow \phi$' is taken to be an axiom of practically all systems of modal logic.

It may be thought that this strange feature of counterpart-theoretic modal logic stems from the fact that we allow every counterpart of 5 at **w*** to be an element of **e**(F^1, **w***) when there are no counterparts of 5 at **w***. That this is not so can be seen by simply specifying an interpretation **J′** which is exactly like **J** except for the fact that **J′** assigns a counterpart to 5 at **w***, namely 2. (The reader can obtain the appropriate diagram for **J′** by simply changing the line from 2 to 4 and 5 so that the end at 2 is made into an arrow pointing at 2.) It is evident that we can obtain this result because we determine whether '$\Box F^1e$' is true under **J′** by seeing whether all the counterparts of **r**(*e*) are in the extension of 'F^1e' at the different worlds, and we determine whether 'F^1e' is true under **J′** by seeing whether **r**(*e*) is in the extension of 'F^1' at the actual world. Since **r**(*e*) need not be identical to a counterpart of **r**(*e*) at the actual world, we can obtain the undesired result.

Perhaps a supporter of counterpart-theoretic modal semantics will argue that this problematic result arises only for cases in which we refer to "non-actual" objects (such as 5 in the above model) that exist in worlds other than the actual world and that such merely possible objects cannot either satisfy a condition or not satisfy a condition. Such an Actualistic position is not, however, open to David Lewis, since it is an important feature of his ontological views that there are non-actual objects (Lewis is a Possibilist).

It might also be replied that the above paradoxical example depends upon having an object (namely, 5) in one world *v* satisfying a condition ($-F^1$) in another world (w^*). But how can that be? Well, Lewis believes that an object in another world can be related to objects in this world. After all, a person in one world can be related by the counterpart relation to a person in another world. Also, Lewis believes that there are people in other worlds that have the same height or the same colour hair as people in this world.

An alternative defence can be mounted by noting that one can only obtain the sort of interpretation under which the sentence is false by assigning to the individual constant '*a*' an object that does not exist in the actual world. Hence, such embarrassing interpretations can be precluded by requiring individual constants to name only actual objects. This is a reasonable restriction on a language, it might be argued, since we cannot in fact name any object that does not exist in this world. This suggestion seems plausible, especially if we suppose that the "merely possible" objects are things that do not really exist. But if we adopt Lewis's metaphysical views according to which these "merely possible" individuals are not only real parts of concrete worlds, but they are just as real as the parts of our world, it is not at all obvious that we cannot name such things.

Another possibility: a counterpart theorist might allow that the above sentence is not valid, but then try to explain why we are inclined to think that '$\Box Fa \rightarrow Fa$' is valid, when it is not. The sentence seems to be valid, he might argue, because we only think of cases in which the name '*a*' denotes something that actually exists; and in all those cases, the sentence would be true. However, I do not think that many would find such defences very persuasive. The sentence '$\Box\phi \rightarrow \phi$' is the most basic of modal laws. If the necessity of ϕ does not imply the truth of ϕ, it may be argued, it is hard to see what meaning we should attach to 'necessity'.[41]

Another problem with the above system of modal logic concerns the existence predicate.[42] The sentence

$$\Diamond{-}E^1a$$

is true at a world under an interpretation iff there is some world w and some counterpart x of $\mathbf{r}(a)$ at w which is such that x is not a member of $\mathbf{d}(w)$. But any counterpart of $\mathbf{r}(a)$ at w must be a member of $\mathbf{d}(w)$, so there cannot be any such counterpart. It follows that the above sentence is unsatisfiable. In short,

$$\Box E^1a$$

is a valid sentence of this system. Everything "named" in the language would have to be, it would seem, a necessary being. Something seems to be drastically amiss.

To avoid this last feature of the semantics, Forbes has suggested that the counterpart function should be defined in such a way that an object x may have a counterpart y in a world w even though y does not exist in w, i.e. even though y is not a member of $\mathbf{d}(w)$. The idea is that for the case of worlds in which no counterpart of x exists, then x is itself a counterpart of x in that world.[43] We can import this feature into our semantics by revising our earlier specification of the **c**-function so as to read:

> **c** is a function whose value, for argument $<x, w>$, where $x \in \mathbf{D}$ and $w \in \mathbf{W}$, is a non-empty set of elements of **D** which is such that:

[41] It should be noted that in the counterpart theory of Lewis 1979, '$\Box\phi \rightarrow \phi$' is a theorem. The counterintuitive features being discussed above should not be attributed to Lewis's version.

[42] There are other counterintuitive features of counterpart-theoretic modal semantics that we should consider. Hazen points out that, in the type of counterpart-theoretic system we are considering, the sentence

$$\Box R^2ab \rightarrow \Box(\exists x)R^2ax$$

is not valid (Hazen 1979: 326–8). This is because there are interpretations in which $\mathbf{r}(a)$ has counterparts in worlds in which $\mathbf{r}(b)$ does not. In those cases, '$\Box(\exists x)R^2ax$' could be false, even when '$\Box R^2ab$' is not. However, it is questionable that this feature of the system should be regarded as counterintuitive, given that the above sentence is also not a valid sentence of **M** (remember, in **M**, $\mathbf{r}(b)$ could be in a relation at a world in which it does not exist).

[43] Forbes 1985: 61.

(i) if an object $y \in \mathbf{c}(x, w)$ and $y \notin \mathbf{d}(w)$, then $y = x$;
(ii) if $x \in \mathbf{d}(w)$, then this set is $\{x\}$.

By means of this expedient, one can indeed avoid the above problem and allow for contingent beings. However, this resolution of the difficulty seems to be deficient for philosophical reasons: it is based on a purely *ad hoc* stipulation that is at odds with Lewis's conception of how worlds represent possibilities. Lewis is definitely not inclined to accept Forbes's definition, writing: "What does it mean to say that the counterpart is 'at W' if not that, at W, the counterpart exists?" He then adds sardonically (in a footnote): "We might just *say* it, and not mean anything by it. That is Forbes's solution to our present difficulty . . ."[44]

A study of these problems suggests that our problems arise from an improper reading of the box operator. We have been giving '□' the strong reading of 'necessary', in which '□*Fa*' is true iff *a* has *F* at every world. Perhaps we should give box the weak reading, so that '□*Fa*' is true iff *a* has *F* at every world in which *a* exists.[45] Such a reading, although not the usual one, would certainly fit Plantinga's understanding of the operator.[46] With such a conception of the box operator, there would be no temptation to defend the validity of the formula '□*Fa* →*Fa*'. Furthermore, if we give the box the essentiality reading, the validity of '□E^1a' will seem perfectly understandable. Translating '□*Fa*' as meaning '*a* is essentially *F*', the sentence in question tells us that, at every world, if *a* exists, then *a* exists—the validity of which is perfectly reasonable.

Looking more closely at the truth definition for *de re* sentences, we see that there are good grounds for taking the box operator to have the essentiality reading. We have:

> '□*Fa*' is true under **I** iff, for every world *u*, and for every object *x*, if *x* is a counterpart of the referent of 'a' at *u*, then $x \in \mathbf{I}(F, u)$.

But we also have:

> '□(*Ea* → *Fa*)' is true under **I** iff, for every world *u*, and for every object *x*, if *x* is a counterpart of the referent of '*a*' at *u*, then if *x* exists in *u*, $x \in \mathbf{I}(F, u)$.

But any counterpart of the referent of '*a*' at *u* exists at *u*. Thus, '□*Fa*' is true under **I** iff '□(*Ea* → *Fa*)' is true under **I**.

It may be thought that giving the box operator this reading for *de re* sentences would result in the operator being ambiguous: it means one thing

[44] Lewis 1986: 10.
[45] This was suggested to me by Jonathan Turkanis.
[46] See Plantinga 1974: 58–9.

for *de re* sentences and something else for *de dicto* sentences. But a defender of Plantinga's way of translating the box can also adopt Plantinga's way of responding to the ambiguity charge: for both kinds of sentences, the box means 'it is essential that'. Thus, for the *de dicto* case, $\Box\phi$, it means that, at every world, if ϕ exists, then ϕ is true. Since, according to Plantinga, every proposition necessarily exists, we can infer that $\Box\phi$ is true iff ϕ is true at every world.[47]

Notice that, however we may explain (or explain away) the seemingly paradoxical nature of the fact that '$\Box Fa \rightarrow Fa$' is not valid, we are left with the fact that the counterpart-theoretic system we have sketched above is not, strictly speaking, an S5 modal logic (although it can still be classified as an "S5-type modal logic"). Once we give up the validity of '$\Box\phi \rightarrow \phi$', we abandon a basic law of S5 modal logic. Of course, this feature of the system need not be considered undesirable.

Problems Involving Identity

There is a decidedly odd consequence of the above counterpart-theoretic semantic definitions concerning identity. Suppose that '$a = a$' is an abbreviation for 'I^2aa'. I shall take the binary predicate 'I^2' to be a logical constant which stands for the identity relation, that is, at every world, 'I^2' must be assigned by the **e** function the extension that consists of the set of all ordered pairs of elements of **D** whose first element is identical to the second. One of the striking characteristics of this counterpart-theoretic modal system is that the sentence '$\Box a = a$' is not valid in it. Figure 4 will make that clear.

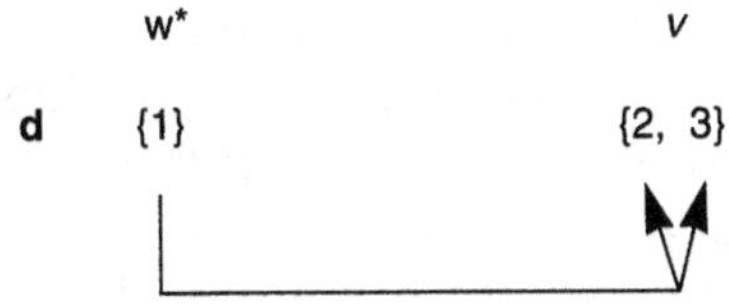

FIGURE 4

Since, according to this interpretation, there is a counterpart of 1 at *v* that is different from some counterpart of 1 at *v*, it is not the case that every counterpart of **r**(*a*) at *v* is identical to every counterpart of **r**(*a*) at *v*. It follows that the sentence '$\Box a = a$' is false under this interpretation.[48] Since

[47] See Jager 1982: 340 for a truth definition in terms of structures that incorporates this feature of Plantinga's views on necessity and essentiality.

[48] This sort of strange feature of this version of counterpart theory has been noted in Forbes 1985: 178. But nowhere in this work can I find any justification for giving the truth conditions as he does in a way that deviates from Lewis's ideas.

most philosophers believe that the above sentence is a necessary sentence of modal logic, many would find counterpart-theoretic modal systems of this sort to be unacceptable.[49]

However, the above difficulty is a result of Forbes's way of giving the truth condition for counterpart-theoretic modal semantics. There are ways of specifying such truth conditions—in a way that would yield a semantics closer in spirit to Lewis's original ideas—so that this result does not obtain: The clause for the case of $\Diamond\phi\alpha_1\alpha_2 \ldots \alpha_n$ can be specified as follows:

> [Revision 1] The sentence
>
> $$\Diamond\phi\alpha_1\alpha_2 \ldots \alpha_n$$
>
> is true at w under **I** iff there is a $w^{\#} \in \mathbf{W}$ and there is some y_1 that is an element of $\mathbf{c}(\mathbf{r}(\alpha_1), w^{\#})$, and some y_2 that is an element of $\mathbf{c}(\mathbf{r}(\alpha_2), w^{\#}), \ldots,$ and some y_n that is an element of $\mathbf{c}(\mathbf{r}(\alpha_n), w^{\#})$, where $y_i = y_k$ if $\alpha_i = \alpha_k$, i, $k = 1, 2, \ldots, n$, such that $<y_1, y_2, \ldots, y_n> \in \mathbf{e}(\phi, w^{\#})$.

Making the same adjustment in the definition for the $\Box\phi\alpha_1\alpha_2 \ldots \alpha_n$, case, one can obviate the above difficulty. With this new definition of 'truth under an interpretation', '$\Box a = a$' will be a valid sentence of the system.

The question arises: why did Forbes give his truth conditions in the way he did, especially in view of the fact that it departs from Lewis's original ideas? I cannot find any explicit explanation in Forbes 1985 for his deviant truth conditions. But evidently the reason is connected with the fact that '$a = b \rightarrow \Box a = b$' is not a valid sentence of counterpart theory. (This feature of the system will be explained shortly.) But Kripke had shown how the validity of the above sentence can be deduced from the validity of '$\Box a = a$' and Leibniz's Law. By giving his truth conditions in the way he does, he can defend the invalidity of the above principle of identity against Kripke's argument by rejecting the premiss that '$\Box a = a$' is valid.[50] In Forbes 1985: ch. 7, Forbes attempts to provide philosophical grounds, from a counterpart theorist's point of view, for rejecting the above law of identity.

Despite Forbes's analysis of identity statements, I shall, in the discussion that follows, assume that the truth conditions of *de re* sentences con-

[49] Alan Gibbard has developed an interesting argument against the necessity of identity in Gibbard 1975. Gupta has enthusiastically seconded Gibbard's argument. He cites Gibbard's case for contingent identity as good grounds for rejecting the Fregean analysis of restricted quantification in terms of unrestricted quantification (Gupta 1980: 93). I personally have not been convinced by Gibbard's argument. For an excellent discussion of Gibbard's views on identity and necessity, see Ray 1992.

[50] See Forbes 1985: 178–9, especially n. 18.

form to those that are in the spirit of Lewis's original article (and hence are those of [Revision 1] given above). With this change, consider an interpretation that yields Figure 5.

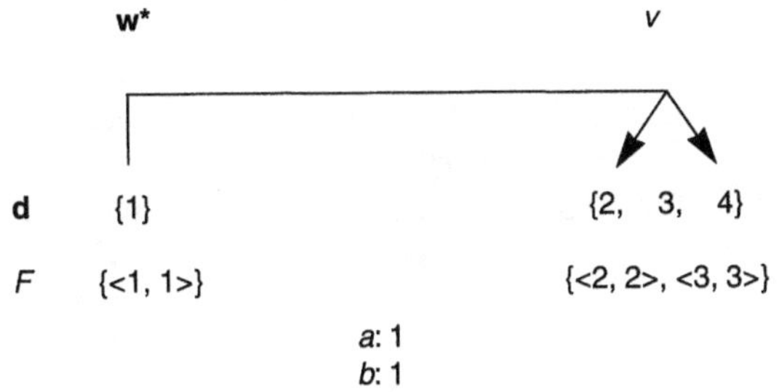

FIGURE 5

It can be seen that the sentence

$$\Box Faa \;\&\; a = b$$

is true under this interpretation, even though '$\Box Fba$' is not. What makes this result especially troubling is the apparent conflict with Leibniz's Law: it would seem that the open-sentence '$\Box Fxa$' can be true of **r**(*a*), when **r**(*a*) is the very same thing as **r**(*b*), even though that same open-sentence is not true of **r**(*b*). How can that be? Surely, this feature of counterpart theory is the most troubling of all.

Notice that this paradoxical result cannot be explained away by construing the box operator as meaning 'essentially'. No matter how one translates the box, one seems to have the flouting of Leibniz's Law. Also, one can see, from the above result, that the sentence '$a = b \rightarrow \Box a = b$' is not valid. This sentence expresses what is widely considered a law of identity, sometimes called "the necessity of identity". Surprisingly, such consequences of his counterpart theory do not seem to bother Lewis: he seems to be quite willing to abandon such a principle of identity.[51]

There is an easier way of responding to the above apparently paradoxical result. Change the definition of 'truth under an interpretation' by "requiring counterparts of identical things to be identical". The words in quotation marks in the preceding sentence require explanation. The idea is to change the definition of 'true under an interpretation' so that the truth of simple modal sentences of the form $\Box\phi_n\alpha_1\alpha_2 \ldots \alpha_n$ will depend upon whether or not all *n*-tuples of counterparts, $<y_1, y_2, \ldots, y_n>$,

[51] Lewis 1979*a*: 121–2. See also Lewis 1986: 259.

satisfying certain conditions to be given, are or are not members of $\mathbf{e}(\phi, u)$, for every $u \in \mathbf{W}$. The certain conditions are that $y_i = y_k$ if $\mathbf{r}(\alpha_i) = \mathbf{r}(\alpha_k)$, $i, k = 1, 2, \ldots, n$. That is, we have:

[Revision 2] The sentence

$$\Box\phi\alpha_1\alpha_2 \ldots \alpha_n$$

is true at w under $\mathbf{I}$ iff for every $w^\# \in \mathbf{W}$ and every y_1 that is a member of $\mathbf{c}(\mathbf{r}(\alpha_1), w^\#)$, and every y_2 that is a member of $\mathbf{c}(\mathbf{r}(\alpha_2), w^\#), \ldots$, and every y_n that is a member of $\mathbf{c}(\mathbf{r}(\alpha_n), w^\#)$, where $y_i = y_k$ if $\mathbf{r}(\alpha_i) = \mathbf{r}(\alpha_k)$, $i, k = 1, 2, \ldots, n$, $<y_1, y_2, \ldots, y_n> \in \mathbf{e}(\phi, w^\#)$.

The sentence

$$\Diamond\phi\alpha_1\alpha_2 \ldots \alpha_n$$

is true at w under $\mathbf{I}$ iff there is a $w^\# \in \mathbf{W}$ and there is some y_1 that is an element of $\mathbf{c}(\mathbf{r}(\alpha_1), w^\#)$, and some y_2 that is an element of $\mathbf{c}(\mathbf{r}(\alpha_2), w^\#), \ldots$, and some y_n that is an element of $\mathbf{c}(\mathbf{r}(\alpha_n), w^\#)$, such that $y_i = y_k$ if $\mathbf{r}(\alpha_i) = \mathbf{r}(\alpha_k)$, $i, k = 1, 2, \ldots, n$, and $<y_1, y_2, \ldots, y_n> \in \mathbf{e}(\phi, w^\#)$.

With these changes, the sentences '$(\Box Faa \,\&\, a = b) \rightarrow \Box Fab$' and '$a = b \rightarrow \Box a = b$' will be valid, and there will be no conflict with Leibniz's Law.

But would Lewis be happy with such changes? Evidently not. As I indicated above, he seems to reject the necessity of identity for philosophical reasons. Ponder the following example. A piece of plastic is synthesized in a mould to produce a dishpan, which is later destroyed by incineration. Thus, at all times t, the plastic exists at t iff the dishpan exists at t. The piece of plastic and the dishpan are, according to Lewis, one and the same thing. The plastic *is* the dishpan; and he tells us that this 'is' is the 'is' of identity. But although the plastic is the dishpan, *it might not have been.* Lewis maintains that the plastic "and the dishpan—it and itself—might have coexisted without being identical!" (Lewis 1986: 253).

Lewis's position is perplexing. If the plastic and the dishpan are one and the same thing, how could they (it) have coexisted without being identical? Lewis attempts to explain his views by saying:

> I only distinguish what everyone must distinguish: two different references, in different words, to the one thing. And I say that these different references tend to evoke two different ways of representing, whereby one world can make conflicting representations *de re* concerning that one thing. (Lewis 1986: 253)

Let us grant Lewis that there are two different ways of referring to one and the same thing, namely via the two descriptions 'the piece of plastic' and 'the dishpan'. And, corresponding to these two different ways of referring to the object before us, we can imagine a world in which two different things are counterparts of the one thing in the actual world that is being referred to by means of the two descriptions: there is one thing that is plastic but not made into a dishpan, and there is another thing that is a dishpan but not made out of plastic. So we can allow Lewis that, given his Modal Realistic framework, the one world can make conflicting representations *de re* concerning the one object. But none of this resolves the apparent conflict with Leibniz's Law. If the plastic is identical to the dishpan, then how can the piece of plastic satisfy an open-sentence that the dishpan does not? Yet it would seem that, by Lewis's own admission, the plastic satisfies the open-sentence 'x might have coexisted with the dishpan without being identical to it', whereas the dishpan does not. How can that be?

Here is one way of making some sense of this view. The descriptions 'the piece of plastic' and 'the dishpan' refer to one and the same thing, but in different ways. The former refers to it by way of one property, say, having such-and-such physical attributes (those that characterize plastic); the latter by way of another property, say, the property of being an article manufactured for use in such-and-such a way in the washing of dishes. These different "modes of reference" tend to evoke different counterpart relations when we think modally about the one object that has both these properties. We might put it this way: the one object, *qua piece of plastic*, has modal property F with respect to counterpart relation α; but this same object, *qua dishpan*, lacks modal property F with respect to a different counterpart relation β. The reference to the one object by the words 'the piece of plastic' gives rise to our using the counterpart relation α when we ask 'Could the piece of plastic have been made into a laundry basket?', whereas reference to this same object by the words 'the dishpan' gives rise to a different counterpart relation β. Thus, the one object, *qua piece of plastic*, could have been made into a laundry basket without being destroyed, but could not have been made out of another piece of plastic. This same object, *qua dishpan*, could have been made out of another piece of plastic, but could not have been made into a laundry basket without being destroyed. There is no conflict with Leibniz's Law since these *de re* modal attributions are relativized, via different modes of reference, to different counterpart relations.

The idea of relativizing *de re* modal attributions to different modes of reference suggests a different form of quantificational modal logic than any that I have discussed up to this point. Such a modal logic has been developed and defended in great detail by Gregory Ray in his Ph.D. dissertation (Ray 1992). From what I have said above, it may come as a surprise to learn that counterpart theory plays no role in the formal developments of Ray's logic: it has a standard Kripkean S5 possible worlds semantics and it does not lead to a rejection of the necessity of identity as appears to be the case, at first sight, in Lewis's counterpart theory. The main feature of the system is the use of descriptors that are subscripted to the standard modal operators to yield the relativization desired. The truth conditions of *de re* modal sentences are given in such a way as to yield the following results:

$$\Box_\delta Fa \text{ is true under } \mathbf{I} \text{ iff, for every } \delta\text{-world } w \in \mathbf{W}, \mathbf{r}(a) \in \mathbf{e}(F,w)$$

where a world v is a δ-world iff (intuitively speaking) $\mathbf{r}(a)$ satisfies the descriptor at v. In terms of Lewis's example, let δ be the descriptor 'x is a dishpan', and let γ be the descriptor 'x is a piece of plastic'. Then interpret F to be the predicate: *is destroyed when its shape is flattened out*. Letting 'a' denote the object that is both a dishpan and a piece of plastic, it would be reasonable to assert '$\Box_\delta Fa \;\&\; -\Box_\gamma Fa$'. What this assertion would come to, roughly, is this: it is necessary that $\mathbf{r}(a)$, qua dishpan, is destroyed when its shape is flattened out; and it is not necessary that $\mathbf{r}(a)$, qua piece of plastic, is destroyed when it is flattened out. Thus, Ray's system yields a "one thing" solution to the problem of the dishpan and the piece of plastic, and it does so without flouting Leibniz's Law. To go into the details of Ray's system further would take us too far afield, but I believe that it has many attractive features.

Let us now use some of Ray's ideas to develop further the ideas underlying Lewis's rejection of the necessity of identity. In particular, let us revise the counterpart-theoretic system we have been discussing so as to accommodate Ray's subscripted modal operators. Within this framework, the subscripts can be used to index different counterpart functions. I shall not bother to give the details of the syntax for the revised language. Since the language is to have modal operators subscripted by finite sequences of natural numbers—the sequence giving the counterpart relations that will be relevant—I shall add arabic numerals and the symbols:

$$< , >$$

to the set of logical constants of the language. The definition of formula can then be given in standard ways. The seventh coordinate $\mathbf{c}$ of an interpretation $\mathbf{I} = <\mathbf{W}, \mathbf{D}, \mathbf{w}^*, \mathbf{d}, \mathbf{r}, \mathbf{e}, \mathbf{c}>$ will satisfy the following:

> $\mathbf{c}$ is a function whose domain is the Cartesian product $\mathbf{D} \times \mathbf{W} \times \mathbf{N}$, where $\mathbf{N}$ is the set of natural numbers, and whose co-domain is the power set of $\mathbf{D}$, subject to the following two constraints:
>
> (a) $\mathbf{c}(x, w, n)$ must be a subset of $\mathbf{d}(w)$,

and

> (b) if $x \in \mathbf{d}(w)$, then $\mathbf{c}(x, w, n)$ must be $\{x\}$.

[*Hence,* $\mathbf{c}$ *must be a function that takes as arguments ordered triples whose first element* x *is a possible object, whose second object* w *is a possible world,* *and whose third object is the index of a counterpart relation; and the value of the function for argument <x, w, n> must be a set of possible objects that exist in w—the "counterparts$_n$ of x at w"—where each possible object that exists in w will have only one counterpart$_n$ in w, namely itself.*]

We can then define 'truth at w under **I**' for *de re* sentences as follows:

First, let ϕ be an n-ary predicate, let $\alpha_1, \alpha_2, \ldots, \alpha_n$ be individual constants (not necessarily distinct), and let s be the sequence $<j_1, \ldots, j_n>$.

The sentence

$$\Diamond_s \phi \alpha_1 \alpha_2 \ldots \alpha_n$$

is true at w under **I** iff there is a $w^\# \in \mathbf{W}$ and there is some y_1 that is an element of $\mathbf{c}(\mathbf{r}(\alpha_1), w^\#, j_1)$, and some y_2 that is an element of $\mathbf{c}(\mathbf{r}(\alpha_2), w^\#, j_2), \ldots,$ and some y_n that is an element of $\mathbf{c}(\mathbf{r}(\alpha_n), w^\#, j_n)$, such that $<y_1, y_2, \ldots, y_n> \in \mathbf{e}(\phi, w^\#)$, where $y_i = y_k$ if $<\mathbf{r}(\alpha_i), j_i> = <\mathbf{r}(\alpha_k), j_k>$, $i, k = 1, 2, \ldots, n$.

The sentence

$$\Box_s \phi \alpha_1 \alpha_2 \ldots \alpha_n$$

is true at w under **I** iff for every $w^\# \in \mathbf{W}$ and every y_1 that is a member of $\mathbf{c}(\mathbf{r}(\alpha_1), w^\#, j_1)$, and every y_2 that is a member of $\mathbf{c}(\mathbf{r}(\alpha_2), w^\#, j_2), \ldots,$ and every y_n that is a member of $\mathbf{c}(\mathbf{r}(\alpha_n), w^\#, j_n)$, $<y_1, y_2, \ldots, y_n> \in \mathbf{e}(\mathbf{r}(\phi, w^\#)$, where $y_i = y_k$ if $<\mathbf{r}(\alpha_i), j_i> = <\mathbf{r}(\alpha_k), j_k>$, $i, k = 1, 2, \ldots, n$.

Let us now analyse the example of the plastic and the dishpan in terms of this revised system of logic. Corresponding to the two ways of referring to the one object, there will be two counterpart relations (we can assume, for purposes of definiteness that the one has index 2 and the other has index 5). Can we conclude in this system that the plastic could not have been different than the dishpan? Well, the sentence

$$a = b \;\&\; \Diamond_{<2,5>} a \neq b$$

is satisfiable. So the sentence

$$a = b \rightarrow \Box_{<2,5>} a = b$$

is not valid.

What about the use of Leibniz's Law to prove the necessity of identity? In this system, the sentence

$$\Box_{<2,5>} a = a$$

is not valid. So one cannot use this sentence and Leibniz's Law, applying it to the open-sentence

$$\Box_{<2,\,5>}x = a$$

to infer that

$$\Box_{<2,\,5>}a = b$$

is a consequence of '$a = b$'. It can seen that Lewis's response to the plastic-dishpan example does not presuppose a rejection of Leibniz's Law after all. Furthermore, one can see that, strictly speaking, Lewis has not rejected the necessity of identity. What he rejects is better expressed by:

$$a = b \rightarrow \Box_{<2,\,5>}a = b$$

In other words, Lewis holds that, even though the plastic and the dishpan are one and the same thing, the plastic *qua plastic* could have differed from the plastic *qua dishpan*.[52]

What is Lewis's response to the various problems with counterpart theoretic semantics for quantificational modal logic that I have discussed above? He writes:

> What is the correct counterpart-theoretic interpretation of the modal formulas of the standard language of quantified modal logic?—Who cares? . . . If this language of boxes and diamonds proves to be a clumsy instrument for talking about matters of essence and potentiality, let it go hang. Use the resources of modal realism *directly* to say what it would mean for Humphrey to be essentially human, or to exist contingently.[53]

I take Lewis to be claiming that we need not worry about the problems arising from our attempts to develop an adequate counterpart-theoretic semantics for modal quantificational logic; he is suggesting that we don't have to express our modal thoughts in a formal modal language at all. With the translation rules of his counterpart theory (for details, see below in Chapter 8), Lewis can explain (in English) what it means to say 'Humphrey is essentially human' in such a way that the truth of the sentence does not imply that Humphrey exists necessarily. The above quotation suggests that Lewis is quite willing to have quantificational modal logic abandoned as a means of formalizing his modal discourse. He does not need quantificational modal logic, since he thinks he can use his first-order version of counterpart theory to formalize modal reasoning.

The above attempts to develop an acceptable version of counterpart-theoretical semantics clearly illustrate how both formal and philosophical considerations can influence the fashioning of a system of modal logic. In

[52] On the other hand, Lewis does note that '$a = b \rightarrow \Box a = b$' is not a theorem of his system, because he allows that something can have more than one counterpart in a world (Lewis 1979*a*: 122). In this case, there is no suggestion that different counterpart relations are involved, so it may be the case that Lewis simply rejects the necessity of identity.

[53] Lewis 1986: 12–13.

the case of counterpart-theoretic semantics, various features of the system are motivated by both mathematical and philosophical considerations, and there are features of the system which can only be properly understood and appreciated when they are viewed within a complex web of philosophical doctrines that support and are supported by the system. It is to these philosophical doctrines that I now turn.

3

Modal Realism

The models studied in modal logic are set theoretical structures. So one can regard the model theory of modal logic as a branch of set theory. The philosophical question arises: what is the connection between the set theoretical structures studied in the model theory and anything objective in virtue of which we can explain how a study of these structures can yield genuine insights into the principles and inferences of modal logic? One answer to this question is provided by the Modal Realist.

1. Lewis's Modal Realism: Exposition

Modal Realism, as I use the term, is the doctrine that, besides the actual world we live in, there exist other possible worlds. Modal Realism, in the full-blown sense of the term, maintains that these other possible worlds are not mere abstractions or descriptions—they contain things that are just as real and concrete as things like tables, chairs, people, and trees in the actual world. According to the Modal Realist (in the full-blown sense), our world—the world we live in—is just one of many that exist; there are other worlds that contain things like tables and chairs that are no less real than the tables and chairs in our world. Thus, as I use the term, Modal Realism is an ontological thesis. It is not the epistemological thesis that we know that such possible worlds exist, even though David Lewis, the Modal Realist *par excellence,* believes that we do know that such possible worlds exist.

It might be thought that, when Lewis proclaims the existence of other possible worlds, he is speaking metaphorically or poetically. But no: he is quite definite about what he is asserting.

When I profess realism about possible worlds, I mean to be taken literally. Possible worlds are what they are, and not some other thing. If asked what sort of thing they are, I cannot give any kind of reply my questioner probably expects: that is, a proposal to reduce possible worlds to something else.

I can only ask him to admit that he knows what sort of thing our actual world is, and then explain that other worlds are more things of *that* sort, differing not in kind but only in what goes on at them. (Lewis 1979*b*: 183–4)

Lewis tells us that possible worlds are like "remote planets; except most of them are much bigger than mere planets and they are not remote" (Lewis 1986: 2). The reason that these worlds are not supposed to be remote from us is this: they are totally isolated—spatially and temporally—from us and from each other. As Lewis puts it: "There are no spatiotemporal relations at all between things that belong to different worlds" (Lewis 1986: 2). Hence, these worlds are neither remote from us nor near to us either in space or in time.

The Indexical Analysis of Actuality

This view of possible worlds gives rise to the following problem. There are these different possible worlds, each one spatiotemporally isolated from all the others; but presumably only one of these is the actual world. What distinguishes the actual world from the other merely possible ones? If all these possible worlds exist, what makes one of them the actual world? Or to put it another way: what does the actual world possess that the merely possible worlds lack?[1] It cannot be existence, since all the possible worlds are supposed to exist.

Lewis has a simple response. What 'actual' means depends upon who the speaker is: if the speaker is α, then 'the actual world' refers to the one and only world in which α exists. This doctrine is called the 'indexical analysis of actuality', since 'actual' is regarded as an indexical term, in the way 'I', 'this', 'here', and 'now' are regarded as indexicals: what these terms refer to depends upon features of the context of utterance.[2] In the case of 'actual', the feature of the context of utterance that is relevant is the world in which the utterance is made. From God's point of view—assuming with Leibniz that God is not an inhabitant of any world—all of the possible worlds are on a par. There is no special world that is singled out as *actual*. Of course, from our point of view, our world is special. It is actual. But that is analogous to saying that, from my point of view, *here* is special: here is where I am.

Possible Objects are Worldbound

I noted above that, according to Lewis, a speaker exists in one and only one world. This may seem paradoxical. After all, according to standard versions of the semantics of modal logic, a person may exist in many pos-

[1] For a general discussion of this problem, see Adams 1979*b*: 190–209.
[2] Lewis 1986: 92.

sible worlds. But in the semantical analysis developed by Lewis, each person is "world-bound": if person α exists in world w, and if world $u \neq w$, then α does not exist in u. From the perspective of Modal Realism, such a doctrine is surely right: if all these possible worlds exist, and I am in one of them, how could I possibly be in any other world? It is easy enough to see how there could be someone very much like me in some other world; but no matter how much like me this person might be, he could not *be* me if he is in another world. That just could not be me, since I am in this world and he is in that world.

Counterpart Theory

The sentence 'Hubert Humphrey could have won the presidential election' is true, according to standard versions of possible worlds semantics, if and only if there is a possible world at which Hubert Humphrey does win the presidential election. Since the above sentence is clearly true, there must be a possible world, according to standard versions of possible worlds semantics, in which Humphrey exists and does things. How, then, does Lewis allow for the truth of such sentences, given his doctrine that every person (indeed, every concrete thing) is world-bound? According to Lewis's account of modality, the above sentence is true if and only if there is a possible world in which there is a *counterpart* of Humphrey who wins the presidential election.

What is a counterpart of Humphrey? A counterpart of Humphrey in world w resembles Humphrey—indeed, resembles Humphrey more closely than does anything else that exists in w—but is not Humphrey. But how does this counterpart of Humphrey represent Humphrey as doing anything? Well, military personnel can use tiny models of soldiers, tanks, trucks, and artillery to represent a battle. Similarly, a possible world can have a real flesh and blood person to represent Humphrey: "[B]y waving his arm, or by winning the presidential election, the other-worldly Humphrey represents the this-worldly Humphrey as waving or as winning" (Lewis 1986: 194). Generally speaking, then, worlds *represent* possibilities; and they do this by having appropriate parts—parts that are similar to things in our world.[3]

The counterpart relation is Lewis's substitute for *transworld identity*.[4] It

[3] Cf. this description of Leibniz's views on counterfactual conditionals: "[W]hen we wonder what would have happened had Caesar crossed the Rubicon, we are thinking about Caesar and our world. But [Leibniz] thought that when we describe possible worlds which fit our counterfactual claims, we are not describing the historical Caesar himself . . . the possible world so described does not contain Julius Caesar himself. It is a world in which a similar Roman general stays in Gaul" (Ishiguro 1990: 177).

[4] Although transworld identity has been discussed in some detail already, the reader may wish to consult Plantinga 1974: ch. 6, for a comparison of transworld identity with Lewis's theory of counterparts.

is not identity; it is a similarity relation. So we cannot assume that this relation is an equivalence relation, as is identity. In general, the counterpart relation is neither symmetric nor transitive. Also, things may be similar in some respects and dissimilar in other respects. And some similarity relations may differ from others because of the priorities placed on the different respects of similarity. Furthermore, two similarity relations may differ from one another, even if they are alike in the respects of comparison, because one can be more stringent than the other (Lewis 1986: 254).

But if we have indefinitely many counterpart relations, how is it determined which counterpart relation is to be the relevant one for a specific statement of possibility, say, "Humphrey could have won the election"? According to Lewis, such determinations are to be made on the basis of contextual factors.[5] The plastic-dishpan example of the previous chapter illustrates the way in which how one refers to an object can influence which counterpart relation is supposed to be selected. Still, if indefinitely many counterpart relations may be presupposed by our modal statements, it seems mysterious to me just how we succeed in communicating which are the relevant ones in the course of making modal statements.

The Mereological Account of Worlds

In the above quotation, a counterpart of Humphrey is said to be a *part* of a world. In Lewis's mereological terminology, Humphrey is not said to be *in* the actual world; he is said to be a *part* of the actual world.[6] The *part of* relation is one of the basic relations of Lewis's ontological theory. So mereology plays a big role in the development of this theory.[7] Any part of a possible world is a "possible individual"; and if two things are parts of the same world, then they are called "worldmates". A possible world is the mereological sum of all the possible individuals that are parts of it (Lewis 1986: 69). It is a *maximal* sum since anything that is a worldmate of any part of a world is a part of that world.

Why can there not be two parts of a world that are not spatiotemporally related? In works of science fiction, two separate "universes", α and β, are sometimes depicted as existing in independent spacetimes. Why could

[5] See Lewis 1986: 254–63. Lewis has suggested to me (in a letter dated 2 Apr. 1996) that "we interpret what's said so as to make the message make sense", adding that we needn't eliminate all the indeterminacy.

[6] I should note, however, that Lewis has indicated in a letter that he is happy to use either expression.

[7] Lewis refers his readers to Leonard and Goodman 1940: 45–55 for the mereology that he uses in Lewis 1986. More recently, in Lewis 1991: sect. 3.4, he has proposed the Transitive Law, Unlimited Composition (or Fusion Existence), and Uniqueness of Composition as the basic axioms of mereology. There, in sect. 3.5, he mounts a defence of mereology against some typical criticisms.

there not be, as parts of some world, such "universes" that are spatiotemporally isolated from each other, as C. D. Broad thought many years ago (Broad 1933: 176–7)? There certainly does not appear to be anything conceptually impossible about such isolated universes. Lewis admits that he has some inclination to agree that a world with two disconnected spacetimes is a possibility. But he feels that he *has* to deny that there is such a possibility:

> Given a choice between rejecting the alleged possibility of disconnected spacetimes within a single world and (what I take to be the alternative) resorting to a primitive worldmate relation, I take the former to be more credible. (Lewis 1986: 72)

Many readers may find this quotation puzzling. Lewis seems to be forced to adopt a counterintuitive position in order to avoid taking the worldmate relation as a primitive. This has struck some commentators as intellectually repugnant.[8] What is so important about avoiding the adoption of another primitive relation? Every theory has some primitive terms. Not every term can be defined. So what is implausible about taking the worldmate relation to be a primitive? Lewis does not explain his thinking on this matter. But at least part of his motivation for his position is to be found in the fact that his justification for postulating and believing in possible worlds is based on the contention that this postulation results in a *reduction in the diversity of primitives* of our "total theory".[9] This suggests that

[8] For example, Gideon Rosen has commented: "Surely the number of 'island universes' ought to be a contingent matter. . . . armchair metaphysics should not pronounce on the number and nature of such things" (Rosen 1995: 71). Phillip Bricker also argues that "island universes" should be deemed possible (in Bricker 1996: 235). See also Bigelow and Pargetter 1987 for several objections to Lewis's position on this matter.

[9] Actually, it is by no means clear that Lewis is able to obtain a reduction in even the number of primitives in the way he advocates. Complications arise because, in a Newtonian world, things are not related, spatiotemporally, in quite the way they are in our Einsteinian world. In a Newtonian world, spacetime points are related by two distance functions, whereas spacetime points in an Einsteinian world are related by a single distance function. Thus, Lewis is prompted to allow that, in some worlds, things are interrelated by a system of relations that are only analogous to the spatiotemporal relations. However, in attempting to spell out the analogy, he introduces a number of new terms: the relations, he tells us, must be *natural*, *pervasive*, *discriminating*, and *external*. (See Lewis 1986: 75–6). Lewis has replied (in his letter of 2 Apr. 1996) that these terms: (1) may not remain primitives; and (2) may serve other purposes than to explain 'worldmate'. But granting Lewis these possibilities, the point remains that just how many primitives will be needed to give the details of Lewis's theory of worldmatehood is not clear. I do not see how one can determine with any degree of confidence whether the total number of primitives needed to spell out Lewis's theory of worldmatehood will be less than or no more than a competing theory that takes the worldmate relation to be primitive, in the absence of anything like a total theory in which these competing theories are to be placed. How can one tell how many new primitives will be needed to accomplish such-and-such, if one has no clear idea of just what primitives one has to begin with? Of course, Lewis is not concerned with the relative number of primitives

he is inferring a metaphysical or conceptual impossibility from a mere methodological principle (to reduce the diversity of primitives). This seems to be a strange way to proceed. Few, if any, mathematicians would infer a mathematical impossibility from such a methodological principle. I shall say much more about this justification later, but for now I merely note the curious fact that Lewis's criterion of "worldmatehood" rests upon a very strange assumption and that it leads to some very counter-intuitive doctrines. For example, as Craig Bach pointed out to me, it implies the falsity of the following statement:

$$\Diamond(\exists x)(\text{I am not spatiotemporally related to } x).$$

And this seems wrong. Surely, there could be something to which you are not spatiotemporally related. I should also emphasize that Lewis presents no *objective* grounds for thinking that a disconnected spacetime is not possible.

Is Lewis Proposing an Analysis of our Modal Concepts?

I now take up the question of just what Lewis is claiming for his realistic account of modality. From what I have said thus far, it is quite possible that Lewis is proposing that we adopt some radically new modal concepts, which are meant to replace the crudely characterized pre-analytic modal concepts[10] that philosophers utilize, just as, according to Quine, Weierstrass and other nineteenth-century analysts developed new epsilon-delta concepts of limits and continuity, to replace the defective intuitively developed concepts of limits and continuity that Newton and other early analysts had utilized.[11]

To investigate this question, I suggest that we turn to Lewis's explanation (in Lewis 1986) of how the model theory of modal logic casts light on the nature of certain controversial principles of modal logic. Lewis follows standard practice in characterizing the semantics of modal logic in terms of *relational frames* and *valuations* over these frames (a relational frame, together with a valuation, corresponds to what I have called an 'interpretation'). He then notes that, instead of trying to answer such baffling questions as "Is it true that whatever is actual is necessarily possible?", we could investigate certain general questions about the relational

of the competing theory, but only with the relative diversity of its primitives; but the same basic point applies. How can one tell if the set of primitives of the new theory is more or less diverse than that of the old theory, if one has no clear idea of just what primitives the old theory has?

[10] I use the expression 'pre-analytic' here to indicate that I have in mind the concepts characterized in Chapter 1 *prior* to their being *analysed* by means of possible worlds semantics.

[11] Quine 1960: sect. 51.

frames, answers to which would also determine the truth value of the controversial modal principle. However, to apply such a result, one would have to adopt some analysis of modality which links the relational frames to modal principles. Lewis unequivocally declares his belief that modal operators are quantifiers over possible worlds, claiming that the metalogical results are applicable, because "there exist frames which afford correct interpretations of the modal operators" (Lewis 1986: 19–20).

Thus, it is abundantly clear that Lewis is not advocating *replacing* our crudely characterized pre-analytic modal concepts by his realistic possible worlds ones. No, he takes himself to be providing *correct analyses* of these concepts and "correct interpretations" of the modal operators. And this is why the metalogical results are applicable.

There is another reason why I interpret Lewis to be advocating a possible worlds *analysis* of our modal concepts (and, in particular, of the modal operators). At one point in Lewis 1986, he takes up the question 'How do we come by the modal opinions that we do in fact hold?' His answer to this question is: "I think our everyday modal opinions are, in large measure, consequences of a principle of recombination" (p. 113). Now this provides us with additional evidence that Lewis regards his possible worlds account as giving us a correct analysis of our modal concepts. For his principle of recombination is a principle that implies that certain kinds of possible worlds exist, given the existence of other worlds.[12] Unless he thought that he was giving an accurate analysis of our modal concepts, it would make no sense to suppose that ordinary people come to their everyday modal opinions as a consequence of such a principle about worlds.

Lewis's Possibilism

It was pointed out in Chapter 1 that Plantinga is an Actualist—someone who holds both that there is no thing that does not exist and also that there could not have been something that did not exist. From the exposition given above, it can be seen that Lewis does not fit Plantinga's characterization of an Actualist. According to Lewis, there are things (in other words) that do not *actually* exist: there are things that exist but that do not exist in our world. To that extent, Lewis can be classified as a Possibilist. A Possibilist, according to this way of viewing the situation, is someone who holds either that there are things that do not (actually) exist or that there could have been things that did not (actually) exist.

[12] Lewis's principle of recombination will be discussed in detail in Chapter 8, Section 4.

2. Is Lewis's Analysis of Modality Correct?

I would like to raise an initial doubt about the correctness of Lewis's analysis of modality. The following considerations are not meant to be decisive. At this point, I only wish to raise a small doubt that the modal notions I discussed in the previous two chapters may not have been correctly analysed by Lewis.

First, it should be noted that there is no logical connection between Lewis's ontology of worlds and his analysis of modal operators, in virtue of which belief in the former commits one to acceptance of the latter. Thus, let us imagine that there is another philosopher—let us call him Lewis#—who has the same many worlds ontology that David Lewis has. We can suppose that Lewis# has not been convinced by Lewis's *analysis* of the modal operators. In particular, he questions the hypothesis that Lewis's possible worlds analysis adequately captures the crudely characterized, pre-analytic concept that was discussed in the previous chapters. Lewis# allows that there is a real advantage to analysing the modal operators in the way Lewis does, especially since they both agree that these many worlds do exist. However, Lewis# is strongly inclined to believe that the other worlds do not exist *necessarily*. To Lewis#, the existence of each of the other worlds appears to be just as contingent a matter as the existence of the parts of the one he inhabits. As he sees it, these other worlds are all parts of one reality; and they share the same contingency that all parts of the actual world possess. In other words, Lewis# allows that, for a great many propositions ϕ that we are apt to consider, ϕ is necessary (in the crudely characterized, pre-analytic sense of the term) iff ϕ is necessary (according to Lewis's analysis of that term). However, since Lewis seems to think that the existence of each of the other worlds is necessary,[13] Lewis# believes that there are some propositions that he himself believes to be contingent (in the crudely characterized sense), which Lewis believes to be necessary (according to his analysis of that term). This leads Lewis# to think that Lewis's analysis may not have completely captured the crudely characterized sense of the term 'necessary'.

3. Lewis's Justification for Possible Worlds

Why does David Lewis believe in possible worlds? It should be noted first that Lewis is a Mathematical Realist, where 'Mathematical Realism' is defined to be the doctrine that mathematical objects, such as numbers and

[13] See Lewis 1986: sect. 2.6.

sets, exist. Mathematical objects are generally held to be causally inaccessible, non-physical objects that do not exist in physical space and time. (Lewis's version differs from the generally held position in so far as it is agnostic about the location of mathematical objects.)[14] It is obvious that Mathematical Realism is an ontological view that goes far beyond the common sensical belief in ordinary objects, such as tables and chairs, with which we come into contact in everyday life.[15] Lewis's reasons for believing in possible worlds are closely related to his reasons for being a Mathematical Realist. Consider Lewis's position on the existence of mathematical objects. Agreeing with Hilbert's well-known statement that the set theoretical universe is a paradise for mathematicians, he goes on to claim:

> We have only to believe in the vast hierarchy of sets, and there we find entities suited to meet the needs of all the branches of mathematics; and we find that the very meagre primitive vocabulary of set theory, definitionally extended, suffices to meet our needs for mathematical predicates; and we find that the meagre axioms of set theory are first principles enough to yield the theorems that are the content of the subject. Set theory offers the mathematician great economy of primitives and premises, in return for accepting rather a lot of entities unknown to *Homo javanensis.* It offers an improvement in what Quine calls ideology, paid for in the coin of ontology. It's an offer you can't refuse. The price is right; the benefits in theoretical unity and economy are well worth the entities. (Lewis 1986: 3–4)

What Lewis is arguing here can be seen to be a development of a line of reasoning that goes back at least as far as *Principia Mathematica*, in which Russell and Whitehead attempted to justify their Axiom of Reducibility on "inductive" or pragmatic grounds. They suggested that the axioms should be accepted because it permits the deduction of mathematical statements (such as the Least Upper Bound Theorem of real analysis) we believe to be true—statements that could not be proved without the axiom.[16] In short, Russell and Whitehead argued that the Axiom of Reducibility should be believed because it is, in some sense, "necessary for mathematics". Gödel took this line of reasoning a step further by arguing that belief in the objects of mathematics could be justified in a similar fashion:[17]

[14] He has written (personal communication): "I have no idea whether mathematical objects are in space and time." In Lewis 1986, he seems to allow that pure sets have no location in space (see p. 83). However, he seems to think that mixed sets (sets of urelements) have a location if its members do. The set is supposed to be located where its members are located. This is a doctrine that Penelope Maddy has developed in Maddy 1990—a doctrine I criticize in Chapter 9.

[15] See Chihara 1990: chs. 1 and 10.

[16] See Chihara 1973: 48–50.

[17] For detailed discussions of Gödel's reasoning, see Chihara 1973: ch. 2, and also

It seems to me that assumption of such objects is quite as legitimate as the assumption of physical bodies and there is quite as much reason to believe in their existence. They are in the same sense necessary to obtain a satisfactory system of mathematics as physical bodies are necessary for a satisfactory theory of our sense perceptions. (Gödel 1964*a*: 220)

Then Quine came along and pushed the envelope a bit further by developing what has come to be known as the "indispensability argument" for the existence of mathematical objects. For Quine, the postulation of mathematical objects was justified by our best theory of the world—best by the light of our present scientific theories. Contemporary mathematics is an essential part of contemporary physics, and mathematics is ontologically committed to the existence of mathematical objects. Thus, in so far as we are justified in believing what is implied by our best theories of physics, so also are we justified in believing in the existence of mathematical objects.[18]

This line of argumentation is carried to new heights by Lewis. If we can justify belief in mathematical objects in this way, perhaps we can also justify belief in other worlds in a similar way. Such a thought no doubt led to Lewis's argument for the existence of other worlds. We find him espousing the following position. Just as the ontology of sets is a paradise for mathematicians, so also is the ontology of possible worlds a paradise for philosophers. We philosophers have only to believe in possible worlds and our task of theory construction is made considerably easier:

We find the wherewithal to *reduce the diversity of notions we must accept as primitive*, and thereby to improve the unity and economy of theory that is our professional concern—total theory, the whole of what we take to be true.[19]

Our total theory includes modal beliefs; and the postulation of possible worlds will allow us to analyse modality as quantification over these possible worlds: possibility, on this analysis, "amounts to existential quantification over the worlds".[20] More specifically, '$\Diamond P$' is true, according to this analysis, iff there is a possible world *w* which is such that '*P*' is true at *w*. Such an analysis has led to fruitful developments in modal logic—fruitful in so far as we can define '$\Diamond$' and '$\Box$' in terms of the one primitive concept, that is, the concept of *possible world*—and this fruitfulness supports the analysis.

Lewis does not believe that possible worlds are only useful in the area of

Chihara 1990: ch. 1, sect. 3. Penelope Maddy give a quite different interpretation of Gödel's thoughts in Maddy 1980. I discuss her interpretation in Chihara 1982.

[18] This description of Quine's reasoning is merely a sketch. For more details, see Chapter 9, Section 7 below. I discuss the indispensability argument in even more detail in Chihara 1973: ch. 3, and Chihara 1990: ch. 1, sect. 2.

[19] Lewis 1986: 4, italics mine.

[20] Lewis 1986: 7.

modal logic. The kind of fruitfulness described above extends to other areas of philosophy. For example, properties can be defined in terms of possible worlds. Properties are just sets of things (both actual and possible): intuitively speaking, the property *F* can be regarded as the set of all things (both actual and possible) that have the property *F*. Propositions are taken to be properties of possible worlds and hence they are regarded as sets of possible worlds: intuitively speaking, the proposition *p* is just the set of all worlds at which *p* is true. Thus, starting with the primitive notions from mereology and set theory, and adding the notion of *possible world,* a significant number of other basic notions can be defined in terms of these primitives.[21] In this way, it is thought, one can reduce the diversity of the primitives one needs in one's total theory. *That* is what possible worlds are good for. Thus, Lewis concludes, "Modal Realism is fruitful; that gives us good reason to believe it is true."[22]

The conclusion of the above argument is so fantastic that it is only if the reasoning is completely rigorous and logically compelling that there could be any real hope of winning the assent of many philosophers. Not surprisingly, Lewis's views have met with a number of objections. In the following section, I shall take up two of the more striking of these objections. These are objections that have been raised by a number of philosophers and to which Lewis has replied in a forceful and convincing fashion.[23] In what follows, I shall be especially concerned to evaluate Lewis's responses to these objections.

4. Two Objections to Lewis's Modal Realism

The Modal Irrelevance Objection

Peter van Inwagen expressed the first of these objections—to be called the 'Modal Irrelevance Objection'—by suggesting that Lewis must "face the

[21] These include the content of a person's belief, the content of someone's knowledge, counterfactuals, and various notions useful in the analyses of languages. (See Lewis 1986: ch. 1, sects. 1.2–1.5.)

[22] Lewis 1986. In Lewis 1979*b*: 182, Lewis puts forward an ordinary language argument to support his belief in the existence of possible worlds. This argument has been criticized by many philosophers, for example, by Robert Stalnaker in Stalnaker 1984: 44–58, and more recently by Michael Jubien in Jubien 1988. Since Lewis does not cite or discuss his ordinary language argument in Lewis 1986 (which is one long defence of his Modal Realism), one might suppose that he no longer accepts this argument. However, I have learned from Lewis (in a letter of 2 Apr. 1996) that he still accepts the soundness of the argument—he does not present it in his Lewis 1986 because it does not address the question of whether the worlds, whose existence is supposedly justified by the argument, are real or merely ersatz. I shall take up one aspect of this argument in Chapter 8.

[23] John Bigelow and Robert Pargetter have judged that "Lewis has been largely successful in his replies to his critics" (Bigelow and Pargetter 1987: 97).

problem of explaining what these things [these possible worlds] would have to do with modality if there were any of them" (van Inwagen 1985: 119). Michael Jubien has put this objection in an especially vivid form:

> [S]uppose for a moment we agree, in the spirit of Lewis, that 'the world' covers only what is in some reasonable and liberal sense, 'spatiotemporally (or perhaps causally), connected' to us. *The world* is then what he would call 'the actual world' (or near enough, for our purposes). And it certainly makes sense to imagine that there *are* entities rather like the world but that lie outside of its spatiotemporal limits. Given this much, it would be very natural, though hardly mandatory, to say that our world could have resembled these entities in various ways. But why should we suppose that for *any* way our world could have been, there is an entity of this special sort that *is* that way? After all, *we* didn't create these special entities or 'abstract' them from our modal talk; they came about in ways similar to the way 'the world' did, however that may have been. They're just there, utterly independent of us . . . What can they have to do with possibility? (Jubien 1988: 305)

Lewis's response to this sort of objection in Lewis 1979*b* is puzzling. He seems not to take it at all seriously and brushes it off: "You might think that I have often explained what these things [these possible worlds] have to do with modality, for instance by saying that the modal operators are quantifiers over them" (Lewis 1979*b*: 98).

This is a puzzling response: Lewis surely must realize that van Inwagen is perfectly aware of the fact that he (Lewis) takes the modal operators to be quantifiers over these worlds. It would thus be reasonable for Lewis to infer that when van Inwagen asks what these worlds have to do with modality, he is not asking a question that can be answered by saying "modal operators are quantifiers over worlds". What is going on here? I shall return to this puzzling response shortly. Let us now consider the second objection.

The Unknowability Objection

This objection, to be called 'The Unknowability Objection', was raised by Tom Richards:

> [W]hile possible-worlds semantics does yield truth-conditions for possibility statements . . . the truth conditions are such that, for any given statement, it is impossible in general to determine whether they are met and hence whether the statement is true.

Lewis rephrases the objection to read: "[I]f modal realism gives the right account of the content of what we know [in believing some modal proposition] then it could not possibly be known at all."[24]

Lewis's strategy for rebutting this objection mirrors Kant's method of

[24] Lewis 1979*b*: 108.

undermining Hume's criticisms of metaphysical philosophers. You will recall that Hume had objected to the claims of metaphysical philosophers to know *a priori* certain *synthetic* truths. For Hume, synthetic propositions can only be known on the basis of sense experience, so if some proposition is truly known a priori, then it would have to be an *analytic* truth. But according to Hume's analysis, the only analytic truths that can be known are those of mathematics and logic. Thus, Hume concludes the *Enquiry*:

> When we run over libraries, persuaded of these principles, what havoc must we make? If we take in our hand any volume; of divinity or school metaphysics, for instance; let us ask, *Does it contain any abstract reasoning concerning quantity or number?* No. *Does it contain any experimental reasoning concerning matter of fact and existence?* No. Commit it then to the flames: for it can contain nothing but sophistry and illusion.[25]

Awakened from his dogmatic slumbers by passages such as the above, Kant responded by arguing, in his *Prolegomena to Any Future Metaphysics*, that the truths of mathematics are *synthetic*, not analytic as Hume had thought. In this way, Kant hoped to show that mathematics could be classified with metaphysics as a science yielding a priori knowledge of synthetic truths. If Hume had seen that mathematics yields a priori knowledge of synthetic truths, Kant suggests, the "good company into which metaphysics would thus have been brought would have saved it from the danger of a contemptuous ill-treatment, for the thrust intended for it must have reached mathematics, which was not and could not have been Hume's intention".[26]

In a similar fashion, Lewis responds to the Unknowability Argument by putting forward mathematical knowledge as a case in which "we can know that there exist countless objects causally isolated from us and unavailable to our inspection" (Lewis 1986: 109). By putting knowledge of possible worlds in the good company of mathematical knowledge, he hoped to save his analysis of modality from a contemptuous ill-treatment, writing: "Causal accounts of knowledge are all very well in their place, but if they are put forward as *general* theories, then mathematics refutes them" (Lewis 1986: 109).

How does Lewis attempt to justify his claim that mathematics gives us knowledge of causally isolated objects? The earlier quotation from Lewis indicates why he thinks we are justified in postulating the existence of sets: it is for the sake of theoretical unity and economy that we are justified in

[25] David Hume, *An Enquiry Concerning Human Understanding* (La Salle: Open Court, 1949): 184.

[26] Immanuel Kant, *Prolegomena to Any Future Metaphysics* (New York: Liberal Arts Press, 1951): 18.

expanding our existential beliefs to include sets. He then puts forward an account of (or thesis about) knowledge according to which, "if we are prepared to expand our existential beliefs for the sake of theoretical unity, and if thereby we come to believe the truth, then we attain knowledge" (Lewis 1986: 109). This is the way, according to Lewis, we have attained knowledge of sets, which are causally inaccessible objects.

Having thus convinced himself that we can have knowledge of causally inaccessible things, Lewis sees no insuperable barrier to obtaining knowledge of the existence and contents of other causally inaccessible things such as possible worlds. Indeed, one might interpret Lewis to be reasoning as follows. The reason Richards and supporters of the Unknowability Objection think we could not have any knowledge of other worlds is because they believe that it is not possible to have any knowledge of causally inaccessible things. But mathematical knowledge of sets refutes this premiss: sets are causally inaccessible, and yet we have knowledge of sets.

Brian Skyrms supports the Unknowability Argument by maintaining that, if possible worlds are supposed to be the same sorts of thing as our actual world, then any claim to know about such worlds should be backed by the same sort of evidence as we would require for any claim to know about some physical constituent of reality.[27] Lewis responds that to know that there are donkeys, we require causal acquaintance with donkeys, but to know that there are possible worlds in which donkeys talk, we do not require causal acquaintance with possible worlds. The difference is supposed to be this: the former is knowledge of a contingent truth, whereas the latter is knowledge of a necessary truth. That donkeys do not talk in the world we happen to inhabit is a contingent truth. To know that donkeys do not talk in the world we inhabit, we need to observe ourselves and our surroundings (Lewis 1986: 112). However, to know that there is a possible world in which donkeys talk is just to know that it is possible that donkeys talk—a necessary truth—and to know this, Lewis believes, it is not necessary that we observe ourselves or our surroundings.

Here again we have a puzzle. On the one hand, Skyrms's position seems perfectly reasonable. On the other hand, Lewis's response, although not without some bite, seems somehow perverse or bizarre. What is going on?

5. A Critique of Lewis's Views on Knowledge

What are we to say about Lewis's replies to the above objections? Much of Lewis's defence rests upon his analysis of mathematical knowledge, so let

[27] Skyrms 1976: 326.

us begin there. Should we postulate the existence of sets merely because we can, as a result, reduce the diversity of primitives and axioms of their total theory? I am deeply sceptical of Lewis's reasoning. But Lewis shows he has a response to this scepticism, when he writes:

> Maybe the very idea of accepting controversial ontology for the sake of [such] theoretical benefits is misguided—so a sceptical epistemologist might say, to which I reply that mathematics is better known than any premise of sceptical epistemology. (Lewis 1986: 4)

Of course, mathematics is more certain than any epistemological theory we now have. But why must anyone sceptical of Lewis's views about ontology either presuppose the falsity of mathematics or question the truth of mathematics? Here, it should be noted that he does not provide proof or even plausible argument that our sceptical doubts must conflict with mathematics; he simply asserts it. It is in this way that Lewis makes it seem as if anyone sceptical about his radical ontological claims is questioning mathematics. But, in fact, it is not mathematics that I oppose but only Lewis's *interpretation* of mathematical theorems and his reasoning in support of believing in the existence of sets. As I view the dispute between the mathematical realist and those sceptical of the existence of mathematical entities, it is a *philosophical* dispute involving two opposing philosophical doctrines—it is not a conflict between a mathematical theorem and some philosophical thesis. To view the dispute as Lewis does, one would have to incur a commitment to a sizeable philosophical analysis of mathematical language and truth.

But my doubts about Lewis's claim to have knowledge of possible worlds extends far beyond his claims about *mathematical* knowledge. Even if we do have knowledge of the existence and properties of sets, I still don't see how we could possibly know that there is a world very much like our own in which someone very much like me gives a lecture on possible worlds. In line with Skyrms's view of the matter, I feel that knowledge of such other worlds is completely beyond our cognitive powers.

As I noted earlier, Lewis has a kind of reply to the sort of doubts just expressed. Recall that, in response to the Unknowability Objection, he advanced an epistemological thesis according to which, *if we come to believe the truth as a result of expanding our existential beliefs for the sake of theoretical unity, we thereby attain knowledge.* This was first put forward as an account of how we attain knowledge of mathematical objects. But he went on to say: "In this way, we can even attain knowledge like that of the mathematician: we can know that there exists countless objects causally isolated from us and unavailable to our inspection" (Lewis 1986: 109). Thus, it is argued, contrary to what is presupposed by the Unknowability Objection, we can attain knowledge of other worlds even though these worlds are causally isolated from us. Lewis's epistemological principle shows how we can have this knowledge. According to that thesis, if we postulate possible worlds for the sake of theoretical unity, and if such a postulation happens to be true (that is, there really are such worlds), then we would have genuine knowledge of other worlds. For purposes of reference, let us label this epistemological thesis (which was italicized above) 'The Theoretical Unity Thesis'.

I do not believe that Lewis arrived at his Theoretical Unity Thesis as the result of a thorough investigation of our uses of the term 'know' and related expressions. I suspect that Lewis accepted this thesis because he could find no other plausible way both of explaining how we could have knowledge of the existence of sets and also of defending his analysis of possibility against the Unknowability Objection.[28] What should be emphasized is that Lewis gives us no justification or grounds for accepting his epistemological thesis: he just asserts it.

Let us suppose that this thesis has been put forward as a serious epistemological thesis involving our ordinary concept of knowledge. Notice that Lewis's thesis is strikingly similar to the *justified true belief* account of knowledge, according to which S knows that ϕ iff S believes ϕ, ϕ is true, and S is justified in believing that ϕ. Replace the third condition of that account of knowledge with Lewis's condition that S has accepted ϕ for the sake of theoretical unity, and one gets, essentially, Lewis's Theoretical Unity Thesis. Of course, the justified true belief account runs afoul of what epistemologists sometimes call 'Gettier examples',[29] so it is reasonable to investigate such examples to see if any casts doubt on Lewis's thesis.

In making the points of the above paragraph, I did not wish to suggest for a moment that Lewis was unaware of the similarities between his account and the justified-true-belief account of knowledge or of the many problems that have plagued that account. Indeed, there is a passage in Lewis 1986 in which he suggests that Gettier-type problems do not arise for the case of "simple non-contingent matters" such as the statements expressed by '2 + 2 = 4' and 'No contradiction is true'. Even if these statements were accepted by some person only because he was told to by a guru, still, Lewis thinks, the person would know that 2 + 2 = 4 and that no contradiction is true (Lewis 1986: 113).

But even if we grant Lewis, for the sake of argument, that his account of knowledge is not undermined by Gettier-type problems *in the case of simple non-contingent matters*, that would not shield his epistemological

[28] This suspicion has been confirmed by Lewis in his letter to me of 2 Apr. 1996.

[29] See Nozick 1981: ch. 3, sect. 1, for a discussion of such examples.

position from such problems, when we are concerned with our supposed knowledge of *other worlds*! There is an enormous difference between the claim to know that 2 + 2 = 4 and the claim to know that there are other possible worlds. Would a person's acceptance of the claim that there are other possible worlds fail to be knowledge if he accepted the claim only because a guru told him to? Absolutely! What epistemologist would allow that this person really has knowledge of other worlds, when his belief is based upon such disreputable grounds?

Let us, then, investigate the plausibility of Lewis's Theoretical Unity Thesis, in the light of the large literature on Gettier-type examples. Let us begin with the following example. Sherlock is to meet the butler in a field in the dark of night. He sees the tall form of the butler coming toward him. Suddenly, the man raises his arm, pointing a pistol at him. A shot rings out. Sherlock is hit in the shoulder. He sees the butler running away. Sherlock is not seriously hurt, but he is initially puzzled when he finds the pistol that the man had dropped and he sees that only one round has been fired. Sherlock is convinced that he heard two shots, not one. He then expands his existential beliefs for the sake of theoretical unity: he can think of only one reasonable way of reconciling what he heard with what he saw, and that is to postulate the existence of another gunman hiding behind the hedge. Suppose that, in fact, the diabolical Dr Moriarty was hiding behind the hedge and that he fired the shot that hit Sherlock. However, Moriarty's gun was equipped with a silencer, so that only one shot could have been heard. It turns out that Sherlock had a bit too much laudanum that evening, so that the noise from the butler's gun produced in his mind the impression that he had heard two shots. Should we say, in this case, that Sherlock *knew* that there was another gunman? Obviously not. He believed that there was another gunman, and, luckily for him, his belief happened to be true. In my opinion, this is not a case of knowledge.

Lewis might respond to this objection by claiming that he had in mind cases in which the agent's existential beliefs are expanded in a more substantial way than is done in the above example: he might specify his Theoretical Unity Thesis in such a way that, if an agent's existential beliefs are expanded to include a belief in a *kind* of entity that the agent had not already believed in, and if this expansion is made for the sake of theoretical unity, and if the agent's new existential belief turns out to be true, then the agent knows that this new kind of entity exists. The above example does not cast doubt on this analysis because Sherlock does not come to believe in the existence of a kind of entity that he had not already believed in.

Well, it is not difficult to come up with examples that cast doubt on this version of Lewis's thesis. Suppose that Tom and Sue are graduate students in physics who room together. One day, Sue discovers that Tom has been reading her diary without asking her permission. So she sets about writing, in her diary, a purely fictional account of the results of an experiment that her group has conducted. The only way of accounting for these results, without substantially revising accepted theory, is to postulate the existence of magnetic monopoles—a type of entity whose existence is suggested by certain theoretical considerations but which has never been experimentally detected. Tom reads Sue's diary one day, and for the sake of theoretical unity, comes to believe in the existence of monopoles. Now suppose that in fact monopoles exist. Should we say that Tom knows that monopoles exist? Again, we have a belief that happens to be true—not a case of knowledge. So there are reasons for doubting the Theoretical Unity Thesis.

I can imagine a Lewis supporter revising or qualifying the thesis in such a way that the above example would no longer be a problem for it. However, I am inclined to think that any such revision or qualification would be open to other damaging examples. The huge literature on the justified true belief analysis of knowledge prompts me to regard this line of analysis as not very promising.

Since I see no good reason for accepting Lewis's epistemological thesis, the Unknowability Objection remains for me a formidable objection to Lewis's analyses of the modal operators. It seems to me preposterous to claim to *know* that there are countless worlds containing planets similar to, but different from, the earth, and on which there live people very similar to us—and also to claim that we know such things not on the basis of scientific studies or experimentation, but rather because accepting such worlds yields a reduction in the diversity of the primitive notions we need for our total theory. Considerations of the sort Lewis puts forward constitute such nebulous grounds for these radical beliefs that I can only wonder at Lewis's ability to claim, seriously, *to know* such things.

6. Additional Doubts about Lewis's Analyses

Reconsider Lewis's thesis that we know there exist possible worlds in which donkeys talk. As Lewis puts it: "If modal knowledge is what I say it is, and if we have the modal knowledge that we think we do, then we have abundant knowledge of the existence of concrete individuals not causally related to us in any way" (Lewis 1986: 110). One is tempted to reply along the following lines: there is a sense in which we know that it is possible that donkeys talk. In this sense of the term 'possible', it does not follow that there are possible worlds in which donkeys talk. There is a sense of the term 'possible' in which, from the fact that it is possible that donkeys talk, it follows that there are possible worlds in which donkeys talk. But in that sense of the term 'possible', we do not know that it is possible that donkeys talk. This seems to constitute good grounds for maintaining that these two senses are different, which agrees with the suggestion in Section 2 that we have grounds for thinking that the modal operators, as Lewis analyses them, do not coincide with the crudely explained pre-possible worlds kinds of modal operators discussed at the beginning. Indeed, it is just the legitimacy of identifying these two senses that the Modal Irrelevance Objection is intended to throw into doubt. Richards, in effect, asks: why should there be any connection between, on the one hand, the possibility (in some commonly used sense of the term) that donkeys talk and, on the other hand, the existence of some possible world in which donkeys talk? Jubien puts it this way: "As long as other worlds are entirely distinct, independent things (at least physically), it is hard to see how they can have anything at all to do with the truth or falsity of simple statements of possibility" (Jubien 1988: 304).

Matthew Henkin has come up with a novel way of developing the same basic point (in a personal communication). According to Lewis, the truth of the statement 'Chihara could have had eggs this morning for breakfast' depends upon whether there is a world in which someone very much like me, at a time that "corresponds" in some appropriate way to this morning, has eggs for breakfast. Imagine that, in a region of our space-time continuum—so remote from us that any light signal emanating now from this region would never reach us before all the stars in the galaxy have been extinguished—there exists a galaxy that is the twin of our own, in which there exists a planet just like the earth, in which there live people just like the ones that live on earth today, and someone, who is so much like me that he could be my identical twin, at a time that "corresponds" in some appropriate way to this morning, has eggs for breakfast. Now Henkin asks: why is the existence of this "world" irrelevant to the truth of the above modal statement? What is so special about Lewis's worlds—worlds that are not causally related to anything in our space-time—that make them just the ones that determine the truth or falsity of our modal statement? In other words, who gave the Lewis "worlds" their special status?

Lewis's suggestion that he has explained the connection between statements of possibility and the existence of his kind of worlds by "saying that the modal operators are quantifiers over" possible worlds simply does not address the concerns that underlie these objections. Few would judge Lewis's reply to these objections to be adequate.

Consider the following example: Barney Bailley advances an analysis of grammaticality, according to which a sequence of English words σ is a grammatical sentence of English iff σ has ever appeared in the imagination of the angel Gabriel. The Grammatical Irrelevance Objection is raised: "You need to explain, Bailley, what a sequence of English words appearing in the imagination of the angel Gabriel has to do with grammaticality." Suppose Bailley responds in the way Lewis did: "You would think that I *have* explained what a sequence of English words appearing in the angel Gabriel's imagination has to do with grammaticality, for instance by saying that a grammatical sequence is one that has appeared in that angel's imagination." The inappropriateness of this reply is obvious: it assumes the very thing that is being challenged, namely, the legitimacy of interpreting or understanding 'grammaticality' in the peculiar way Bailley advocates. Similarly, in responding in the way he does, Lewis has simply assumed that what is being challenged is the logical connection between modality *as he analyses it* and the existence of possible worlds. When Richards speaks of possibility, he clearly does not have in mind Lewis's sense of that term, according to which possibility presupposes the existence of possible worlds. But when Lewis replies that he has explained the connection, he is thinking in terms of his own sense of 'possibility'.

It may be objected, in response to the above passages, that Bailley's analysis of grammaticality differs from Lewis's analysis of modality in so far as Bailley does not provide any details of how what appears in the angel's imagination is supposed to connect with the semantics of grammaticality, as Lewis does for the case of possible worlds and the semantics of possibility. Lewis shows, modulo certain assumptions about possible worlds, that many intuitively acceptable modal principles can be derived from his possible worlds analysis—nothing comparable has been shown for Bailley's analysis.[30] True enough. But the point of my example is simply to make perspicuous the logic of the Unknowability and Modal Irrelevance Objections and also the inappropriateness of Lewis's reply to these objections. There is little point in supplying such additional details.

These considerations point to a diagnosis of Lewis's puzzling replies to the objections I have been discussing. The Modal Irrelevance and the Skyrms Objections are directed at throwing into doubt Lewis's *analysis* of modal statements. But Lewis responds to these objections from the perspective of one for whom the analysis is already beyond question. Recall that his response to the Modal Irrelevance Objection was that he had already explained what worlds have to do with modality by saying that the modal operators are quantifications over them. But that is just what was being disputed: the objector was claiming that modal operators cannot reasonably be taken to be quantifications over worlds, since these worlds are irrelevant to modality. So, from the point of view of his disputers, Lewis was simply begging the question against his opponents.

Take Skyrms's objection that if possible worlds are real worlds, with

[30] This sort of objection was raised at my lectures at both the University of Notre Dame and the University of California, Davis.

real flesh and blood people in them, just like the actual world, then any claim to know about them and their contents should be backed by the same sort of evidence as we would require for any claim to know about some physical constituent of reality. Skyrms was disputing Lewis's analysis of modality, suggesting that, since we obviously cannot obtain any evidence of the appropriate sort supporting significant hypotheses regarding the existence and contents of such other worlds, if Lewis's analysis were correct, we would not have any knowledge of modality—an absurdity. Again, Lewis's reply can be seen to be coming from the position of one for whom the analysis in question has already been adopted. He simply takes it for granted, in his reply, that modality is quantification over worlds, so he can assert that to know that there exists a world of the appropriate sort, all that is needed is that we know such-and-such is possible. And it is easy to know that. Again, we have a reply based upon the position that is being disputed—from Skyrms's point of view, a kind of begging the question.

Returning to the Bailley example, imagine that an Unknowability Objection is raised: "If your analysis were correct, Bailley, we would not know if any sequence of English words were grammatical." We can imagine Bailley responding in the Lewis fashion:

> You probably think that discovering what has appeared in the angel Gabriel's imagination is impossible because you think it is impossible to discover facts about causally inaccessible things. But mathematics refutes such an impossibility thesis: mathematics is more certain than any epistemological theory that implies the impossibility of learning anything about the causally inaccessible. Since we do have knowledge of what is grammatical, we do have knowledge of what has appeared in the angel Gabriel's imagination. Nothing is easier.

I should now like to examine Lewis's justifications for his radical ontological theses.

7. Doubts about Lewis's Justification of his Mathematical Realism

The earlier quotation from Lewis, in which he explained why we are justified in believing in the existence of sets, contains some amusing rhetoric—Mafia talk about "an offer you can't refuse"—but apart from such rhetoric, the case really comes down to the simple idea that, if we postulate the hierarchy of sets described in standard versions of set theory, we can reduce the diversity of the primitives and axioms of our total theory, thus contributing to the overall unity of our total theory; and *that* makes it rational to postulate sets.[31] It is this simple idea that I wish to investigate.

Is it obvious that we can reduce the diversity of the axioms of our "total theory" by postulating the existence of sets, as Lewis claims? It is hard to see how we can answer this question when we have only the vaguest of ideas of what our "total theory" is supposed to look like. How do we know that we can reduce even the number of axioms of our total theory, if we do not know what the theory is? Besides, why should we suppose that our total theory *has axioms*? Indeed, what is an axiom of our total theory? Is our total theory supposed to be a deductive theory of some sort? In the next section, I shall give some reasons for questioning this way of regarding the "whole of what we take to be true". But for now, let us just accept Lewis's way of regarding the situation.

I can imagine Lewis responding that, whatever our "total theory" may be, it would have to include mathematics. Since mathematics can be reduced to set theory, he may have reasoned that all the axioms we would need for mathematics would be the "meagre axioms of set theory", and these would surely be fewer than all the axioms we would need for mathematics.

But suppose that our "total theory" already contains a recursive (infinite) set of axioms. Then replacing some finite number of them with some small finite number will not reduce the number of axioms. And even if our total theory contains only finitely many axioms, it is not obvious that eliminating all of the axioms of mathematics for those of set theory would result in a reduction in the total number of axioms. It surely would depend upon the system of set theory we decided to accept. If we opted for Zermelo-Fraenkel set theory, for example, we would be taking on an infinite number of axioms, so we might end up with even more axioms than we had originally.[32]

Still, the axioms of set theory are all *set theoretical* axioms; so even if the *number* of axioms is not reduced, still it could be argued that a unification of sorts would be achieved by the reduction: all of the axioms would be about just one sort of thing, namely, sets. But even this view needs further investigation. The usual reduction of all mathematical notions to set theoretical ones requires many definitions. In typical formalized theories, definitions are object-language sentences that have the status of axioms. So even if one started with only set theoretical axioms,

[31] Lewis's reasoning bears an obvious similarity to Quine's reasons for believing in sets. For a detailed critical discussion of Quine's Mathematical Realism, see Chihara 1990: ch. 1, sect. 2.

[32] Lewis has responded that counting axioms is not a good measure of theoretical disunity, when most of the axioms are instances of a single axioms (personal communication).

one might well have to add a huge number of definitional axioms about all sorts of mathematical entities, such as relations, functions, numbers, vectors, spaces, and the like. At this point, the difference between this way of proceeding and the unreduced way may not appear to be all that great. From the perspective of the set theoretical reducer, we start with the domain of sets and then break up this domain into relations, functions, numbers, spaces, etc. From the alternative perspective of most practising mathematicians, we start with the domain of mathematical objects and then break this domain into sets, relations, functions, numbers, vectors, spaces, etc. Is there all that much difference to these contrasting procedures? We can grant that there are advantages to the set theoretical approach to mathematics. We do get a nice unified approach to mathematics, where the rules of what is and what is not allowed are made clearer. But I think the set theoretical realists have grossly exaggerated these advantages. Is it obvious that the set theorist's way carries all the advantages? After all, there may be a steep price to pay for the set theorist's way of proceeding: the resulting theory may be more complex, trickier to teach, harder to work with, or more difficult to comprehend. To the physicist, all the added details required to work entirely within set theory may seem unnecessary, confusing, and physically meaningless. Is it obvious that achieving such a reduction would be all that theoretically significant? (It is noteworthy that physicists do not construct their scientific theories using the sort of mathematical system that Lewis advocates: their theories are not formulated in terms of Zermelo-Fraenkel set theory.) Surely, a detailed and convincing argument would be required, if such a view is to support our belief in sets.

Even more fundamentally, what is so important about reducing the diversity in the axioms and primitives of one's "total theory"? I know of no widely accepted general principles of confirmation theory or justification theory from which one could deduce an explanation of Lewis's doctrine. Consider the case of the sentential calculus. In their introduction to the second edition of *Principia Mathematica*, Bertrand Russell and Alfred North Whitehead assert that the "most definite improvement resulting from work in mathematical logic during the past fourteen years" is the replacement of the two primitive logical connectives '∨' and '–' with the one connective '|' known as the Sheffer stroke (named after the logician D. H. Sheffer, who defined it). Consequently, they tell us, Jean Nicod showed that the five axioms of the sentential calculus of *Principia Mathematica* could be replaced by a single axiom formulated in terms of the Sheffer stroke. The first of the two connectives replaced is a binary connective; the second is a unary connective. The replacing system has only the one binary connective. And the one axiom of the resulting sentential calculus is about this one connective. The whole of the sentential calculus could then be regarded as being concerned with only one binary primitive. So one obtains, by this replacement, a reduction (of sorts) in the diversity in the primitives and axioms of the sentential calculus. However, I know of no logician today who thinks that this reduction is extremely important.[33] Indeed, one would think that if reducing the diversity of primitives and axioms of the sentential calculus in this way is all that important, then we would all be using a Sheffer stroke sentential calculus or at least teaching such a system in our logic courses. But in fact, no one that I know of either uses the stroke or teaches the stroke in his or her courses.

8. Doubts about Lewis's Justification of his Modal Realism

Lewis's case for believing in possible worlds is much weaker than his case for believing in sets. One can at least give a plausible argument for claiming that the set theoretical reduction of all of mathematics accomplishes a reduction in the kind of axioms one needs in mathematics. Lewis does not even pretend to show that belief in possible worlds yields any sort of reduction of the *axioms* of our total theory. So *the whole justification for belief in possible worlds rests upon the supposed reduction in the diversity of the primitives of our total theory*.

All this weight Lewis puts on reducing the diversity of the primitives of one's total theory is highly suspect right from the start. After all, what *is* a primitive of one's total theory? And how is one to tell what is, and what is not, a primitive? There is no difficulty in picking out the primitives of, say, Frege's formal system of logic of his *Fundamental Laws of Arithmetic*: that system is a formalized axiomatized deductive theory. Such a theory has starting-points, namely, the axioms, and a recursive set of primitive symbols, out of which all the sentences of the theory are constructed. The primitives of such a theory will simply be the non-logical constants occurring in the axioms. But how does one find the primitives of someone's total theory? Does the whole of what someone takes to be true even have axioms? If so, how does one pick them out?

To think that the whole of what one takes to be true can be axiomatized in some language is questionable for a variety of reasons. Let me enumerate just a few of these reasons. (1) One's total theory—the whole of what one takes to be true—involves a great many beliefs that are just too vague and amorphous to be accurately expressed in sentences. (I have in mind

[33] It should be pointed out that Jean van Heijenoort's *Sourcebook* (van Heijenoort 1967) contains fourteen works that were published during the fourteen years between 1913 and 1927 when the first and second editions of *Principia Mathematica* were published, and neither Sheffer's article nor Nicod's work is included.

such things as hunches and belief attitudes tied to feelings of foreboding, familiarity, or *déjà vu*.) (2) Besides, it is misleading and unrealistic to suppose that the totality of what we believe can be expressed as a set of sentences, since this suggests that these sentences are just believed, when in fact, as Bayesians have emphasized, we believe most, if not all, things to varying degrees—we are more or less confident that Homer was one person, that Shakespeare wrote *Hamlet*, that Fermat's Last Theorem has been definitely proved.[34] (3) Our beliefs are undoubtedly linked in all sorts of ways (e.g. by association, similarities of content, and memories) that simply won't be captured by the deductive logic of the language one uses. (4) One's total theory, when regarded as an axiomatized theory, will very likely be *inconsistent*—the various paradoxes and antinomies that have been uncovered over the years provide us with some indications of how this could be[35]—so it is likely that an axiomatized deductively closed version would have to assert everything. (5) If one's total theory were not inconsistent, it would probably be incoherent in various ways—the history of philosophy shows us how the most brilliant of thinkers can have fundamental beliefs that are incoherent.

Let us, however, subdue our doubts about this way of regarding the whole of what we take to be true. I intend to evaluate Lewis's justification (of his belief in possible worlds) by means of an investigation of two questions: (i) Is it obvious that we *can* reduce the diversity of the primitives of our total theory by postulating possible worlds, as Lewis claims? (ii) If the postulation of possible worlds does effect a reduction in the diversity of the primitives of our total theory, does that make it rational to believe in possible worlds?

Let us first consider question (i). (I shall take up question (ii) later in this section.) It may seem that it is obvious that we do get a reduction in primitives from the postulation of possible worlds and possible objects. We have seen how Lewis is able to define *possibility* and *necessity* in terms of possible objects. As I said earlier, Lewis takes *properties* to be sets of things (both actual and possible), and he takes *propositions* to be properties of possible worlds. So it would appear that starting with the primitives of Lewis's basic ontology of sets, one can obtain a theory of sets and possible worlds; one can then go on to eliminate *possibility*, *necessity*, *properties*, and *propositions* from the list of primitives that are apparently needed in our total theory. So it would appear that a reduction in primitives can be achieved.

Unfortunately, it is hard, if not impossible, to evaluate such claims when we have no clear idea of how one should determine what is and what

[34] See Chihara 1981, where this is argued in detail.
[35] See, for example, Chihara 1979.

is not a primitive of our "total theory". However, there are some reasons for questioning that one does get a reduction in the primitives of one's total theory in the above way. I do not believe in possible worlds and counterpart relationships. So there are grounds for thinking that I do not mean by 'possible' what Lewis does: I do not intend 'possible' to involve quantification over possible worlds. Now suppose that I become convinced by Lewis that I should adopt his way of understanding modal operators, property talk, and sentences containing the term 'proposition'. Then we need to distinguish at least two senses of the term 'possible': the old pre-Lewisian sense, and the new possible worlds version. Similarly, we need to distinguish at least two senses of the terms 'property' and 'proposition'. Now, by adopting Lewis's possible worlds versions of these terms, we could, it would seem, simply drop from the totality of truths we hold to be true those that are expressed using the old senses. For example, *whereas, before, I believed that my Acura has the property (in the old sense) of being red, I now believe that it has the property (in the new sense) of being red.* Since the new sense of property is a defined term, I can drop the term 'property' from the list of primitives of my total theory. In this way, it would seem, the adoption of Lewis's version of these notions will allow me to reduce the primitives of my total theory.

But this scenario is much too simple to be acceptable. Consider the italicized sentence in the previous paragraph. Won't that very sentence express one of my beliefs? And doesn't the first occurrence of the term 'property' which occurs in that sentence have the old sense? Or consider the belief I have which is expressed by the sentence: 'My wife believes that it is possible that donkeys talk.' The term 'possible' in that sentence has the old sense, since I know that she thinks belief in possible worlds and counterparts is fantastic. Indeed, won't there be indefinitely many beliefs of this sort remaining in my total theory—beliefs expressed by sentences containing occurrences of 'possibility', 'property', or 'proposition', where the terms have their old sense? Thus, there are grounds for questioning whether postulating possible worlds does allow me to economize on the diversity of primitives of "my total theory" after all.

Turning to question (ii), even if one can achieve a more parsimonious set of primitives by postulating possible worlds, why should that give us good reason for thinking there are such possible worlds? Even if it is agreed that a reduction in the diversity of primitives is theoretically beneficial, and even if it is allowed that postulating possible worlds does yield a genuine reduction in the diversity of primitives, do the benefits outweigh the considerable ontological costs? Is the price right?[36]

[36] Bas van Fraassen has argued that the kind of highly theoretical benefits promised by Lewis's sort of ontological postulation pale besides the potential costs of believing in unreal entities. See van Fraassen 1995: 147.

Lewis seems to think so, because he does not believe that the ontological costs are very weighty. He distinguishes two types of *ontological parsimony*: quantitative parsimony and qualitative parsimony. A theory or doctrine is *quantitatively parsimonious* if it requires a relatively small number of items of the kind of entity it postulates. A theory or doctrine is *qualitatively parsimonious* if it requires a relatively small number of "fundamentally different *kinds* of entity" (Lewis 1979*b*: 185). Lewis agrees that qualitative parsimony has theoretic value, but he sees no such value in quantitative parsimony. He allows that his Modal Realism is quantitatively unparsimonious, since he populates his ontology with so many possible worlds. Surprisingly, however, he does not believe that his modal views are qualitatively unparsimonious, since we believe in our actual world already. "I ask you to believe in more things of that kind," he says, "not in things of some new kind" (Lewis 1979*b*: 185).

Despite Lewis's clever attempt to deflect the charge of ontological excessiveness, Lewis's ontological doctrines can be seen to be qualitatively unparsimonious. After all, as Joseph Melia has pointed out, Lewis's version of Modal Realism requires an ontology containing unicorns, spirits, gods, trolls, golden mountains, parallel universes—indeed, every possible kind of thing (Melia 1992*b*: 192).

"Oh no," it might be replied, "Lewis does not have an excessive ontology: He is not ontologically committed to the existence of *actual* unicorns, spirits, etc., but only *possible* unicorns, spirits, and the like." But this clever manœuvre should not distract us from the justness of Melia's objection. Recall that possible people are held to be just as real and full-blooded as actual ones. Also, reconsider the example of Lewis#. He accepts, you will recall, exactly the same ontology that Lewis does. And clearly Lewis# is ontologically committed to the existence of unicorns, spirits, etc. Therefore, Lewis is too. Indeed, Lewis is quite explicit about the existential commitments of his views. He makes it quite clear that he uses the existential quantifier in a standard way. Responding to an objection of Lycan's, he writes: "I quantify just the way he or anyone else does: over all the entities I think there are" (Lewis 1986: 98). So Lewis does think there are unicorns, spirits, and all manner of strange creatures. *The ontological commitments of his doctrines are as heavy as any theory any philosopher I know has ever espoused.* This is a very high theoretical price to pay for something as nebulous and questionable as a parsimonious set of primitives.

Up to this point in my analysis of Lewis's justification for his Modal Realism, I have been accepting, for the sake of argument, a big assumption of Lewis's reasoning, namely that the "benefits in theoretical unity and economy" that the postulation of theoretical entities supposedly brings to our total theory provide us with good grounds for believing that there really are such theoretical entities. This is an assumption that is far from obvious and it should certainly be questioned. Gideon Rosen has expressed wonderment that such considerations of utility "should be even prima facie compelling for Lewis given his other views" (Rosen 1990: 338). The reason is that Lewis allows that there is no necessary connection between a theory's being unified, economical, elegant, etc. and its being true. But Lewis's argument for his Modal Realism, at best, shows that the postulation of possible worlds allows us to have a more unified, economical, elegant, etc. total theory. And that may provide us with reason to hope the theory is true.

The trouble is that reason to hope is not in general reason to believe. To suppose otherwise is to countenance wishful thinking as a sound policy for fixing belief in metaphysics.[37]

9. Magical Ersatzism

Part of Lewis's defence of his Modal Realism consists in attacking alternatives that have been proposed to his own view. In particular, he delivers an impressive array of objections against what he calls 'ersatz modal realism' or 'ersatzism'—the view that there is only one world, the actual one, but an infinity of abstract entities that go proxy for the possible worlds that Lewis postulates by somehow representing ways that this one world could have been (Lewis 1986: ch. 3). Since these representing entities are supposed to be actual entities, ersatzism is a version of Actualism. As Lewis views these alternative views of modality, the abstract entities that supposedly do the representing he calls "ersatz worlds" and the proponents of these alternatives are called "ersatzers".

One form of ersatzism he takes up is *linguistic ersatzism*. This is the view that the things that go proxy for possible worlds are sets of sentences. One form of linguistic ersatzism takes the sentences in question to be sentences of some natural language. According to this view, a set that functions as a possible world would have members that literally say what would have happened or been the case had the world been as it is represented as being. Lewis can easily show that such a form of ersatzism is untenable. For one thing, there are just not enough "worlds" to represent all the possibilities we want: there is at most a continuum of such "worlds", whereas there are clearly more possibilities than that (Lewis 1986: 143).

In this section, I shall concentrate on Lewis's objections to "magical ersatzism", since these objections are directed at Plantinga's notion of

[37] Rosen 1990: 338 n. 18. For a quite different kind of critical response to Lewis's justification for believing in possible worlds, see van Fraassen 1995.

possible worlds, which I shall be taking up next. Enough has been said about Plantinga's views already for the reader to see why Lewis classifies Plantinga as an ersatzer. From Lewis's point of view, Plantinga's possible worlds—his maximal possible states of affairs—are not genuine worlds at all: they are artificial worlds that somehow "represent" ways the world might have been.

The crucial question for the ersatzer to answer, according to Lewis, is this: how is this representing done? Let me clarify this question. One of the principal goals of Lewis's analysis of modality is to give the *truth conditions* of modal statements in terms of the existence or non-existence of the appropriate possible worlds. For the ersatzer, one wants to give the truth conditions of modal statements in terms of the existence or non-existence of the things that go proxy for the possible worlds. Thus, as Lewis analyses modality, Humphrey could have had a wooden leg iff there is world that represents Humphrey as having a wooden leg. So the question arises: how is this representing done? For Lewis, the worlds represent possibilities via their parts: a world may represent Humphrey as having a wooden leg by having a part that is sufficiently similar to our Humphrey for it to be a counterpart of Humphrey—the counterpart of Humphrey having a wooden leg.

Now how do Plantinga's "possible worlds" represent? Plantinga is silent about this matter: Plantinga does not even attempt to explain how the maximal state of affairs represents Humphrey as having a wooden leg. This is why Lewis thinks of Plantinga as a "magical ersatzer", that is, one who, in effect, holds that the representing is done by magic (Lewis 1986: 183).

To clarify further the problem Lewis is posing for the ersatzer, let us call the one concrete (actual) world we live in 'the C-world'.[38] Now there are infinitely many maximal possible states of affairs, but almost all of these states of affairs do not obtain. Somehow, the C-world *makes obtain* some maximal possible state of affairs: this state of affairs is the one that "represents" this C-world. But what is it about that maximal state of affairs that makes it the one that "represents" this particular C-world? In virtue of what properties does that maximal possible state of affairs represent this C-world and not some other one? In other words: why does this C-world make just that maximal state of affairs obtain?

Lewis does not demand that the ersatzer define the *makes obtain* relation. He is willing to allow that the ersatzer may take the relation to be a primitive of her theory. But Lewis feels that it is legitimate to ask for, at least, a classification of the relation (Lewis 1986: 176). He thus asks: is this

[38] I here follow van Inwagen's usage in van Inwagen 1986.

relation intrinsic or extrinsic? If the former, is it internal or external? To understand these questions, we will need some definitions. First of all, we need to distinguish what Lewis calls 'intrinsic properties' from what he calls 'extrinsic properties'. An *intrinsic property* is one that a thing possesses in virtue of how that thing itself is, whereas an *extrinsic property* is possessed by the thing in virtue of its relation or lack of relation to other things (Lewis 1986: 61).

Consider now Michael Jubien's explication of what *intrinsic* relations are (in Jubien 1991). There are two kinds of intrinsic relations: internal and external.

> A relation is *internal* if its holding *depends only on the intrinsic nature of the relata taken by themselves.*

Thus, whether or not *A* is taller than *B* depends upon the nature of *A* and *B* (what they are like in and of themselves, without reference to their relationships to other things).

> A relation is *external* if its holding *depends on the intrinsic nature of the composite of the relata*, but not just of the relata themselves.

For example, the relationships of distance holding between the electron orbiting the proton of a classical hydrogen atom are not internal, since these relations do not just depend upon the intrinsic natures of the electron and the proton taken separately. But if we take the composite—the hydrogen atom—then the relations holding do depend upon the intrinsic nature of the composite.

> A relation that is not intrinsic is said to be *extrinsic*. An extrinsic relation, then, is one that is neither internal nor external.[39]

An example of an extrinsic relation is that of being *an owner of*: a person is related to a piece of property by this relation. (I shall return to this example shortly.)

Lewis is thus asking the ersatzer to classify the *makes obtain* relation as internal, external, or extrinsic. Suppose that Plantinga decides that the relation is internal. Then Lewis will want to be told what it is about just that maximal possible state of affairs in virtue of which the C-world *makes it (and not the others) obtain.* The problem, of course, is that Plantinga knows nothing of the intrinsic nature of maximal possible

[39] Lewis has pointed out to me (personal communication) that calling a relation 'extrinsic' can produce confusion since it might suggest merely that the relation is external, or it might suggest that the relation is not even intrinsic to the sum of the relata: some third thing, distinct from *a* and *b* is involved in making it so that *a* and *b* are related (as for example in the case of the relation of *having the same owner*).

states of affairs. They are abstract entities, with which we are not in any sort of causal relationships. We have no understanding of or insight into the differences among these abstract entities in virtue of which some one of them is in the *makes obtain* relation with the C-world. Then, says Lewis, it is a mystery how any one could have grasped this relationship. "If the ersatzer has understood his own primitive, he must have done it by magic" (Lewis 1986: 178).

Perhaps, then, the relation is an external one. Here, the difficulty is in seeing how the composite natures of the concrete world and the various abstract entities in question can account for the relationships that are thought to obtain. Since one term of the relationship is entirely beyond our apprehension, it is hard to see how we could have grasped this relation.[40] After all, the composite in question consists of something concrete with something abstract. It is a mystery how one would have more insight into just what this relation is when one takes into account not just the intrinsic natures of the relata, but also the intrinsic nature of the composite. What more is there to grasp when one has the composite?

Since Lewis does not think there is any real possibility that the *makes obtain* relation could be extrinsic, he concludes that, whether the relation of selection is internal or external, there is an insoluble mystery of how we could have grasped the relation. Therefore, says Lewis, "ersatzism is to be rejected" (Lewis 1986: 182).

Peter van Inwagen, who has defended Plantinga's views on possible worlds on several occasions, is impressed by Lewis's objection. He gives it the following form: suppose that a philosopher were to put forward a theory of *typosynthesis* that makes the following assertions:

[1] There are exactly ten cherubim.
[2] Each human being bears a certain relation, typosynthesis, to some but not all cherubim.
[3] The only things in the domain of this relation are human beings.
[4] The only things in the converse domain of this relation are cherubim.

The philosopher admits that she cannot single out or refer to individual cherubim except by means of the term 'typosynthesis' (e.g. 'the cherub that all Greeks bear typosynthesis to'). Nor can she single out even one non-empty set of cherubim without bringing in the typosynthesis relation. In that case, says van Inwagen, if it is admitted that no one knows any of the relevant intrinsic properties of a cherub in virtue of which the relation should hold, it must be by magic that we understand what 'typosynthesis'

[40] Lewis 1986: 179. This is not Lewis's only reason for being sceptical of classifying the relation as external, but it is not necessary for my purposes to go into that.

means. Such a theory must be rejected as philosophically unsatisfactory (van Inwagen 1986: 206).

Having thus characterized Lewis's objection to Plantinga's possible worlds theory, van Inwagen is forced to admit that he can find nothing wrong with the reasoning. Yet, he feels that there must be something wrong, since it seems to prove too much. It seems to show, he argues, that we do not understand set theory. If so, says van Inwagen, we would not understand mathematics—which is absurd. So there must be something wrong with the argument.

Now how can this reasoning be used to show that we do not understand set theory? van Inwagen asks the set theorist to explain just what this relation of membership is that obtains between certain things and certain sets. Is this relation internal, external, or extrinsic? Suppose that it is internal. Consider the various unit sets. What are the intrinsic properties of some unit set in virtue of which just that one set is the one that has its only member Bill Clinton? We haven't the vaguest idea. We are not in any sort of causal relation with sets. So we don't know anything about the intrinsic natures of sets. Evidently, it is a mystery how anyone could have understood this relation. It can be seen that the mystery is not solved by supposing the membership relation to be external or extrinsic.

Michael Jubien takes up the dispute in Jubien 1991. Unlike van Inwagen, Jubien thinks he knows what is wrong with Lewis's argument. Lewis argues that, if the *makes obtain* relation is internal, it is a mystery how we could have grasped the relevant intrinsic properties of the maximal possible states of affairs in virtue of which the C-world makes obtain just the right candidate. Jubien claims that Lewis's objection to the ersatzer presupposes that "we have to know about the relevant intrinsic features of the *relata* in order to "grasp" an internal relation" (Jubien 1991: 251). This, for Jubien, is an unreasonable demand. He gives an example to support his objection. Take a cube of sugar that is soluble in the water found in some particular glass. Then the sugar is related by the *soluble in* relation to this particular mass of water. But a person could understand this internal relation without knowing anything about the chemistry of solutions and hence anything about the relevant features of the intrinsic natures of the relata:

> I think we can and do grasp internal relations without understanding their inner workings. In a Kripkean spirit, I maintain that we can do this in much the way that we can pick out something like *heat* by means of its outward manifestations, without knowing that it happens to be the motion of molecules. (Jubien 1991: 251)

What is wrong with Lewis's objection to the supposition that the relation is internal, according to Jubien, is the unreasonable demand that Lewis places on understanding an intrinsic relation.

But has Jubien solved the mystery of how we could grasp the *makes obtain* relation? In order to evaluate Jubien's criticisms of Lewis's argument, we need to re-examine what is meant by 'internal', 'external' and 'extrinsic' relations. To do this with any confidence, we need to examine the definitions of these terms given by Lewis.[41] (Recall that I used the explications of these terms given by Jubien.) Here's how Lewis defines 'internal relation':

An internal relation is one that supervenes on the intrinsic natures of its relata. (Lewis 1986: 62)

He then explains what is meant by 'supervenience' in the following way:

Suppose that objects X^1 and Y^1 stand in the relation. If objects X^2 and Y^2 do not stand in this relation, then there *must* be a difference in the intrinsic nature of either the Xs or the Ys. And if there is no difference in the intrinsic nature of either the Xs or the Ys, then X^2 and Y^2 *must* stand in this relation.

It is clear from this explanation that there is a modal element implied by the characterization of a relation as internal. This modal element, which may not be evident from the definition given by Jubien, is something we need to investigate if we are to evaluate Jubien's example with accuracy.

Some insight into this modal element can be obtained from Lewis's discussion of the notion of *supervenience*. He tells us that "we have supervenience when there could be no difference of one sort without differences of another" (Lewis 1986: 14). And the 'could' here is clearly stronger than mere physical possibility. Indeed, his comments on p. 17 indicate that he has in mind the 'could' of what I have been calling 'broadly logical possibility'.[42]

In this light, reconsider Jubien's supposed example of an internal relation that we can grasp even if we do not know anything about the intrinsic natures of the relata that accounts for the relation *S* obtaining. Suppose that there is a possible world *v* in which there is a cube of sugar and some other glass of water, and these are such that the cube of sugar of world *v* has the same intrinsic nature as the cube in the actual world and the water of world *v* has the same intrinsic nature as the water in the actual world. Might it not be the case that, in *v*, the cube of sugar is not soluble in the

[41] In particular, we need to examine the definitions of the relevant terms given in Lewis 1986, where Lewis gives his objections to ersatzism. He gives slightly different definitions of these terms in Lewis 1983: 356 n. 16.

[42] See also his discussion of the dot-matrix picture (Lewis 1986: 14–15)—something that has global properties that supervene on the dots. No two such pictures could differ in their global properties without there being somewhere a dot in one where there is no dot in the other. This 'could' is not mere physical could. It has to be something like metaphysical or conceptual possibility.

water? After all, the laws of physics could be different in world *v*. If so, *being soluble in* is not an internal relation. It would seem, then, that there is some doubt that Jubien has come up with an appropriate counterexample to what Lewis is presupposing.

Besides, the *being soluble in* example differs in crucial ways from the *makes obtain* case. In Jubien's case (unlike Lewis's), we are in causal contact with both of the relata: this allows us to infer, from past experience with cubes of sugar dissolving in water, that the one before us would also dissolve if we dropped it into this water. And we would be justified in making such an inference, even if we do not know anything about the chemical composition of the sugar or the water. In the case of Plantinga's worlds, we have no such causal contact with any of them, and we have no past experiences upon which to make the sort of inference we could make in the sugar case. Thus, even if solubility is an internal relation, it is by no means obvious that we can clear up the mystery of how we could have grasped the *makes obtain* relation.[43]

One can also make the case that *being soluble in* just is a modal relation that obtains in virtue of what the relata would do in such-and-such circumstances, that is, the sugar would dissolve in the water (at normal temperature and pressure) if the cube were immersed in the water. And we know what it would be for the cube to be immersed in the water and for the cube to dissolve in the water. This makes the *being soluble in* relation very different from the *makes obtain* relation.

What, then, should we make of the Lewis–van Inwagen reasoning? Are we forced by this argumentation to conclude that no one understands mathematics? Consider how Lewis has responded to van Inwagen's *ad hominem*. Surprisingly, he does not, as one might expect, attack van Inwagen's conclusion that set theory is in the same predicament as magical ersatzism. Lewis accepts much of the force of van Inwagen's reasoning,[44] writing: "His point is well taken, and my present complaints against membership . . . amount to my endorsement of it" (Lewis 1991: 36). So Lewis is forced to admit that he is in a fix.

[43] I have been informed by Jubien (letter dated 15 Aug. 1994) that part of his motivation for questioning Lewis's argument against Magical Ersatzism was to defend his belief in propositions against such an argument. In Pt. III of Jubien 1991, Jubien argues that we could have indirect evidence for the existence of propositions that represent by virtue of their inner structure. He then argues that if there really are such propositions, they must be complex. Thus, he attempts to provide additional grounds for rejecting Lewis's objection to Magical Ersatzism.

[44] Lewis is not willing to allow that van Inwagen's *ad hominem* is exact; however, he does admit that he is left in a very unpleasant situation.

It's a nasty predicament to claim that you somehow understand a primitive notion, although you have no idea how you could possibly understand it. That's the predicament I'm in. (Lewis 1991: 36)

But Lewis goes on to say (in apparent agreement with van Inwagen's sentiments) that he cannot accept the conclusion that he does not grasp the primitive relation of set theory, for that implies, he believes, that he should reject "present-day set-theoretical mathematics" (Lewis 1991: 36)—something he cannot do. He concludes:

If there are no classes, then our mathematical textbooks are works of fiction, full of false 'theorems'. Renouncing classes means rejecting mathematics. That will not do. (Lewis 1991: 58)

"Hot though it is in the frying pan, that fire is worse," he says (Lewis 1991: 36). So, whereas he had argued earlier that his arguments show that the ersatzer's theory must be rejected (Lewis 1986), his later view is that these same arguments do not show that set theory must be rejected.

Lewis's attitude on this matter is worth exploring. In Lewis 1991, the concept of singleton (or unit set) is taken as the one primitive of set theory. The rest, he tells us, is "mereology" (Lewis 1991: vii). Unfortunately, Lewis admits, we know nothing about "the nature of the primitive relation between things and their singletons" (Lewis 1991: 45). A number of mysteries about singletons bedevil us. Where are they? What is their intrinsic nature? Do they differ qualitatively from one another?[45] However, he feels that he cannot continue this train of argumentation to its conclusion that set theory is a philosophically unsatisfactory theory that should be rejected. He tells us that he is moved to laughter "at the thought of how *presumptuous* it would be to reject mathematics for philosophical reasons" (Lewis 1991: 59). "How would *you* like the job of telling the mathematicians," he continues, "that they must change their ways, and abjure countless errors, now that *philosophy* has discovered that there are no classes?" "Not me," says Lewis:

And so I have to say, gritting my teeth, that somehow, I know not how, we do understand what it means to speak of singletons. (Lewis 1991: 59)

So he continues to maintain his belief that he somehow grasps the fundamental primitive of set theory, on the grounds that giving up that belief will commit him to giving up set theory. And giving up set theory, he believes, amounts to giving up mathematics.

I believe Lewis has moved much too fast to his conclusion that the fire would be worse than the pan; a number of steps have been skipped over

[45] He goes on to say that singletons, and hence all classes, are "profoundly mysterious". See Lewis 1991: 57.

much too quickly. What, after all, does van Inwagen's *ad hominem* really show? Evidently, van Inwagen thought that Lewis's reasoning, applied to set theory, would show (if correct) that we do not understand mathematics. But what is it to understand mathematics? Let us restrict ourselves to just some area of mathematics, say analysis. By the usual criteria that we use to determine if someone understands a substantial amount of analysis, we see if the person can explain the principal concepts of analysis (by giving the relevant definitions, applying the definitions, and explaining their implications), knows the axioms or fundamental assumptions of the area, and also can cite, prove, explain, and apply (in a variety of contexts) the principal theorems of the area (both basic and advanced). Nothing in the Lewis–van Inwagen arguments shows that mathematicians do not understand mathematics *according to these criteria*. Thus, I agree with van Inwagen's reasoning to this extent: if his *ad hominem* showed that no one understands mathematics according to the above criteria, then we really would have an absurdity. But it doesn't.

This is not to deny, of course, that there are many things at the very foundations of mathematics or set theory that are not clearly understood. It has been my experience that a great many mathematicians would allow that there is much about the fundamental nature of mathematics (and set theory) that is not understood. Should the existential theorems of mathematics be taken at face value to be assertions of the existence of abstract mathematical entities? Should the axioms of, say, Zermelo-Fraenkel set theory be taken to be intuitively obvious truths? Should they be taken to be true statements of fact? Do the objects spoken of in set theory truly exist? Are they related to one another by the membership relation in the way apparently described by set theory? About such matters, there is no consensus. Most mathematicians I have known are very undogmatic. I suspect that many mathematicians would not claim to *know* the answers to any of the above questions. Certainly, many do not claim to know that the axioms of set theory are true statements about some actually existing entities. Nor would they claim to have knowledge of the existence and properties of causally inaccessible entities as Lewis claims for himself.

Mathematics is very useful. About this, there can be no doubt. But its usefulness is something in need of philosophical analysis and explanation. Is mathematics useful because there really are entities that we call 'sets', which are related by something we call the 'membership relation' in just the way set theory tells us they are? The answer to this question is by no means obvious. Mathematics has been described by Wittgenstein as a "motley of techniques". Certainly, one can regard the history of mathematics as the development of a whole array of ever more useful and increasingly sophisticated mathematical procedures, calculation methods, and techniques. I see no reason at all to suppose that "grasping" (in the way these philosophers require) the supposed relation that obtains between a set and its members is required to understand these procedures, methods, and techniques, or to have a real feeling for applying and exploiting these methods and techniques to solve sophisticated problems in science and engineering.

From the Lewis–van Inwagen point of view, set theory is a theory about certain abstract entities, the sets, that truly exist, and a particular relation, membership, that obtains among these entities. For this reason, it is easy to come to believe that, if we do not "grasp" this membership relation, we do not understand set theory. But this is a *philosophical* point of view (an extreme form of Set Theoretic Realism); it is not, itself, a piece of mathematics. Abandoning such a point of view is not tantamount to abandoning mathematics.

If the Lewis–van Inwagen arguments do not establish that we do not *understand* mathematics (in some appropriate sense of that term), they provide even less proof that mathematics is false. Hence, even if the arguments are completely sound, there is no need to tell mathematicians that they must "change their ways and abjure countless errors". There is no need to reject mathematics for philosophical reasons.

But these arguments do seem to raise some doubt about a certain philosophical view of the nature of mathematics and set theory—a view that Lewis and van Inwagen both share. The arguments also constitute grounds for questioning the sort of Set Theoretical Realism espoused and defended by Penelope Maddy—a view I shall discuss in more detail in Chapter 9.[46] But these are views I am more than happy to reject anyway. Indeed, I feel that there are independent grounds for rejecting such views.

What about Plantinga's Modal Realism? Does Lewis's "magical ersatzism" objection undermine Plantinga's analysis of modality in terms of his maximal possible states of affairs? Certainly, it raises doubts that need to be answered before the view can be judged to be acceptable.

10. Plantinga's Actualistic Modal Realism

I have been claiming that Lewis is a Modal Realist. But Alvin Plantinga has argued that Lewis is not a Modal Realist at all.[47] On the other hand, Plantinga considers himself to be a Modal Realist, whereas I do not classify Plantinga as a Modal Realist in the full-blown sense in which I have been using that term in this chapter. For me, Plantinga is only an ersatz Modal Realist. Who is right?

[46] See Maddy 1980 and Maddy 1990.
[47] Van Inwagen 1986: 189.

Let us examine Plantinga's reasons for refusing to classify Lewis as a Modal Realist. First, we need to become clear about how Plantinga defines some crucial terms. As he defines 'Modal Realism', the modal realist "asserts that there are such things as *possible worlds* and that for any (temporally invariant) state of affairs or proposition *S*, *S* is possible if and only if there is a possible world that includes or entails it".[48] But, as was noted in Chapter 1, Section 2, a possible world, according to Plantinga, is a "maximal possible state of affairs". So a possible world, as Plantinga uses the term, is not at all the sort of thing that Lewis would classify as a possible world.[49]

Given the above definitions, we can now see why Plantinga does not consider Lewis a Modal Realist. Lewis asserts the existence of "possible worlds", in one sense of that term, but these worlds are not maximal possible states of affairs. Lewis's possible worlds do not entail propositions. They are not abstract entities. So in Plantinga's sense, Lewis is not a Modal Realist. On the other hand, it is also clear that Plantinga is not a Modal Realist in my sense. For Plantinga does not assert the existence of possible worlds that are as real and concrete as the actual world. Plantinga's possible worlds are abstract. They do not contain, as genuine parts, real flesh and blood people.

We can now see why Plantinga can accept, and Lewis cannot, a person's being in more than one possible world. Using Plantinga's notion of possible world, Socrates being in two distinct possible worlds is regarded as comparable to (and no more paradoxical than) there being two distinct possible states of affairs about Socrates: one actual—say, *Socrates being shorter than Plato*—and one not—say, *Socrates being able to play the violin*. And this is similar to having two distinct descriptions of Socrates: one true—say, 'Socrates was shorter than Plato'—and one false—say, 'Socrates knew how to play the violin'. However, if we restrict ourselves to Lewis's notion of possible world, Socrates being in two distinct possible worlds would be comparable to Socrates being a part of two mereological sums α and β, where no part of α is spatiotemporally related to any part of β—something that just wouldn't fit our conception of Socrates.[50]

By now, it should be apparent that Lewis and Plantinga choose different sides on the Actualism–Possibilism dispute. Possibilism is the doctrine that *there are objects that do not exist (in the actual world) but that do exist*

[48] Van Inwagen 1986: 191.

[49] Cf. van Inwagen 1986, where one can find a detailed discussion of the differences between the sort of view of possible worlds that Plantinga advocates and Lewis's kind of view.

[50] Of course, there is more to Lewis's rejection of "overlap" of two worlds than I have indicated here. For details, the reader should study Lewis 1986: sect. 4.2. See also Armstrong 1989: 18–19.

in other possible worlds. Plantinga is quite explicit about his espousal of Actualism. Lewis, on the other hand, can be classified as a Possibilist since he maintains that there are merely possible objects that have properties and are related to other merely possible objects. I shall have more to say about Actualism and Serious Actualism later in Chapter 8.

Another difference between Plantinga's view of possible worlds and Lewis's concerns the actual world. Plantinga does not regard 'the actual world' as an indexical. Instead, Plantinga asserts that, among the possible worlds, there is at least one that includes only actual states of affairs.[51] He believes that, just as there are logically equivalent, but distinct, states of affairs, there may well be more than one such possible world. Hence, according to this conception, there may not be just one actual world. Such features of Plantinga's conception are indications that his possible worlds are not genuine worlds at all: they are more like descriptions of possible worlds than the worlds themselves. And this is one more reason why I do not classify Plantinga as a true Modal Realist.

I find it curious that Plantinga wants to assert the existence of merely possible worlds (possible worlds that are not actual), but not merely possible individual objects. As I noted earlier, Plantinga is an "Actualist" who holds that *there are no objects that could have existed but in fact do not*. Of course, he does not deny that, *in some sense*, there is some possible world in which there are people who do not exist in the actual world. But what this means, roughly, is only that these people would have existed had this possible world been actual. These possible people do not really exist—they are not actual. What does exist, according to Plantinga, are the *essences* of these people. However, it seems to me that what Plantinga should say, to be at all consistent, is that merely possible worlds do not exist either. That is, he should be a similar sort of Actualist with respect to possible worlds. What exists, according to his theory, is not what I would call a possible world but only a kind of maximally specific proposition describing what such a world would be like: the world itself, as I see it, simply does not exist. The merely possible worlds (in this sense) are no more actual than the merely possible individuals discussed above. Thus, there is a kind of inconsistency in Plantinga's views.

The inconsistency I mention here is not a strict inconsistency in the doctrines Plantinga espouses: I am not suggesting that Plantinga asserts both ϕ and not-ϕ. Rather, I am pointing to a kind of inconsistency in treatment. When Plantinga says that a possible world exists, different from the actual world, he means only that there is some maximal state of affairs which would have obtained iff there had been a concrete world, different from the one that actually exists, of the appropriate sort—there is no sug-

[51] Plantinga 1985*b*: 89.

gestion that any such concrete world actually exists. So one would think that Plantinga would allow that some possible man different from all actual men exists, meaning by this only that there exists a state of affairs that would have obtained iff there had been a man, different from all actual men, of the appropriate sort—without any suggestion that any such concrete man actually exists. But he doesn't.

There are suggestions in Plantinga's writings that the very notion of a merely possible object is incomprehensible or incoherent. For example, in contrasting the notion of a possible state of affairs with that of a possible object, he notes that a merely possible state of affairs is not something that does not exist—it does not obtain but it does exist.

> But a possible object that does not exist is a horse of a different colour. *This is a thing such that there exists no such thing*, to paraphrase Meinong—a monumentally perplexing idea. What could such a thing be like? (Plantinga 1974: 132, italics mine)

Of course, what I am suggesting as a possible object for Plantinga is just an appropriate state of affairs: say the state of affairs of *O having obtained,* where O is an essence that obtains in at least one possible world. Such a state of affairs could function in place of (concrete) possible objects much in the way maximal possible states of affairs function in place of (concrete) possible worlds.

11. Jager's Applied Semantics

It is time to re-examine Thomas Jager's version of quantificational modal logic that was developed to model the sort of "applied semantics" advocated by Plantinga[52]—a logical system, you will recall, that was discussed earlier in Chapter 1. Jager's possible worlds semantics for his system of modal logic (in Jager 1982) was devised to conform to both Actualism and Serious Actualism—doctrines that Plantinga explicitly espouses. The main ideas are these: in the formal semantics, an interpretation of the language **I** = <**W**, **D**, **w***, **d**, **r**, **e**>, is the same as an interpretation of **M** except for this one condition:

> For any *n*-ary predicate θ, **e**(θ, *w*) is a subset of the *n*th Cartesian product of **d**(*w*).

[52] After describing the essentials of his applied semantics for quantificational modal logic, he wrote: "Thomas Jager has worked out a pure semantics for quantified modal logic that appropriately reflects this way of viewing the matter" (Plantinga 1985*b*: 92).

This condition guarantees that interpretations will conform to Serious Actualism. In order that identity should conform to Serious Actualism, the following definition is given:

> The binary relation symbol 'I^2' is a logical constant. In any interpretation **I**, $\mathbf{e}(I^2, w)$ must be the set of all ordered pairs $<\alpha, \alpha>$ such that $\alpha \in \mathbf{d}(w)$. (In other words, 'I^2' is interpreted to be the identity relation on $\mathbf{d}(w)$.)
>
> For all individual symbols α and β, '$I^2\alpha\beta$' will be abbreviated: '$\alpha = \beta$'.

It is the *applied semantics* of this modal logic that embodies the distinctive views of Plantinga. In the applied semantics, one is required to take **W** to be a set of "possible worlds" in Plantinga's sense of that term, i.e. **W** must be a set of maximal possible states of affairs. **D** must be a set of essences; **d** must be a function which is such that, for any maximal possible state of affairs w, $\mathbf{d}(w)$ is the set of all essences exemplified in w; and $\mathbf{e}(\theta, w)$ must be the set of n-tuple of essences which are coexemplified with θ in w.[53] Since possible worlds and essences actually exist, according to Plantinga, such an interpretation will be Actualistic.

The definition of truth at a world under an interpretation needs to be changed slightly in order to give '$\Box$' the essentiality reading (as discussed in Section 2 of Chapter 1). The main idea is to change the definition of truth for the case of *de re* sentences in the following way:

> In what follows, let ϕ be a sentence and let $\alpha_1, \alpha_2, \ldots, \alpha_n$ be the individual constants (not necessarily distinct) that occur in ϕ.
>
> The sentence
>
> $$\Diamond\phi$$
>
> is true at w under **I** iff there is a $w^{\#} \in \mathbf{W}$ such that $\mathbf{r}(\alpha_1) \in \mathbf{d}(w^{\#})$, $\mathbf{r}(\alpha_2) \in \mathbf{d}(w^{\#}), \ldots, \mathbf{r}(\alpha_n) \in \mathbf{d}(w^{\#})$, and ϕ is true at $w^{\#}$ under **I**.
>
> The sentence
>
> $$\Box\phi$$
>
> is true at w under **I** iff for every $w^{\#} \in \mathbf{W}$ which is such that $\mathbf{r}(\alpha_1) \in \mathbf{d}(w^{\#})$, $\mathbf{r}(\alpha_2) \in \mathbf{d}(w^{\#}), \ldots, \mathbf{r}(\alpha_n) \in \mathbf{d}(w^{\#})$, it is the case that ϕ is true at $w^{\#}$ under **I**. (In both cases, I omit the requirement that the $\mathbf{r}(\alpha_i)$ be in $\mathbf{d}(w)$.)

I would now like to take up the question of whether all the fundamen-

[53] Jager 1982: 339.

tal notions of Plantinga's applied semantics have been adequately defined. In particular, I would like to investigate the question of whether the framework of notions Plantinga uses to define the terms of his applied semantics is adequate to define satisfactorily the notion of *essence* which, as is obvious from the above account, is so fundamental to his Actualism.

At first glance, Plantinga's definition of 'essence' seems to require quantifying over the set of possible objects—something that would clash with his avowed Actualism. Recall that Plantinga said that a property F is an essence iff there is some world w and some individual x in w which is such that x has F essentially and there is no world $w^{\#}$ in which there exists an object distinct from x that has F. But the phrase 'some individual x' occurring in this definition seems to involve quantification over possible objects.

Plantinga seems to sense this problem in his book. Thus, he writes:

> Objects or individuals *exist in* possible worlds . . . To say that an object x exists in a world W is to say that if W had been actual, x would have existed; more exactly, x exists in W if it is impossible that W obtain and x fail to exist. (Plantinga 1974: 46)

Extrapolating from this passage, we can understand apparent quantification over possible objects to involve *intraworld* quantification over possible individuals. To say that there exists a possible object x that has property F is just to say that there exists an individual x in world w which is such that x has property F; and this is just to say that if w had obtained, then there would have been an individual that had property F.

Reconsider Plantinga's procedure in explicating the notion of essence. Plantinga first explains what it is for a property to be an essence of Socrates. The following explication, he tells us, is not satisfactory:

> F is an essence of Socrates iff F is an essential property of Socrates and everything that is not identical to Socrates lacks F. (Plantinga 1974: 70)

The reason this explanation fails is because the quantifier 'everything', which occurs in this sentence, only ranges over actual objects. We want something that functions like a universal quantifier that ranges over *all possible objects*. So he gives the following explanation:

> F is an essence of Socrates iff F is an essential property of Socrates and there is no possible world in which there exists an object that is both not identical to Socrates and also has F. (Plantinga 1974: 70)

Thus, Plantinga tries to get the effect of universally quantifying over all possible objects by saying: "There is no possible world in which there exists an object which is not . . ." In effect, he is replacing the Possibilist's universal quantifier 'Every possible object x is such that . . .' with 'Every world w is such that every object x that exists in w is such that . . .' Then, he gives a definition of 'essence', which I paraphrase as follows:

> A property F is an essence iff there is a world w in which there exists an individual x such that x has F essentially, and there is no world $w^{\#}$ in which there exists an individual y which is distinct from x and which has F.

Now this definition would be easy to understand, given Plantinga's explanation of intraworld quantification, if the definition stopped immediately after the occurrence of the world 'essentially'. In that case, we could take the definition to be: a property F is an essence if there is a world w such that had it obtained, there would have been an object that had F. Unfortunately, the definition continues beyond that occurrence of the word to bring in a second world quantifier and another individual quantifier. Now what does it mean to say that, in a world w, there is some possible individual x that has some property and that, in any other world $w^{\#}$, if an object y exists in $w^{\#}$ that is different from x, then y lacks property F in $w^{\#}$? We have a sentence in which the variable 'x' occurs within the scope of two worlds w and $w^{\#}$, which seems to presuppose the intelligibility of quantification (over possible individuals) into distinct worlds in a way that would require *making sense of the transworld identity of possible individuals.*

One might suppose that the meaning of the above definition of 'essence' can be given the following (ghastly) explanation: Had world w obtained, then there would have been some individual that had property F essentially, and had any other world $w^{\#}$ obtained, then any individual y that would have existed had $w^{\#}$ obtained would have been such that, had it been distinct from that very individual of world w—the one that would have existed and that would have had property F essentially had w obtained—it would have lacked property F. But can we speak of "that very individual" in this way? Since, according to Plantinga, merely possible individuals do not really exist, how can we refer to "it" if it is a merely possible individual? If all talk of such things is a mere *façon de parler*, then it would seem that we cannot really refer to "that very individual" in world w. This problem is disguised by the fact that, when Plantinga defined what an essence of an individual is, he picked an individual, Socrates, who is actual. There is, of course, no problem with then saying that had $w^{\#}$ obtained, no object different from Socrates would have had F.

It might be thought that Plantinga could simply say that F is an essence iff (i) $\Diamond(\exists x)Fx$ and (ii) $\Box(x)\Box(y)((Fx \,\&\, Fy) \rightarrow x = y)$. There are two problems with this suggestion: (1) the meaning of the quantifiers '$(\exists x)$' and '(x)' are defined by Plantinga in terms of essences and hence the suggested explanation would be circular; and (2) the meaning of the modal operators is explained in terms of possible worlds, so it would seem that we would end up with the same sort of difficulty described above.

Perhaps Plantinga will retreat to the position of taking the notion of *essence* as one of the undefined primitives of his theory. He could say that, although he cannot define the notion within the framework he uses to ground the other fundamental notions of his applied semantics, he can give a partial explanation of the notion by explaining what the essences of actual objects are; and he can then hope that his readers will meet him halfway and grasp the idea. This is not the sort of move that Lewis would find theoretically attractive, but Plantinga may find it less troubling.

Another problem with Plantinga's applied semantics has been uncovered by several philosophers, who have expressed doubts that there are enough essences to do the job being assigned to them. The question arises: *are there any essences that are not exemplified in the actual world?* If not, quantifying over essences will be tantamount to quantifying over actual objects—which would clearly be unsatisfactory for what Plantinga hopes to accomplish with essences.[54]

Alan McMichael has argued in support of a negative answer to the above question (in McMichael 1983*b*). The argument has two parts: the first attempts to justify the claim that an essence cannot be a purely "qualitative" property—a property that is a compound of "general properties" ("properties that do not themselves involve either individuals or their essences"). He gives two arguments for concluding that essences are not purely qualitative (only one of which I take up here). There are, it is claimed, symmetrical possible worlds in which one individual on one side of the axis of symmetry has exactly the general properties that its twin on the other side of the axis has. So if essences were qualitative, then both twins would have the same essence. But that is impossible. So essences are not qualitative.

It seems to me that even if the modal premiss is accepted, the above argument, at best, shows that not all essences are qualitative—not that no essences are qualitative. However, McMichael attempts to generalize the reasoning, first, by claiming that "the theoretical impossibility of adequate interpretations in the case of symmetrical worlds does suggest that the essences view yields the wrong truth-conditions for modal sentences in general" (McMichael 1983*b*: 57), and, second, by suggesting that any individual could have existed in a symmetrical world.[55] McMichael concludes

[54] Besides McMichael (whose reasoning I discuss below in some detail), Fine has also argued strenuously that there are no unexemplified essences (Fine 1985, especially sect. 1). For an interesting response to McMichael's objections from a radically different Actualistic perspective, see Jubien 1996. Unfortunately, it would take me too far afield to give an evaluation of Jubien's Actualistic response here.

[55] Plantinga seems to agree with McMichael to this extent: he tells us that he is dubious of the claim that every object has a purely qualitative essence (Plantinga 1985*a*: 333).

that essences are *Haecceities*, that is, "properties which necessarily characterize single individuals but which are not reducible to general properties".[56] We can express McMichael's thesis using Adams's terminology by saying: every essence of an individual is a thisness which is not a suchness.

If essences are *Haecceities*, then in order that there be essences of merely possible individuals, which Plantinga clearly needs in order to develop a possible worlds semantics of the sort that he advocates, there must be unexemplified *Haecceities*. But that is something that McMichael cannot accept. He is willing to allow that there is a *Haecceity* corresponding to the expression 'is identical with Socrates'. This is because Socrates is an actual individual to which one can make reference. But he cannot see how there could be a *Haecceity* corresponding to 'is identical to *X*' if there is no actual individual *X* to which we can refer. Thus, McMichael comes to hold that there are no unexemplified *Haecceities*.[57]

If McMichael is correct, then Plantinga's Actualistic version of applied semantics simply cannot do what has been claimed for it. We cannot replace the "possible objects" of the canonical conception with Plantinga's "essences" if we are to retain anything like the sort of quantificational modal logic that has been proposed by modal logicians.

12. Paradoxes in Plantinga's Ontological Theory

Why should we believe that there are "possible worlds" in Plantinga's sense of that term? What justification does *he* give for his belief in these strange abstract entities? Initially (in Plantinga 1974), he gave nothing that could be considered a justification or proof of the existence of possible worlds. This last claim may be surprising to those who recall that, in the same work, he wrote: "I have argued that there are possible worlds" (1974: 121). But the closest that I can find in that work to an *argument* for the existence of possible worlds is to be found in the following quotation:

[56] McMichael 1983*b*: 59. I have capitalized the term '*Haecceities*' to distinguish McMichael's use of the term from Adams's, which was discussed earlier. The reader should be warned that there are several closely related uses of the word 'haecceity' in the philosophical literature.

[57] McMichael is by no means alone in adopting such a position. Fine gave various arguments in support of the view that *Haecceities* ontologically presuppose the objects they involve (Fine 1985). But see Plantinga's responses to Fine's arguments (Plantinga 1985*a*: 338–40). Hazen not only supports McMichael's arguments against Plantinga's Actualism, he advances additional considerations that tell against the Plantinga position (Hazen 1996). In this article, Hazen also replies to two responses to McMichael's article that have attempted to resuscitate some form of Actualism.

> Of course *the actual world* is one of the possible worlds; it is the maximal possible state of affairs that is actual . . . Obviously *at least* one possible world obtains. (Plantinga 1974: 45)

This reasoning does not justify belief in possible worlds—it only leads to the conclusion that at least one possible world obtains, on the assumption that there are possible worlds. I do not know if he failed to give a defence of his postulation of possible worlds because he believed that no defence was necessary—he might have believed that it was self-evident that such entities exist—or because it just did not occur to him that a defence of his belief in possible worlds was required.

I shall argue that, in the absence of any solid justification, reasonable people should withhold their acceptance of the hypothesis that these maximal possible states of affairs truly exist. In the case of mathematical entities—such things as numbers and sets—I can at least see a *prima facie* case for thinking there are such things. After all, we do talk in mathematics as if there were such things. But possible worlds, in Plantinga's sense, have no such initial plausibility. So far as I can see, it is only philosophers who believe in such things, and this belief seems to arise primarily because of questionable philosophical considerations. My scepticism is partly a matter of being doubtful about a philosopher's postulations of strange entities—postulations that serve as part of the foundations of a highly theoretical and abstract ontological theory that is supposed to take in essentially everything that exists. I believe the history of philosophy gives us some reason for being sceptical about any such postulations involved in an all-encompassing ontological theory. But my scepticism is also due partly to the strangeness of the entities being postulated in this particular case.

Consider what properties these "possible worlds" are supposed to have. The amount of "structure" that is postulated for such states of affairs is mind-boggling. Let α be the "actual world". Everything that exists (people, the earth, particles of sand, atoms, electrons, properties, numbers, sets, etc.) is in α; and every fact about all these entities is also included in α. But that is not all: every possible fact about every possible thing is also included in α. That is because every possible world is in α. Indeed, Plantinga tells us that, since possible worlds necessarily exist, every possible world is in every possible world.[58] Thus, α contains not only all the structure of property theory, set theory, and mathematics in general, but also all the structure of every other possible world, and these other structures contain all the structure of every other possible world, including α's, and so on ad infinitum. It is as if we had these incredibly complex structures reflected in mirrors that also reflect infinitely many other of these

[58] Van Inwagen 1986: 193.

structure-containing mirrors, each of which, in turn, reflects all the other mirrors, etc. Why should we suppose that such fantastic things truly exist?

We have seen that these possible worlds are very strange entities, postulated with a mind-boggling amount of structure. Reasonable people, I believe, will not accept such things just because Plantinga maintains that there are such things: good grounds should be demanded. John Pollock, responding to such a need (I suspect), has given a proof of the existence of possible worlds.[59] This has prompted Plantinga to admit that "Pollock is right in pointing out the non-triviality of the claim that there is at least one possible world (even given that there are states of affairs and that every state of affairs has a complement)".[60]

Later in his paper, Plantinga does give a proof for the existence of at least one possible world. He does so in response to a proof that Pollock set forth for the principle:

> [IPW] For any state of affairs *S*, *S* is possible iff there is a possible world in which *S* obtains.

Although he believes all the strong assumptions of Pollock's proof of [IPW], Plantinga gives a proof of [IPW] that requires what he takes to be weaker premisses. Since I shall be referring to certain features of Plantinga's proof, I give below a sketch of the only-if direction (which is the non-trivial one).

Plantinga's reasoning proceeds by way of a principle that, when conjoined with both the hypothesis that at least one possible state of affairs exists and the hypothesis that every state of affairs has a complement, implies the existence of possible worlds.[61] The principle in question is a sort of expansion principle that allows one to "expand" a possible state of affairs to a maximal possible state of affairs:

> [Expansion] For any possible state of affairs *S*, there exists a maximal possible state of affairs that includes *S*.

Given the existence of at least one state of affairs *S*, we could conclude from the hypothesis that every state of affairs has a complement that *S* has a complement. It is obvious that either *S* or *S*- complement must be possible. Hence there must exist at least one possible state of affairs. Then, from [Expansion], we could infer that at least one possible world exists. Of course, if *S* is possible, then there must exist one possible world that includes *S*, so we would have the only-if direction of [IPW].

How then does Plantinga attempt to prove [Expansion]? The reasoning

[59] Pollock 1985: 123–4.
[60] Plantinga 1985*a*: 327.
[61] Plantinga 1985*a*: 328–9.

seems to be patterned after Lindenbaum's Lemma (frequently used in proving the completeness of the first-order predicate calculus).[62] The reasoning depends upon another principle that Plantinga advances:

> [Quasi-compactness] For any set of possible states of affairs α, if α has a maximal linearly ordered subset, then α has a maximal linearly ordered subset that has the quasi-compactness property.

Some definitions are required to understand this principle. When Plantinga speaks of a 'maximally linearly ordered subset' here, he has in mind a subset that is *linearly ordered by the proper inclusion relation* and which is maximal in the sense that there is no linearly ordered subset that properly includes it (in the set theoretical sense).[63] A maximal linearly ordered set *A* has the *quasi-compactness property* iff *A* is possible if every finite subset of *A* is possible. A set of states of affairs is *possible* iff it is possible that all of its members obtain. In addition, Plantinga's proof requires an appeal to a state of affairs version of a fundamental axiom of set theory—a form of the Axiom of Choice known as 'Hausdorff's maximal principle'.[64]

Plantinga's proof proceeds as follows. Let *S* be a possible state of affairs. Then, he concludes, there exists a set α that is the set of all possible states of affairs that *include S* (in the sense defined above for states of affairs). It can be seen that Plantinga accepts another principle:

> Given any possible state of affairs *S*, there exists a set whose members are all those possible states of affairs that include *S*.

The reasoning seems to be this. If *S* is a possible state of affairs, then there is the property of *being a state of affairs that includes S*. Since, according to Plantinga, the set of all states of affairs exists, it follows by the Subset Axiom that there is a set whose members are all those possible states of affairs that include *S*.

Now every set of states of affairs is partially ordered by the proper inclusion relation (that is, the relation *properly includes* defined above).[65] So by Plantinga's state of affairs version of Hausdorff's maximal principle,

[62] See, for example, Mendelson 1987: 68–9.

[63] Plantinga 1985*a*: 328.

[64] See Pinter 1971: 119.

[65] There is, however, some room for confusion here. The trouble arises because of different definitions of 'partial ordering' in the literature. For example, Patrick Suppes defines a partial ordering of a set *A* as a relation in *A* that is reflexive, anti-symmetric, and transitive. See Suppes 1960: 72. Using this definition, the proper inclusion relation is not a partial ordering since it is not reflexive. The relevant notion of partial ordering is to be found, for example, in Enderton 1977: 168. There, a partial ordering is defined to be a relation that is transitive and irreflexive.

α has a maximal subset β that is linearly ordered (by this proper inclusion relation). By Quasi-compactness, it follows that α has a maximal linear ordered subset δ which is possible if every one of its finite subsets is possible. At this point, Plantinga appeals to another existential principle:

> [Conj] For any set of states of affairs θ, there is a state of affairs that is the *conjunction* of θ, that is, there is a state of affairs that obtains if and only if every member of θ obtains.

Such a principle is accepted by Plantinga, I believe, because if θ is a set of states of affairs, the gerund phrase 'all the members of θ having obtained' seems to refer to just such a conjunction. The conjunction of δ includes *S*, and it is easy to show that this conjunction is a possible world. It only remains to show that this conjunction is possible. To show this, it is sufficient to show that δ is possible. But δ is possible if every finite subset of δ is possible. Since δ is linearly ordered by proper inclusion, every finite subset of δ has a last or greatest member that is possible and that properly includes all the other members of the subset. So every finite subset of δ is possible.

It needs to be pointed out that this wide-ranging proof makes use of principles involving more than just states of affairs; Plantinga is quite willing to bring in set theoretical principles, when it suits his purpose. On the other hand, as will become clear, he is quite willing to reject principles of set theory when he feels that they raise problems with his body of beliefs about states of affairs or essences.

The reasoning of Plantinga's "proof" would be acceptable if the premisses were all acceptable. But the principles Plantinga accepts in this proof hardly seem to be anything like "self-evident truths". Indeed, Christopher Menzel has given a convincing argument for rejecting the Quasi-compactness principle.[66] So Plantinga's "proof" does not amount to an improvement on Pollock's. But more importantly, we still do not have a convincing justification for belief in possible worlds.

Before leaving Plantinga's attempted proof of the existence of possible worlds, a few observations should be articulated. Plantinga's reasoning proceeds from principles that look a lot like principles of set theory. Principles like the Hausdorff or Quasi-compactness may seem to be acceptable because of their similarity to widely accepted principles of set theory. What is clear is that Plantinga is operating with an underlying ontological theory encompassing states of affairs, properties, and sets; and this ontological theory has some very strong existential principles. So the question arises: can one prove the existence of a possible world in Plantinga's ontological theory, without appealing to such strong prin-

[66] See Menzel 1988: 600.

ciples as Quasi-compactness? It turns out that one can, indeed, construct an acceptable proof of the existence of at least one possible world from the principles implicit in the above reasoning, without appealing to Quasi-compactness, so long as one is allowed to use some uncontroversial principles of set theory.

A Proof for the Existence of at least One Possible World

I first state five principles that Plantinga accepted in the above.

[1] States of affairs exist.
[2] Every state of affairs S has a complement S'.
[3] If S is a state of affairs, then necessarily either S or S' obtains.
[4] Given any possible state of affairs S, there exists a set whose members are all those possible states of affairs that include S.
[5] Given any set β of states of affairs, the conjunction of that set, *all the members of* β *having obtained*, exists.

From these principles, I argue as follows. From [1], I infer that there is at least one state of affairs, say A. From [2] and [3], I infer that necessarily, either A or A' obtains. Let B denote the state of affairs *either A or A' having obtained.* (That there is such a state of affairs is routinely taken for granted by Plantinga.) It follows from the above that B is a possible state of affairs. Hence, using [4], I infer that there is a set β which is the set of all possible states of affairs that include B. From the Subset Axiom,[67] I infer that there is a subset γ of β which is the set of all members of β that obtain. Since B necessarily obtains, γ is just the set of all states of affairs that obtain. From [5], I conclude that there is a state of affairs, say D, that is the conjunction of γ. D is the state of affairs *all the members of* β *having obtained.* Since D obtains, D *can* obtain, and hence is a *possible* state of affairs. Let S be any state of affairs. If S obtains, then S is a member of γ, and hence D includes S. If S does not obtain, then by [2], there is a complement of S and this complement is included in D. Hence, D is a possible world. Ergo, there exists at least one possible world.[68]

One step in the above proof might appear dubious: I argued that if S obtains, then D includes S. Now D includes S only if *it is not possible that* D obtain and S not obtain. But the conjunction principle [5] above does not seem to provide the modal element that is needed to make the above inference. Here, we need to take account of the specific state of affairs that is the conjunction of γ. I claim that it is not (conceptually) possible that D (*all the members of* β *having obtained*) obtain and S not obtain. To see this,

[67] This axiom is also called the 'Aussonderung' or 'Separation' axiom. See Enderton 1977: 21 for a discussion of this axiom.

[68] This argument is close to Pollock's.

let us analyse the situation using, for heuristic purposes, Kripke's (in contradistinction to Plantinga's) possible worlds; and suppose that there is a Kripkean possible world *w* that makes *D* obtain (to use the terminology of section 5), but that does not make *S* obtain. (What we are imagining is that the Kripkean world had been such that *D* obtained and that *S* not obtained.) Then *w* makes obtain all the members of β and, in particular, *S*. But that is impossible. So there is no such Kripkean possible world. We can conclude that it is impossible that *D* obtain and *S* not obtain.

A Contradiction in Plantinga's Ontological Theory

We now have a proof of the existence of at least one possible world from principles accepted by Plantinga. But this gives rise to the question: should *we* accept these principles? One reason we should not is this: *using these principles and generally accepted principles of set theory, we can derive a contradiction.* Here is how. As shown above, we can conclude from Plantinga's principles that there is a set α whose elements are *all the possible states of affairs that obtain.* Then, according to Cantor's Theorem, the power set of α, 2^α, has cardinality greater than the cardinality of α. Hence, there cannot be a one–one correspondence between 2^α and any subset of α. But, as will be shown below, there is.

Suppose that β is an element of 2^α. Let *s*(β) be the conjunction of β guaranteed by [4] (that is, *s*(β) is the state of affairs: *all the members of* β *having obtained*). Then, given any element β of 2^α, there is a state of affairs, *s*(β), which is the conjunction of β. It is obvious that, since β is a set of elements of α (all of whose elements obtain), *s*(β) must itself obtain. Since *s*(β) obtains, it must be a member of α.

I shall first argue that, if σ and τ are members of 2^α, and if $\sigma \neq \tau$, then $s(\sigma) \neq s(\tau)$. However, before I begin this argument, it will be helpful to recall that states of affairs are closely related to propositions, so much so, indeed, that some philosophers believe that states of affairs just are propositions.[69] At one time, Plantinga was willing to allow that such philosophers may be right,[70] but later he came to believe that states of affairs are different from propositions, on the grounds that propositions have a property, namely, truth or falsehood, that states of affairs lack.[71] In any case, the close relationship between these two types of entities will be heuristically useful in what follows, since most philosophers have clearer intuitions about propositions than they do about states of affairs. Now suppose that $\sigma \neq \tau$, and consider the propositions:

[69] See Plantinga 1974: 45 for references.

[70] Plantinga 1974: 45. In conversation, I have found Plantinga to pass freely from talk about states of affairs to talk about propositions in a way that suggests that Plantinga does not regard the distinction between states of affairs and propositions to be very important.

[71] Plantinga 1979: 258.

P_1: All the members of σ are true
P_2: All the members of τ are true

Even if P_1 is logically equivalent to P_2, still $P_1 \neq P_2$, for they are about different things: P_1 is about σ, whereas P_2 is about τ. For a similar reason, if $\sigma \neq \tau$, then $s(\sigma) \neq s(\tau)$.

If it could be proven that, whenever $s(\sigma) \neq s(\tau)$, $\sigma \neq \tau$, then we would have a proof that there is a one–one correspondence between 2^α and a subset of α—a contradiction of Cantor's Theorem. There is some doubt, however, that the needed conditional can be proven, the reason being that states of affairs are more fine-grained than sets. It seems to be possible that state of affairs $s(\sigma)$ may be different from $s(\tau)$, even though σ is the same set as τ. Thus, the proposition expressed by 'All the members of σ obtain' may differ from the proposition expressed by 'All the members of τ obtain', even though the set σ is identical to the set τ (since τ turns out to be the set $\sigma \cap \alpha$). So perhaps we can have states of affairs $s(\sigma)$ and $s(\tau)$ which are different, even though σ is the same set as τ.

Let S be the set of all states of affairs x such that $x = s(\sigma)$, for some $\sigma \in 2^\alpha$. I then define a relation E on S as follows:

$$x\,E\,y \text{ iff for some } \sigma, \tau \in 2^\alpha, x = s(\sigma), y = s(\tau), \text{ and } \sigma = \tau.$$

In other words, x and y are in the relation E iff they are conjunctions of the same set. It is easy to see that E is an equivalence relation. E is reflexive since, by definition, every member of S is $s(\sigma)$ for some $\sigma \in 2^\alpha$. That E is symmetric and transitive follows trivially from the fact that identity of sets is symmetric and transitive. E then partitions S into mutually exclusive and exhaustive subsets. From the Axiom of Choice—which, it should be recalled, Plantinga in effect accepts, since he accepts Hausdorff's maximal principle[72]—there is a set C which consists of exactly one element from each of these equivalence classes. C clearly is a subset of α. Now given any set $\sigma \in 2^\alpha$, let $s^*(\sigma)$ be that member of C that is the conjunction of σ. Then, to each member θ of 2^α, there corresponds a member of C, $s^*(\theta)$. This can easily be shown to be a one–one correspondence. As before, if $\sigma \neq \tau$, then $s^*(\sigma) \neq s^*(\tau)$. But now, we can conclude that if $s^*(\sigma) \neq s^*(\tau)$, then $\sigma \neq \tau$. Thus, 2^α is equinumerous with C, contrary to Cantor's Theorem.[73]

[72] Plantinga includes his version of Hausdorff's principle among the premisses he uses in his argument for [IPW]—an argument which he takes to have weaker assumptions than the one Pollock gives for [IPW]; and Plantinga tells us that he believes that Pollock's assumptions are true (Plantinga 1985*a*: 328). Thus, we certainly have grounds for concluding that Plantinga accepts his version of Hausdorff's principle.

[73] The kind of difficulty I raise here is similar to that set forth in Bringsjord 1985: 64. Bringsjord's argument is critically discussed and, in some significant respects, amplified in Menzel 1986*a*: 68–72. See also Grim 1986: 186–91. Cf. also Jubien 1988.

Plantinga's Options

There are various options open to Plantinga. He could take the position, for example, that there is no such state of affairs as $s(\beta)$. This would be hard to maintain, however, since not only would he have to give up [5], but the existence of $s(\beta)$ seems to follow from the intuitive principle of state of affairs existence—a sort of Abstraction Axiom—that Plantinga makes use of throughout his discussion of possible worlds.

Another way of blocking the above reasoning would be to reject some part of classical set theory. At one time, this was the option that Plantinga evidently favoured. In his introduction to *The Possible and the Actual*, Michael Loux relates that Plantinga has, in conversation, "indicated doubts about Cantor's proof".[74] Doubts of this sort are also expressed by Plantinga (in Plantinga 1993) regarding a property version of Cantor's Theorem. Although such a position may be logically consistent, there are reasons for being sceptical of it. There are, of course, set theories within which Cantor's Theorem cannot be proved. But such set theories are generally motivated by constructivistic or predicativistic scruples: the diagonalization reasoning that forms the heart of Cantor's Theorem involves the use of impredicative specifications—specifications rejected by predicativists and certain constructivists. But I find it impossible to reconcile such scruples with the ontological views accepted by Plantinga (recall that Plantinga was willing to accept a state-of-affairs version of Hausdorff's principle—a set theoretical principle that is equivalent to the Axiom of Choice, hardly an axiom that constructivists would find appealing). In any case, when faced with a choice between Cantor's Theorem and the axioms of Plantinga's theory of states of affairs, I believe most philosophers will opt for Cantor's Theorem. Cantor's Theorem has been a theorem of an established part of mathematics for almost a century. It has been thoroughly studied by a large number of expert mathematicians and applied within mathematics in a variety of ways and in many different areas of mathematics. Given that Plantinga has not given us the slightest reason to believe in possible worlds, if Cantor's Theorem conflicts with Plantinga's ontological beliefs, it seems to me much more likely that one of Plantinga's principles is false than is Cantor's Theorem.

I recently learned from Plantinga himself that he has become convinced from some work of Menzel's that there is no such thing as *the set of all states of affairs* or *the set of all states of affairs that include a given state of affairs*.[75] This would allow him to deny that there is such a set as α and, in this way, avoid the contradiction I have just derived.[76]

[74] Loux 1979: 53 n. Such a response to the paradox is also suggested in Menzel 1986*a*.

[75] From a letter dated 11 Sept. 1991.

[76] It would also provide him with the means of avoiding the paradox in Plantinga's metaphysical theory that Jubien has derived in Jubien 1988.

Why he denies the existence of such sets was not made clear to me. It is possible that the denial is simply an ad hoc manœuvre designed to salvage as much of his system as possible in the face of contradiction. In any case, with this new position, Plantinga can avoid the above contradiction. Premiss [5] can be rejected. However, he can no longer claim to have justified belief in the existence of possible worlds, by giving a kind of "proof" from intuitively acceptable principles. Nor can he even claim to have given a plausible argument for the existence of possible worlds. But he now thinks that no such argument is needed!

Plantinga's attitude toward the postulation of abstract entities is one of extreme tolerance. No Ockham's Razor here. Plantinga sees no reason not to believe in the existence of states of affairs. So barring any special difficulty with the supposition that such things exist, he believes that it is rational to accept the supposition. *A possible world is just a state of affairs of a special kind.* So, he believes, it is rational to assert that there are such things, unless there is some difficulty in the hypothesis that such things exist.[77] With such an attitude, it is not surprising that Plantinga sometimes comes up with ontological theories that are paradox-ridden. I wonder if he believes that it is rational to believe in the existence in our atmosphere of alien spaceships, unless there is some difficulty in the hypothesis that there are such things? After all, an alien spaceship is just a special kind of vehicle—certainly, there seems to be no more a priori reason for being doubtful of the existence of such a special kind of vehicle than there is for being doubtful of the existence of possible worlds. Yet most people will not think it reasonable to assert that there exist such spaceships, simply because they see no special problem with the hypothesis that there exist such things. Scientists do not think it is reasonable to believe in the existence of special kinds of subatomic entities just because they can think of no difficulty in the supposition that entities of this sort exist; they demand evidence for an existential belief of this kind. Others, more cautious than Plantinga, will have doubts about an ontological theory having no more solid connection to evidence and rational inference than is suggested by Plantinga's tolerant attitude towards existence. Besides, as the earlier discussion of Lewis's Magical Ersatzism objection has shown, there are grounds for thinking Plantinga's notion of possible worlds has serious problems.

Is Plantinga's Revised Ontological Theory Consistent?

Let us call Plantinga's original ontological theory '$\mathbb{O}1$'. Then, when we drop the axiom of this theory that there is a set of all states of affairs (as Plantinga has suggested), we will have the theory $\mathbb{O}2$. The question arises: Is $\mathbb{O}2$ free of contradiction? I shall argue it is not.

[77] This was communicated to me in a letter of 11 Sept. 1991.

I begin by noting that Plantinga has not abandoned his belief in the existence of the set *PE* of all possible essences, that is, all essences exemplified in at least one world. After all, his actualistic interpretation of quantificational modal logic requires such a set—that is what **D** is supposed to denote. As we shall see, this belief will also lead to contradictions.

Plantinga holds that each existing individual thing has an *essence*, that is, a property that is an essential property of the individual and that no other thing can possess. In particular, every state of affairs has an essence. Plantinga also holds that there is a particular kind of essence that each thing *x* has that is easily described: it is the property of being identical to *x*. Socrates, for example, is said to have the property of being identical to Socrates.[78] Let us call such essences: *Identity Essences*. Obviously, every possible state of affairs also has such an essence. Let *PSAIE* be the set of all essences that are Identity Essences of possible states of affairs, that is, *PSAIE* is the set of *possible states-of-affairs Identity Essences*. That there is such a set can be inferred (using the Subset Axiom) from the fact that *PE* exists. Clearly, there is a one–one relation that correlates each member of *PSAIE* to that possible state of affairs of which the property in question is an Identity Essence.[79] This one–one correlation has, in effect, a domain that is a set—the set *PSAIE*. (All the elements of this set are urelements.) Corresponding to this domain, there should be, one would think, a range whose elements are all possible states of affairs. After all, we have two properties *F* and *G* which are such that there is a one–one relation that correlates one to the other, and one of these properties determines a set of urelements as its extension. Surely the other property must also determine a set of urelements as its extension. Then, what would this range be? It would be none other than the set of all possible states of affairs. Of course, it was just this set that Plantinga singled out as leading to the previous contradiction. It seems likely, then, that *PSAIE* would also lead to some sort of contradiction. As will become clear, it does. I shall derive a contradiction in Plantinga's "revised" ontological theory (the theory minus the assumptions that there is a set of all states of affairs or that there is a set of all possible worlds).

The main idea is to convert the previous derivation of the contradiction in $\mathbb{C}1$ into a derivation of a contradiction in $\mathbb{C}2$, by shifting to talk about identity essences of states of affairs. Thus, whereas before it was shown

[78] Plantinga 1974: 62.

[79] Notice that I say 'relation' and not 'function'. It might be argued that one cannot maintain that there is a one–one *function* of the above sort if it is questionable that there is a set of all states of affairs, since it then would be questionable that there is a codomain of the sort that is required. But one can use a notion of one–one correlation used in Frege 1959.

how the power set of the set of all obtaining states of affairs is equinumerous with a subset of the set of all obtaining states of affairs, this time it will be shown that the power set of all identity essences of obtaining states of affairs is equinumerous with a subset of the set of identity essences of obtaining states of affairs.

> [Def] *OSAIE* is the set of all *obtaining states-of-affairs Identity Essences,* that is, *OSAIE* is to be the set of all *Identity Essences of states of affairs that obtain.*

The power set of *OSAIE*, 2^{OSAIE}, must have cardinality greater than that of *OSAIE*. This means that it cannot be equinumerous with a subset of *OSAIE*. However, as I shall now show, within the framework of Plantinga's revised ontological theory, 2^{OSAIE} must be equinumerous with a subset of *OSAIE*.

To each member α of 2^{OSAIE} there corresponds a proposition that asserts: *each member of* α *is an essence of a state of affairs that obtains.* Corresponding to such a proposition, there is a state of affairs: *each member of* α *being an essence of a state of affairs that obtains.* I shall say that any such state of affairs is a "*members obtaining state of affairs*" or "*mosa*" of α. Now, for each *mosa*, there exists a unique Identity Essence.

> [Def] *MOSAIE* is the set of all such Identity Essences of *mosas*.

Now the underlying idea of the paradox is that 2^{OSAIE} is equinumerous with *MOSAIE*. That would give us a contradiction with Cantor's Theorem, since *MOSAIE* is a subset of *OSAIE*. Unfortunately, there is a small complication, so we cannot proceed quite so easily. But the rest is just a matter of working out some details.

Is 2^{OSAIE} equinumerous with *MOSAIE*? Suppose that α and $\beta \in 2^{OSAIE}$, and suppose that x is the identity essence of a *mosa* of α and that y is the identity essence of a *mosa* of β. Clearly, if $\alpha \neq \beta$, then $x \neq y$. But can we affirm the converse? Again, we run into the problem that states of affairs are more fine-grained than sets. So I shall again resort to the mathematical trick I used earlier. The relation **E** among the elements of *MOSAIE* defined as follows:

> For any elements x and y of *MOSAIE*, x **E** y iff there are sets α and β that are elements of 2^{OSAIE} and which are such that x is an identity essence of a *mosa* of α and y is an identity essence of a *mosa* of β and $\alpha = \beta$.

In other words, x and y are in the relation **E** iff they are both identity essences of the same set.

An example will clarify this definition. Let $\alpha \in 2^{OSAIE}$. Let **s** be the state of affairs *every member of* α *being an essence of a state of affairs that* *obtains*. Clearly, **s** is a *mosa* of α. Let β be defined to be the union of α with the set of all unicorns that exist in the actual world; and let **t** be the state of affairs *every member of* β *being an essence of a state of affairs that obtains.* Then, **t** is a *mosa* of β. I cannot say for sure that **s** and **t** are different states of affairs, but it is clear, in any case, that the Identity Essence of **s** is in the relation **E** to the Identity Essence of **t,** since $\alpha = \beta$.

It is easy to see that **E** is an equivalence relation among the elements of *MOSAIE.* **E** then partitions *MOSAIE* into subsets that are equivalence classes. These classes are mutually exclusive and they exhaust *MOSAIE.* From the Axiom of Choice, there exists a set C which consists of exactly one element from each of these equivalence classes. C is clearly a subset of *MOSAIE*, and hence it is a subset of *OSAIE*. Now I claim that there is a one–one correspondence between C and 2^{OSAIE}.

I first define the relation **R**.

> For every Identity Essence x in C, and every set $\alpha \in 2^{OSAIE}$, x **R** α iff x is an essence of a *mosa* of α.

I then show that **R** is a one–one relation correlating C to 2^{OSAIE}. Thus, suppose that $x \in C$, and that α and β are elements of 2^{OSAIE}. If x **R** α and x **R** β, it follows that x is an Identity Essence of some *mosa*, say **s**, of both α and β. I claim that, in that case, $\alpha = \beta$; for the state of affairs *every member of* α *obtaining* cannot be identical to the state of affairs *every member of β obtaining* if α is a different set from β.

Suppose this time that x and $y \in C$, and that α is an element of 2^{OSAIE}. If x **R** α and y **R** α, then x is an Identity Essence of some *mosa* **s** of α, and y is an Identity Essence of some *mosa* **t** of α. It follows that x **E** y. Since C is the choice set specified above, which contains exactly one element from each of the equivalence classes, $x = y$.

We can thus conclude that **R** is a one–one relation that correlates C, a subset of *OSAIE*, to 2^{OSAIE}. This contradicts the implication of Cantor's Theorem that there cannot be such a one–one relation. Thus, we see that ©2 is also paradox-ridden.

One reason I provided the details of this contradiction was to show how the various ontological elements in Plantinga's theory are interrelated. Implicit in Plantinga's theory is a set theory, a property theory, and a state of affairs theory. These theories are not independent of one another. The existential principles of one theory are related to the existential principles of the others. Given the existence of some state of affairs S, we are allowed to infer the existence of a property (the identity essence of S) and a set (the unit set whose only element is S). Given the existence of a property P, we are allowed to infer the existence of a state of affairs, *P having been instantiated,* and a set, the set of things that have P. Given the existence of a set α, we are allowed to infer the existence of a state of affairs, *all the members of α having existed*, and a property, the property of *being a member of* α. The principles of one of the theories may have far reaching implications because of these interrelations. Where do we go from here?

An Investigation of 𝒪3

Several people have suggested to me that Plantinga could avoid the above paradox by taking the class of all essences to be, not a set, but a proper class.[80] This would certainly block the above derivation of a paradox. But what would motivate such a move? Since it can be proved that, in Zermelo-Fraenkel set theory with the Axiom of Choice (henceforth ZFC), every set must be equinumerous with some von Neumann ordinal, it could be argued, with some plausibility, that the totality of all essences is simply too large to be a set. Would such a position yield an ontological theory 𝒪3 which is free from paradox?

Lewis considered such a move in connection with an apparent paradox in his own theory of possible worlds (Lewis 1986: 104). However, he rejected the suggested position because he felt that the sheer size of a class could not be a barrier to sethood, arguing that what prevents a class from being a set is that "the members of the class are not yet all present at any rank of the iterative hierarchy". He went on to claim:

> But all the individuals, no matter how many there may be, get in already on the ground floor. So, after all, we have no notion what could stop any class of individuals . . . from comprising a set.[81]

If Lewis is right, the suggested defence of Plantinga will be seen to be not only unmotivated, but unacceptable from a set theoretical point of view. However, what justifies Lewis's position? He does not provide his readers with any justification; he merely states his position. And justification is surely required, since, in rejecting the cardinal size criterion of sethood, Lewis is setting himself against the views of a rather formidable group of set theorists, which includes Cantor, Jourdain, Mirimanoff, and von Neumann.[82] For example, von Neumann advocated the cardinality criterion, asserting that a class is a proper class iff it is equinumerous with the class of all things.[83]

[80] This was suggested to me at a talk I gave at the Philosophy Colloquium of the University of Minnesota, and also Jonathan Turkanis has argued this with me at Berkeley.

[81] Lewis 1986: 104. However, unlike Menzel, Lewis gives no justification for his claims about proper classes.

[82] For an enlightening discussion of the views of this group, see Hallett 1986, especially ch. 8. I have also consulted a number of set theorists on this question of whether cardinal size or bounded rank should be taken as the criterion of proper classhood; and the responses I have received strongly suggest that there is no generally accepted answer to this question.

[83] Hallett 1986: 288.

Menzel has given an argument for the view that the true boundary between sets and proper classes should be taken to be position in the hierarchy. For Menzel, as for Lewis, it is not the fact that a class has too many members that makes it a proper class; rather, it is the fact that the class cannot occur at some stage in the set theoretical hierarchy (Menzel 1986*b*: 70). Thus, since the totality of all essences has only urelements as members, it would appear at the very bottom of the hierarchy. If we accept the Menzel–Lewis view, then, we would have to allow that the totality of all essences could be a set.[84] I shall explain shortly Menzel's reasons for adopting *position in the hierarchy*, as opposed to excessive size, as the criterion of proper classhood.

One reason I took pains to discuss the details of Plantinga's alleged proof of [IPW] was to show how very much like set theory Plantinga's ontological theory is: we saw how it included among its postulates very strong existence assumptions—it even asserted a principle very much like Hausdorff's maximal principle. But if this ontological theory is like set theory, it is even more like naive set theory before the discovery of Russell's Paradox: the strong existence assumptions used are not only unformalized—many are not even explicitly formulated, and there is nothing like a complete listing of its postulates. Given this similarity to naive set theory, even to the point of being inconsistent, it may be enlightening to review some of the historically significant reactions to the discovery of the paradoxes in naive set theory. In particular, let us consider the reactions of two of the most important of the philosophical logicians who attempted to repair the damage.

Frege did not accept set theory as Cantor developed it. However, he did have a theory of extensions (of concepts). And his logical system implicitly had a very strong axiom of extension existence. It was this axiom that Russell exploited in deriving his paradox in Frege's system. In the appendix to the *Grundgesetze*, Frege sketched one possible way of blocking the derivation of the paradox: he saw that, by weakening his infamous Fundamental Law V, the specific proof Russell gave could no longer be given. But he was by no means confident that he had adequately repaired his system—and rightly so, since his revised system was shown to be inconsistent.[85] As time went on, Frege abandoned his attempt to take principles about classes or extensions of concepts as "Basic Laws of Logic", writing:

The difficulties which are bound up with the use of classes vanish if we only deal

[84] This option is discussed in some detail in Menzel 1986*b* and Grim 1986. It should be noted, as I will explain in detail shortly, that taking such a large totality as a set will require changing somewhat the Replacement Axiom.

[85] See Resnik 1980: 215 for a discussion of how this was done.

with objects, concepts, and relations, and this is possible in the fundamental part of Logic. The class, namely, is something derived, whereas in the concept—as I understand the word—we have something primitive. Accordingly, also, the laws of classes are less primitive than those of concepts, and it is not suitable to found Logic on the laws of classes. (Jourdain 1912: 251)

Eventually, Frege came to believe that there are no extensions of concepts at all. We are, he claimed, misled by certain idiosyncrasies of language into forming proper names to which no object corresponds.

A particularly noteworthy example of this is the formation of a proper name after the pattern of 'the extension of the concept *a*', e.g. 'the extension of the concept star'. Because of the definite article, this expression appears to designate an object; but there is no object for which this phrase could be a linguistically appropriate designation. From this has arisen the paradoxes of set theory which have dealt the death blow to set theory itself. (Frege 1979: 269)

Russell did not think that the paradoxes had dealt a death blow to set theory. He felt that the paradoxes were due to some sort of fallacy. But after a great many years of arduous work, he too came to believe that there are no such things as sets, classes, or extensions, developing what he called a "no-class" theory.[86]

If one studies the history of foundational studies after the discovery of the paradoxes, it is clear that, to put naive set theory on a relatively secure foundation, an enormous amount of systematic and detailed work was required: the illness was too serious to be cured by band-aids or pills; radical surgery was called for. This suggests that to put Plantinga's ontological theory on a sound foundation much more is required than the sort of ad hoc moves that Plantinga seems to be making. After all, Plantinga's theory involves much more than just classes: its ontology includes, besides classes and objects, states of affairs, propositions, properties, and even propositional functions.[87] And all these different sorts of abstract entities are intimately related, so that the existence of entities of one sort may entail the existence of infinitely many entities of another sort. If many years of hard work by many of the best minds in logic and mathematics were required to put set theory on the relatively firm foundation it now has (a suggestion that would be disputed by few savants), how much more

[86] See Chihara 1973: ch. 1 for a detailed discussion of Russell's attempt to solve the paradoxes.

[87] That Plantinga's ontology includes propositional functions can be seen from the following quotation: "Indeed there are conditions; conditions are just propositional functions (functions in intension) from individuals to propositions" (Plantinga 1985*a*: 320). Plantinga does not say what properties are and seems content to give examples of properties (Plantinga 1974: 46–8). Presumably, properties are not the same thing as propositional functions, although these are closely related entities (compare the pair: *states of affairs* and *propositions*).

would be required to develop a secure ontological theory of the sort that Plantinga has in mind? Seeing how radically the ontological views of Frege and Russell were revised as a result of their investigations into the paradoxes, one wonders how much of Plantinga's original theory would remain after such an investigation.

Consider, now, Zermelo's groundbreaking work in the foundations of mathematics. It would seem that Zermelo presented his axiomatization of set theory in 1908, at least in part, in response to the paradoxes. Describing set theory as that branch of mathematics whose task it is to develop "the logical foundations of all of arithmetic and analysis", he goes on to say that "the very existence of this discipline seems to be threatened by certain contradictions or "antinomies", that can be derived from its principles" (Zermelo 1967: 200). Zermelo's strategy for dealing with the paradoxes is to give an axiomatization of set theory which is, on the one hand, sufficiently restricted to eliminate the paradoxes, and, on the other hand, sufficiently strong to establish the foundations of mathematics (Zermelo 1967: 200).

Notice that, at that time, Zermelo made no claim to "solve" the paradoxes, as did Russell with this theory of types. It was sufficient for Zermelo that the paradoxes could not be derived in his system. Menzel suggests, however, that Zermelo's investigations into the model theory of his set theory provided philosophers with the central idea of what is widely thought to be a solution to the paradoxes. Zermelo found that the models of his second-order set theory are constructible according to what is now called "the iterative conception of set": a conception according to which sets are built up in stages by certain set-building operations. One begins with some initial set of *urelements*, which make up the things of the bottom level, say H_0. Letting $\mathcal{P}(\alpha)$ be the power set of α, we obtain the things of the next level, H_1, by setting it equal to $H_0 \cup \mathcal{P}(H_0)$. Repeating this process, we obtain H_2, H_3, H_4, . . . Then, taking the union of the finite levels, we obtain H_ω. The power set and union operations can be employed to obtain levels $H_{\omega+1}$, $H_{\omega+2}$, $H_{\omega+3}$, . . . And the union operation can be repeated at the next limit ordinal. Repeating this process without end gives rise to a hierarchy of sets, which yields a criterion of sethood: a collection is a set iff it appears at some level in this hierarchy. Since each level contains all the things in the previous levels, we have what is called a "cumulative hierarchy".

The rank of a collection is defined to be the least ordinal greater than the ranks of all its members.[88] It follows that a collection is a set iff it has a rank. Since a proper class is a class that is not a set, we can say, with

[88] This definition is due to D. Mirimanoff (Hallett 1986: 188). Mirimanoff is also credited by Hallett with developing the idea of bounded rank as a criterion of sethood.

Menzel, that a proper class C is a class of unbounded rank, that is, there is no ordinal α such that all the members of C have rank less than α.

Now how does this conception of set give rise to an explanation of the paradoxes that many find convincing? Take the Russell paradox. The paradox begins with the assumption that there is a set of all those sets that are not members of themselves. But for every rank α, there are sets of rank α (and hence sets of rank α that are not members of themselves). Hence, the Russell class is a class of unbounded rank, and there is no such set as the set of all those sets that are not members of themselves. The derivation of the paradox is thereby blocked.

The Burali-Forti paradox can be handled in a similar fashion. The Burali-Forti proceeds by assuming that there is a set of all ordinals. Taking the ordinals to be the von Neumann ordinals, it is easy to show that the class of all ordinals is a class of unbounded rank. Hence, it is a proper class, and the paradox never gets off the ground.

Are these explanations correct? Menzel cannot accept the above explanation of the Burali-Forti. This is because Menzel rejects the assumption that the ordinal numbers are the von Neumann ordinals. I see no reason for discussing all the reasons he gives for this rejection. His strongest argument, in my opinion, consists in applying the multiple structure argument (which had been used by Paul Benacerraf to reject the identification of the natural numbers with the von Neumann finite ordinals)[89] against the identification of the ordinal numbers with the von Neumann ordinals. And just as Benacerraf concluded that natural numbers are not collections at all, so Menzel concludes that ordinal numbers are not collections either.[90] But if the ordinals are not collections, then they must be urelements; so there should be a set of ordinals.[91] Far from justifying the thesis that there is no set of all ordinals, "the iterative conception provides positive reason for thinking there *is* such a set" (Menzel 1986*b*: 43).

This prompts Menzel to construct a formal set theory (to be called 'CZF') that embodies the iterative conception, while treating the ordinals as urelements.[92] The intended model of CZF is the cumulative hierarchy that one obtains when one begins at level 0 with set of ordinals as

[89] Benacerraf's argument is to be found in Benacerraf 1965.

[90] Cf. Hallett's comment that "if set theory is to be the ultimate framework one is still left with the conceptual mystery of why all mathematical objects *should* be sets" (Hallett 1986: 305).

[91] Charles Parsons had suggested just such a possibility some time earlier, in Parsons 1983: 505, writing: "Why should it not be that all ordinals are individuals and therefore 'prior' to all sets, so that there is no obstacle of this kind to the existence of a set of all ordinals?"

[92] 'CZF' evidently stands for Cantorian Zermelo-Fraenkel set theory. Actually, Menzel uses the letters 'CZF' with a star appended, but a discussion of the significance of the star is not needed for my purposes.

urelements. This set is taken to be equinumerous with the class of von Neumann ordinals. However, since it can be proved in ZFC that there cannot be a set that is equinumerous with the class of all von Neumann ordinals, Menzel has to make some significant adjustments to the Zermelo-Fraenkel axioms to avoid contradictions in his set theory. This is done primarily by restricting the replacement axiom to what he calls "determinable sets"—which turn out to be just the pure sets of ZFC (Menzel 1986*b*: 48–50). That CZF is not paradox-ridden is supported by a relative consistency proof of CZF—a consistency proof relative to ZFC, supplemented with an axiom asserting the existence of an inaccessible cardinal (Menzel 1986*b*: 55–8). Thus, we obtain strong support for his contention that allowing there to be sets to be as large as proper classes does not necessarily generate a contradiction.

In Menzel's set theory, then, one cannot use cardinal size to determine that a collection is a proper class. The collection of ordinals is a set (of rank 1), even though it is equinumerous with a proper class (the class of von Neumann ordinals of unbounded rank). It is unbounded rank that is the criterion of proper classhood. But this feature of the set theory is taken to be perfectly acceptable: "For all the iterative conception says, size has nothing whatever to do with whether or not a given collection is a set; the only relevant factor is unboundedness" (Menzel 1986*b*: 44).

I have presented Menzel's views here not because I agree with everything he says on this matter (I don't), nor because I believe that his views on the divide between sets and proper classes are without problems, but in order to indicate some of the deep issues that one must face in undertaking the task of revising Plantinga's ontological theory in response to the paradoxes that undermine it. The foregoing discussion will also indicate why I am sceptical of "off the top of one's head" solutions to the above paradoxes or to ad hoc moves that have not percolated out of a thorough investigation of the principles that gave rise to the paradoxes in question. The claim that the contradiction I derived in Plantinga's theory is due to the assumption that the collection of all essences is a set has far-reaching implications that involve some of the most basic questions regarding the foundation of set theory. Further, such a claim is not at all easy to assess because, as we saw earlier, the system is so complicated—with interlocking principles from set theory, property theory, and a state-of-affairs theory—that selecting any one principle or assumption as the source of the contradiction would seem to require a thorough investigation into the nature of sets, properties, states of affairs, and propositions. A defence of any such claim should not be undertaken lightly.

What Happens to Plantinga's Applied Semantics?

There is another problem that arises when one takes the position described above in response to the Cantorian contradictions. Recall that

Plantinga advocates taking applied possible worlds semantics to give a "sober and literal account of modality" (Plantinga 1974: 126). Supposedly, the applied semantical theory provides a "literal explanation and articulation of our modal notions" (Plantinga 1974: 126). Recall also that Jager's system of quantificational modal logic was hailed by Plantinga himself as appropriately reflecting his Actualistic theory of modality—a view supported by Konyndyk, who calls the Actualistic system of modal logic axiomatized by Jager "the best system of quantified modal logic relative to a possible worlds interpretation" (Konyndyk 1986: 77). Now, as we saw earlier in Chapter 3 (Section 10), the intended applied semantics of Jager's formal language requires **W** to be the set of all possible worlds and **D** to be the set of all possible essences. But if one were to take the position described above that there is no such sets as the set of all possible worlds or the set of all possible essences, then one could no longer maintain that Jager's possible worlds semantics gives a "sober and literal account of modality". In short, the position described above would require Plantinga and his followers to significantly revise their views on what constitutes an acceptable applied semantical system.

Plantinga's Property Theory

There is another reason why I am sceptical of the suggestion that Plantinga's ontological theory can be made paradox-free by simply taking the collection of all essences to be a proper class. As I mentioned earlier, implicit in Plantinga's ontology is a very strong property theory; and as Frege showed (Frege 1959), one can develop cardinality theory in such a theory. Thus, using the sorts of intuitive property-theoretic principles Plantinga makes use of in his theorizing, I was able to construct, in property theory, the sort of paradox I constructed earlier in $\mathcal{O}$1. One can model such a derivation on the earlier one using, instead of the assumption that there is a set of all states of affairs, the assumption that all states of affairs have a property, namely the property of *being a state of affairs.* Of course, it is not surprising that such a paradox can be constructed, since Plantinga operates with something like a property version of the unlimited Abstraction Axiom of naive set theory. It is well known that analogues of the paradoxes of set theory can be constructed in property theories with such strong existential axioms.[93] To block such paradoxes, both Frege and Russell adopted theories of types, according to which the variables of their propositions were restricted in certain ways. Unfortunately for Frege, his type theoretical restrictions were not enough, because (in a way) the extensions of concepts allowed one to circumvent the restrictions.[94]

[93] This was stressed in Carnap 1959. See also Plantinga 1993: sects. 4 and 5.
[94] See Carnap 1959: 138 for a discussion of this point.

Russell's type theory blocked this route to the paradoxes by eliminating extensions of concepts entirely.[95] Plantinga's ontological theory, however, cannot block the paradoxes in the way Russell's theory did. Plantinga's system is no austere "no-class" theory: there are classes, propositions and states of affairs that allow one to move up and down the orders or "levels", so that for any property θ of order *n*, one could find (it would seem) a property of lower (or higher) order that is extensionally identical θ.[96] It is thus not surprising that one can construct, within a plausible formalization of Plantinga's theory of properties, a version of Cantor's Theorem. Then, making use of his theory of states of affairs and classes, one can prove a proposition that contradicts this version of Cantor's Theorem.

Of course, there are many ways that Plantinga could block the derivation of the paradox. In his response to an argument of Grim's, Plantinga suggests that he would reject the reasoning that gives us the property analogue of Cantor's Theorem, by abandoning the sort of Abstraction Principle that gives us the crucial "subproperty".[97] Given that he is willing to make such drastic revisions in the sort of property theory he seemed to be presupposing in his theorizing, one wonders what kind of property theory he would end up with. As Frege found when he tried to revise his logical theory in response to Russell's paradox, it is no simple matter to pick and choose principles for one's fundamental logical theory, when one is trying to obtain a theory that is consistent, intuitively plausible, and also strong enough to yield the sort of ontological theory Plantinga needs for his applied semantics. How is one to do property theory within Plantinga's revised ontological framework? Here, one is left in the dark.

One way out of this mess, it might be thought, is for Plantinga to adopt a version of type theory according to which there are type restrictions that govern not only the properties and propositional functions, but also classes and propositions. Since states of affairs are so similar to propositions, it would be a simple matter to extend, say, the Russellian type restrictions on his propositions to cover states of affairs as well. This could be done in a way that would block the derivation of the kind of paradox described above. Unfortunately, this solution would make it doubtful that there are states of affairs of the sort that Plantinga needs in his theory of possible worlds. Plantinga needs to have a state of affairs that includes all states of affairs that obtain (a maximal possible state of affairs that Plantinga calls 'the actual world'). That would be analogous to having a

[95] For more on this point, see Carnap 1959: 139, and Chihara 1980.

[96] In effect, the system incorporates a feature that would justify a kind of Axiom of Reducibility—an axiom that received much criticism when it was first put forward as a postulate of Russell's theory of types. See Chihara 1973: ch. 1 for more on this axiom.

[97] Plantinga 1993: 275.

proposition that entailed all truths; and there just could not be such a proposition, according to the Russellian system, since no proposition in the system can talk about all the propositional functions that there are. Of course, one could take the position that Russell's type restrictions are too severe. So far as I can see, there is nothing to block the development of a theory of propositions that is, on the one hand, more liberal than Russell's, in a way that permits the existence of propositions with variables that range over all propositions (of all types), variables that range over all states of affairs, variables that range over all sets, and variables that range over all properties; and, on the other hand, is not open to the sort of paradoxes and conflicts with set theory that Russell sought to avoid. But constructing such a theory is no trivial task.

What lessons can we draw from this investigation? First, it is clear that Plantinga's ontological theory, in its present state, is deeply unsatisfactory. Second, we should take Plantinga's many ontological theses—especially those that are based solely upon his intuitions—with a certain dose of scepticism. We have seen that his willingness to assert the existence of all sorts of abstract entities, based upon these intuitions, has led to contradiction. I believe we should be much more cautious about making such existential assertions—especially when they are so far-reaching. Finally, it is by no means clear how Plantinga's ontological view can be revised so as to be, on the one hand, strong enough to yield the sort of possible worlds semantics he advocated, and on the other hand, weak enough to be consistent. Only a systematic and thoroughgoing investigation will yield anything like a consistent principled ontological theory which would do what Plantinga espouses. Seeing how many years of arduous research were required to put the foundations of set theory on a relatively satisfactory footing (and there are some who would argue that the current state of set theory is far from satisfactory),[98] I do not expect to see Plantinga's ontological theory on a secure and firm foundation in the near future.

[98] See the conclusion to Hallett 1986.

4

Anti-Modal Realism

Given the many problems and implausibilities that infect Modal Realism, some form of anti-realism may appear at this point to be irresistible. In his book, Graeme Forbes adopts an anti-realist view of modality, which can be classified as a form of instrumentalism.[1] His main reasons for rejecting modal realism are, basically, epistemological.[2] On the other hand, Gideon Rosen's modal fictionalism—another version of anti-realism—seems to be motivated more by ontological concerns. In this chapter, I shall present an analysis and evaluation of these two versions of anti-Modal Realism in order to set the stage for the following chapters, in which my alternative anti-realistic view will be developed.

1. Forbes's Anti-Realism: Exposition

It is widely believed that possible worlds semantics provides us with an accurate analysis of modal statements. Some philosophers would claim that this kind of semantics provides us with accurate and fundamental analyses of the nature of modal concepts.[3] But as we saw earlier, it has seemed to many philosophers that the acceptance of such semantical analyses commits one to the existence of possible worlds of some sort—concrete ones, according to Lewis, or abstract ones, according to Plantinga. So an important goal of Forbes's anti-realism is to give analyses of the appealing features of possible worlds semantics that show how these features can arise even though there are no such things as possible worlds. However, Forbes is no ersatzist: he does not attempt to give the truth conditions of modal sentences in terms of things that go proxy for possible worlds. In fact, he does not attempt to analyse the truth conditions of modal sentences at all. His principal goal is to explain how possible worlds semantics can be used to determine the validity or invalidity of modal

[1] Forbes 1985. It needs to be emphasized that the anti-realism developed by Forbes in this work should not be identified with the anti-realistic view expressed in Forbes 1989.

[2] Forbes 1985: 79–80.

[3] See, for example, Plantinga 1974: 126.

arguments, without presupposing the existence of any possible worlds; and he hopes to do this in a way that explains the "naturalness" of the possible worlds analysis.

Another feature of Forbes's anti-realistic analysis should be noted: there is no attempt, on the part of the analysis, to assign any significant role to the individual structures (or the set theoretical sextuples that I have called 'interpretations'). They seem to drop out of sight completely in Forbes's analysis. What he does instead is to use the definition of 'true at *w* under an interpretation' to come up with sentences of a two-sorted first-order language, with one sort of variable for "worlds" and one sort of variable for "individuals", which are generally thought to be the first-order translations of the modal sentences of the quantificational modal logic. He thus has two types of sentences: the *modal sentences* with boxes and diamonds and the *world sentences* of the first-order language. The main problem for the anti-realist, according to Forbes, is to explain how the sentences from these two languages are related.

Thus, to take a very simple example, possible worlds semantics enables us to show that, and explain why, the modal inference schema

$$\frac{\Diamond \mathrm{P} \quad \Diamond \mathrm{Q}}{\Diamond(\mathrm{P} \,\&\, \mathrm{Q})}$$

is not valid. Its rules for truth under an interpretation are such as to allow us to express the above modal sentences in the following way:[4]

$$\frac{(\exists w)P(w) \quad (\exists w)Q(w)}{(\exists w)(P(w) \,\&\, Q(w))}$$

where the quantifiers are regarded as ranging over worlds and '$P(w)$' is interpreted as saying: P is true at w. Since our prior understanding of simple quantificational logic allows us to see that the latter inference schema is not valid, we can see that the former inference is not valid. Now the Modal Realist can account for this use of possible worlds semantics by maintaining that each possible worlds statement in the latter inference schema exhibits the real meaning of the corresponding modal statement in the former (recall Lewis's claim that possibility "amounts to existential quantification over the worlds"). It can thus be argued that each possible worlds statement in the latter inference schema provides us with an *analysis* of the corresponding modal statement in the former schema, in terms of which one can understand why it is not valid.

[4] This way of translating sentences of modal sentential logic into sentences of a first-order logic is well known. See, for example, Fine 1975: sect. 1.4.

Forbes asks whether there could be any *relation weaker than synonymy* that would enable the anti-realist to account for the above use of possible worlds semantics. He insists that any candidate relation must satisfy the following condition:

> If σ is a sentence of modal language and σ' its rendering in possible worlds language according to the quantifier treatment of modal operators, then the hypothesis that σ stands in the candidate relation to σ' should be sufficient to guarantee that σ behaves that same way in a modal argument as σ' does in the rendering of that argument in possible worlds language. (Forbes 1985: 73)

But after an investigation of this topic, Forbes concludes that "it is hard to think of any candidate for this relationship other than synonymy" (Forbes 1985: 80).

The Reverse Translation Thesis

Hemmed in as he is by the above reasoning, Forbes comes up with a novel way of attacking the Modal Realist's position: he rejects the realist's claim that the possible worlds sentences provide *analyses* of their corresponding modal sentences. The Modal Realist treats the modal sentence as the *analysandum* and the possible worlds sentence as the *analysans*. Forbes says that it is the other way round: he puts forward the surprising thesis that it is the modal sentence—not the possible worlds sentence—that is the *analysans*. Indeed, for Forbes, the possible worlds sentences "do not mean what they appear to mean" (Forbes 1985: 80). The possible worlds sentence '$(\exists w)P(w)$' in the above inference does not exhibit the real meaning of the corresponding modal sentences '$\Diamond P$'; instead, the possible worlds sentence just means '$\Diamond P$'. This is Forbes's *reverse translation doctrine*: the meaning of a possible worlds sentence is given by its rendering in (or reverse translation into) the modal language.

An example will clarify this doctrine. According to Russell's Theory of Descriptions, the statement

[A] The present King of France is bald

has the meaning:

[B] There is a person x such that, for every person y, y is presently King of France iff $y = x$; and x is bald.

[A] is the *analysandum*; [B] is the *analysans*. In other words, [A] gets analysed as meaning what [B] does. According to Russell's theory,[5] the "surface structure" or the apparent grammatical form of [A] is mislead-

[5] An excellent discussion of Russell's Theory of Descriptions, bringing to bear current work in the theory of reference and linguistics, is Neale 1990.

ing. The sentence [A] is not really a sentence of the form '*Fa*'; it has the meaning of an existential sentence (Forbes 1985: 72). Similarly, it is widely believed that the world sentence expresses the real meaning of its corresponding modal sentence. But this view has gotten things backwards, as Forbes sees it. For Forbes, it is the "surface structure" of the possible worlds sentence

$$(\exists w)(P(w) \,\&\, Q(w))$$

that is misleading: it is not really an existential sentence at all; its meaning is simply that of the modal sentence

$$\Diamond(P \,\&\, Q)$$

What makes this view plausible is the thought that, for a great many years, philosophers (and non-philosophers) have reasoned modally without any talk of possible worlds. It is reasonable to suppose that all this talk about possible worlds did not proceed from the *discovery* of new universes so much as from the *invention* of a new way of thinking about possibility and necessity; perhaps, what underlies this new way of talking is merely a new metaphor. If so, it might be thought, all of our meaningful talk about possible worlds should be reducible to straightforwardly modal talk, involving just the modal operators and such words as 'can', 'could', and 'might'.

In an appendix, Forbes tries to bolster his reverse-translation case by defining an effective function, *Trans*, that supplies the possible worlds translation of each sentence of a standard modal quantificational language, which we can take to be the language **M** presented in the first chapter.[6] *Trans* takes a sentence of **M** and gives as its value a sentence of the world language *W*, a two-sorted first-order language that involves quantification over worlds and individuals.[7] As an example, suppose that the sentence '$\Diamond(\exists x)F^1x$' is taken to be an argument of *Trans*; then the value will be '$(\exists w)(\exists x)(E^2xw \,\&\, F^2xw)$'. He also defines another function, $Trans^{-1}$, that supplies the reverse modal translation of possible worlds sentences. Thus, $Trans^{-1}$ is a function from a subset of the sentences of *W* onto the sentences of **M**.

But this doctrine raises a problem for Forbes: since it is maintained that possible worlds sentences do not have their standard quantificational meaning, the standard possible worlds account of validity in modal logic

[6] Although Forbes does not supply the syntax of his modal quantificational language or formulate its derivational system, he does say that its semantics is due to Kit Fine (Fine 1978: 125–56). Since the language **M** is similar to Fine's modal language, I shall (for purposes of specificity) take Forbes's quantificational S5 language to be **M**.

[7] This sort of translation is well known. See, for example, van Benthem 1983: 137–41.

must be abandoned. Validity must be given an alternative account. This, in turn, requires Forbes to explain why possible worlds semantical methods work, when validity is given this alternative account.

Forbes's Proof-theoretical Account of Validity

Forbes's response to this problem is to adopt a proof-theoretic approach to validity. Instead of starting with truth conditions and then specifying valid inference rules that will preserve truth values, Forbes suggests that competent English speakers, in fact, begin with certain "bedrock 'pre-theoretic' intuitions" of what follows from what (Forbes 1985: 82). He suggests, in particular, that the introduction and elimination natural deduction rules, which tell us by means of the symbol '$\vdash$' what can be deduced from what, give the meanings of the logical connectives. For example, he suggests that the introduction and elimination rules

&-I If $X \vdash \sigma$ and $Y \vdash \tau$, then $X \cup Y \vdash (\sigma \,\&\, \tau)$
&-E If $X \vdash (\sigma \,\&\, \tau)$, then $X \vdash \sigma$ and $X \vdash \tau$

(where X and Y are sets of sentences and σ and τ are sentences) give the meaning of '&', since the usual truth table assignment for '$(\sigma \,\&\, \tau)$' can be inferred from the above rules when it is assumed that σ and τ each have definite truth values (Forbes 1985: 83). This view, then, "looks to the 'natural deduction' rules governing a connective for the embodiment of the essence of a native speaker's mastery of the connective" (Forbes 1985: 82). In this way, Forbes adopts a view of validity of the sort that A. N. Prior attacked in his classic paper (Prior 1960) and that Nuel Belnap defended in (Belnap 1962). It is a view that was especially popular during the forties and fifties, when Wittgenstein's talk of language games and the analogy between semantical rules of language and rules of games were extremely influential in Anglo-American philosophy. A distinguished group of philosophers has been attracted to such a view, so it should not be dismissed out of hand.[8]

What, according to Forbes, are the introduction and elimination rules for '$\Box$' that supposedly give the meaning of that modal operator? Here they are:

$\Box$-I If $X \vdash \sigma$ and for every sentence τ in X, each occurrence of a sentential letter in τ is within the scope of some occurrence of '$\Box$', then $X \vdash \Box\sigma$.[9]

[8] As Prior points out, in his paper, Karl Popper was an early defender of this sort of view.

[9] Rule $\Box$-I is supposed to capture the intuitive idea that if ϕ can be inferred from a set of sentences all of which are necessary, then ϕ itself will be necessary. The actual form of the rule will undoubtedly appear strange to the reader unfamiliar with Forbes's book. Such

$\Box$-E If $X \vdash \Box\sigma$, then $X \vdash \sigma$.

Forbes then claims that when '$\Diamond$' is introduced by definition as '$-\Box-$' and the above rules are added to the usual natural deduction rules for the sentential calculus, one gets an S5 modal sentential calculus.[10]

To see how this could be so, note that the standard axioms of S5 sentential logic are:

[A1] $\Box\phi \rightarrow \phi$
[A2] $\Box(\phi \rightarrow \psi) \rightarrow (\Box\phi \rightarrow \psi)$
[A3] $\Diamond\phi \rightarrow \Box\Diamond\phi$.

Now in Forbes's natural deduction system, we have, for any sentence ϕ,

$$\phi \vdash \phi.$$

Hence, $\Box\phi \vdash \Box\phi$. By $\Box$-E, we obtain $\Box\phi \vdash \phi$, which is Forbes's proof-theoretic version of [A1].

We also have

$$\begin{array}{lll} \Box(\phi \rightarrow \psi) \,\&\, \Box\phi & \vdash \Box(\phi \rightarrow \psi) \,\&\, \Box\phi & \\ & \vdash \Box(\phi \rightarrow \psi) & \text{By \&-E} \\ & \vdash \phi \rightarrow \psi & \text{By } \Box\text{-E} \\ & \vdash \Box\phi & \text{By \&-E} \\ & \vdash \phi & \text{By } \Box\text{-E} \\ & \vdash \psi & \text{By } \rightarrow\text{-E } (\textit{modus ponens}) \\ & \vdash \Box\psi & \text{By } \Box\text{-I} \end{array}$$

Thus, we obtain Forbes's proof-theoretic version of [A2].

For [A3], we have

$$\begin{array}{lll} -\Box-\phi & \vdash -\Box-\phi & \\ & \vdash \Box-\Box-\phi & \text{By } \Box\text{-I.} \\ & \vdash \Box\Diamond\phi & \text{By definition of `}\Diamond\text{'.} \end{array}$$

The Problem of the Naturalness of Possible Worlds Semantics

According to this account of the modal operators, their "use" is given by the natural deduction rules, and the possible worlds semantical account of their meaning is illusory. A big problem for Forbes, at this point, is to ex-

readers may wish to consult Forbes 1985: 84 for an explanation of this rule. I should add that what I am calling '$\Box$-I' is what Forbes calls '$\Box$-I*'.

[10] Since the modal operator '$\Box$' that concerns Forbes is the broadly logical necessity operator, Forbes takes S5 to be the appropriate modal logic. In this respect, Forbes goes along with the majority opinion. See Loux's Introduction to Loux 1979: 28 for one explanation of why S5 has been accepted as the right modal logic for metaphysical modality. Some philosophers, however, disagree. See, for example, Salmon 1989.

plain the naturalness and intuitive correctness that possible worlds semantics possesses, which other purely algebraic methods do not. Forbes attacks this problem by trying to show that there is an important connection between the "use" of the modal operators, given by his natural deduction rules, and sentences containing quantifiers over worlds. He tries to do this in two ways.

(1) He purports to show that it is legitimate, within his framework, to infer:

[PW] In every possible world, '*P*' is true

from

[Box] '□*P*' is true.

This is supposed to show that possible worlds semantics "can be related to our use of modal sentences in a way that others cannot be" (Forbes 1985: 85).

(2) He attempts to establish a technical result that will allow us to utilize first-order logic in assessing modal arguments, when modal arguments are translated into possible worlds language according to his reverse translation scheme. Thus, Forbes hopes to explain why possible worlds semantics works, even though validity is not defined within the possible worlds semantical framework. To this end, Forbes makes the following claim:

> [A]ny modal argument is correct according to the inference rules (say, for S5) iff its translation into possible worlds language is valid according to ordinary first-order semantics. This result establishes that even though world quantifiers do not mean what quantifiers usually mean in first-order languages, the difference is not sufficient to prevent us from employing our familiarity with first-order logic in assessing modal arguments. (Forbes 1985: 86)

He then claims that the argument needed here is trivial; and, in a footnote, he goes on to assert that what has to be shown is that if *X* is a set of modal sentences and σ is a modal sentence, then there is a proof of σ from *X* in a natural deduction system with □-E and □-I iff there is a first-order proof of $Trans(\sigma)$ from $Trans(X)$, where $Trans(X) = \{Trans(\sigma) \mid \sigma \in X\}$.

The Problem of Reverse Translating all Meaningful Worlds Sentences

Another problem Forbes must face is the problem of finding a reverse translation for *all* meaningful possible worlds sentences. Forbes characterizes his anti-realism as resting on "an elimination by paraphrase of the ontology of possible worlds", which he feels is defensible only if "every meaningful possible worlds sentence is a rendering of some modal sentence" (Forbes 1985: 81). But Alan Hazen has argued that there are meaningful worlds sentences that cannot be reverse translated by Forbes's rules (Hazen 1976). Thus, consider:

$$(\exists w)(\exists x)(E^2xw \,\&\, -E^2x\mathbf{w}^*)$$

which we can translate into English: *there could have been something that does not actually exist*. The above worlds sentence is clearly meaningful; yet it cannot be translated into **M**.

Forbes's response to this problem is enlightening, not so much for what he says, as for what he does not say. One might suppose that he would have responded to this example of Hazen's by maintaining that the supposed problematic sentence has already been reverse translated into a modal sentence that makes no reference to worlds, namely, by the above italicized sentence which was given as its English translation. But he does not. It is clear that he feels he must find a *formal* sentence of his modal language which is the reverse translation of the problematic sentence. The reverse translation doctrine is evidently a thesis asserting that possible worlds sentences are reverse translatable into sentences of his *formal* modal language. This is somewhat surprising, since one would think reverse translating worlds sentences into the modal sentences of a natural language would also accomplish his goal of removing by paraphrase the apparent ontological commitments of the worlds sentence.

This gives rise to the following enquiry. Having concluded that the reverse translation doctrine implies that the worlds sentences are reverse translatable into Forbes's formal modal language, we may now wish to know if the worlds sentences that are reverse translatable include only the *formal sentences* of some two-sorted first-order language. There are a number of reasons for concluding that Forbes intends his reverse translation doctrine to cover worlds sentences from natural languages as well as formal sentences. For example, in trying to show how possible worlds semantics is related to the use of modal operators, he asserts that the English sentence

In every possible world, '*P*' is true

has, according to his anti-realist thesis, the same meaning as

$\Box$'*P*' is true.

It can be seen that his reverse translation doctrine applies to worlds sentences of natural languages, as well as to sentences of formal languages.

In any case, in order to save his reverse translation thesis from Hazen's counterexample, Forbes follows a suggestion of Hazen's and expands the vocabulary of his S5 modal language (which can be taken to be our language **M**) to include an actuality operator '$\mathcal{A}$'. This can be done in the way described above at the end of Chapter 1, resulting in the expanded language $\mathbf{M}^{\#}$. Hazen's sentence can then be expressed in $\mathbf{M}^{\#}$ as:

$$\Diamond(\exists x)-\mathcal{A}Ex.$$

This expansion of the modal language, however, invites a search for still other examples of meaningful worlds sentences that cannot be reverse translated to the new formal language. Hazen finds some. For example, the worlds sentence

$$(\exists u)(v)(\exists x)(E^2xv \;\&\; E^2xu)$$

seems to be perfectly meaningful; yet it is not translatable into Forbes's expanded modal language. In response to such examples, Forbes expands the vocabulary of his formal language even further to include subscripted modal operators (which were devised by Christopher Peacocke) that allow one to refer back to states of affairs introduced by other modal operators. This requires a number of revisions of the possible worlds semantical definitions. Thus, when the new actuality operator has the same subscript as a diamond operator, what is said to be actually the case refers back to the possibility covered by the indexed diamond operator. The above problematic worlds sentence is then translated into the expanded language by the sentence

$$\Diamond_1\Box(\exists x)\mathscr{A}_1E^1x.$$

The matter does not end there, since there are still other examples that force Forbes to introduce still other types of operators. The sentence

My car **a** could have been the same colour as your car **b** is actually

can be expressed in the worlds language as

$$(\exists w)C^4awb\mathbf{w}^*$$

where 'C^4' is interpreted as: ① in world ② is equicoloured with ③ in world ④. To deal with such examples, Forbes introduces, in his modal language, actuality operators that are both subscripted and superscripted (the superscripts being individual constants), the superscripts being used to associate a particular object with a given subscripted modal operator.[11] The preceding worlds sentence can then be reverse translated into:

$$\Diamond_1\mathscr{A}^a{}_1\mathscr{A}^bCab$$

Nevertheless, Forbes has to admit that there are worlds sentences that cannot be reverse translated into even his various expanded modal languages. He can find no plausible way of reverse translating sentences, such as '$(w)(w = w)$' and '$(w)(y)w \neq y$', that seem to be just about worlds (Forbes 1985: 94). Forbes suggests that we should regard these sentences "in the same way as certain mathematical sentences were regarded by Hilbert,

[11] Later (Forbes 1985: ch. 7) Forbes introduces an Actuality operator that can have variables as superscripts.

that is, as instruments or uninterpreted stipulations which enable us to establish facts of interest about the interpreted sentences" (Forbes 1985: 94). The "certain mathematical sentences" referred to in this quotation are the "ideal statements" that Hilbert classified as meaningless.[12] Thus, Forbes in effect classifies possible worlds sentences into two sorts: those that can be reverse translated into his modal language, and those cannot be so translated. The former are meaningful; the latter are meaningless instruments or uninterpreted stipulations.

Forbes's instrumentalist attitude toward possible worlds sentences is extended to cover possible worlds semantics as well: we are to regard possible worlds semantics merely as a "tool for fixing the powers of the logical operators, for determining which modal logical arguments are valid and which invalid" (Forbes 1985: 70). As for the set theoretical structures used in the model theory, they are simply part of the instrumentally useful apparatus for fixing which modal arguments are valid (Forbes 1985: 189).

2. Forbes's Anti-Realism: Some Doubts

Doubts about the Proof Theoretic Analysis

Let us begin with an examination of Forbes's proof-theoretic analysis of the meaning of the modal operators. Recall that, according to Forbes's analysis, "the fundamental method of encapsulating the meaning of a logical constant is to give a rule for when a sentence with that constant as its main connective follows from other sentences, and also a rule which determines what other sentences follow from it" (Forbes 1985: 82). In the following, I intend to raise some doubts about this doctrine—nothing like a decisive objection, mind you, but, as a group, these considerations raise considerable doubts about Forbes's analysis.

Is it really reasonable to conclude, from what Forbes has shown, that the introduction and elimination rules give one the meaning of '□'? In the case of the connectives, one can derive, from the natural deduction rules, what are essentially the truth conditions of molecular sentences, relative to the truth values of their atomic components. However, for the case of the box operator, Forbes has not supplied us with anything like a derivation of such relative truth conditions from the two rules. Can we reasonably suppose that our grasp of the meaning of the modal operator is obtained just from Forbes's rules? One indication that Forbes's position is questionable is the fact that the two natural deduction rules would be equally valid for strict logical necessity. But there is a significant difference

[12] See Hilbert 1983.

in meaning between the strict logical necessity operator and the broadly logical necessity operator, as can be seen from the fact that, when given a list of typical examples of broadly logically necessary and strictly logically necessary truths, most philosophers can sort the list into the appropriate two sublists. Since this difference in meaning cannot be obtained from the introduction and elimination rules, we have some grounds for thinking that there must be more to the meanings of these operators than is given by these two rules.

A closely related reason for suspecting that the two rules do not succeed in giving the meaning of the broadly logical box operator is to be found in the fact that native speakers of English seem to be able to determine the truth of all sorts of statements of broadly logical necessity that appear to require more understanding than just knowing the above two rules of inference. It is hard to see how native speakers could conclude that the sentence

$$\Box(\text{All bachelors are unmarried})$$

is true, if their knowledge of the meaning of '□' was restricted to just the introduction and elimination rules. Surely, Forbes must do a lot more explaining before many philosophers would feel comfortable with his thesis that the above natural deduction rules really give the meaning of the box operator.

These doubts about Forbes's proof-theoretical account of the meaning of the modal operators deepen when one considers how textbooks usually explain to students what these operators mean. The introductory logic books that take up S5 modal logic that I have examined do not define or specify what these operators mean in terms of Forbes's introduction and elimination rules. In many of these books, examples are given of broadly logically or conceptually necessary truths, some kind of explanation is given of what 'necessary' means, and the connection to "possible worlds" (of some sort) is discussed; rules of derivation are not even taken up until the semantics is presented. What this way of explaining the operators suggests is that Forbes's proof-theoretic account of the meaning of the operators does not gybe with the type of explanations of their meaning generally given.

Now consider Forbes's responses to Hazen's examples that raise problems for the reverse translation doctrine. It will be recalled that Forbes felt it necessary to introduce a number of new operators into his modal language. Subscripted modal operators were introduced, as well as actuality operators that had both subscripts and superscripts. Given Forbes's proof-theoretic account of the meaning of the modal operators, one would think that he would introduce these operators with introduction and elimination natural deductions rules. But he doesn't. He explains their meaning in terms of possible worlds. (It is as if he had completely forgotten his own position on the meaning of the operators.) Furthermore, it is not at all clear that one can explain their meanings proof-theoretically in the way he advocated for the case of the simple operators.

Another problem with this proof-theoretic analysis is noted by Forbes himself: there is, perhaps, some plausibility to the suggestion that the introduction and elimination rules for the connective '&' "embody" the meaning of the connective, since we can explain how a speaker can come to understand the meaning of the connective as a result of being given such rules:

> [I]f it is introduced into an already understood language, then one can straight away understand sentences with one occurrence of the new connective, since one grasps the operation (knowing the rules) and also grasps the meanings operated upon (since these are from the already-understood language in this case); and one's understanding of sentences with more than one occurrence of the new connective is built up from there. (Forbes 1985: 87)

Unfortunately, the "natural deduction" rules Forbes gives for the operator '□' do not allow such an explanation. Rule □-I tells us how to infer sentences beginning with '□' from sets of sentences whose vocabulary already contains the operator. We cannot understand the rule to be introducing a new operator into an "already understood language", as we could in the case of the "natural deduction" rules for '&'.

Forbes suggests, in response to the problem, that perhaps "this difficulty can be overcome recursively" (Forbes 1985: 87). But he does not develop this idea. I am not confident that I have grasped his suggestion, but he may have had the following idea in mind: given any valid sentence ϕ of first-order free logic, one has

$$\vdash \phi$$

Then one can use □-I to conclude

$$\vdash \Box \phi$$

And from such sentences, one can use the introduction and elimination rules to conclude that still other sentences involving the box operator are valid. So in this way, one can infer the validity of a large number of modal sentences.[13]

Of course, the question is: will the fund of modal theorems obtained in the above way be sufficient to pin down the meaning of the modal operator? In particular, will one obtain the set of all valid sentences of S5 quantificational modal logic? Evidently, Forbes had doubts that this process

[13] I owe this suggestion to Josh Dever.

would yield the appropriate set of theorems, since, at this point in the chapter, he turns to a new approach to validity: the substitutional account. According to this second approach, a modal argument schema is valid iff uniform substitutions into the schema of expressions with *n* free variables for the *n*-ary predicates, and of names for individual constants, always yield arguments with true conclusions if the premisses are all true (Forbes 1985: 88). Noting that this definition of validity is relativized to the language from which the substituends are chosen, Forbes revises his definition to read: a schema is valid for language **L** iff it is not possible for there to be some extension of **L** from which substituends are chosen in such a way that it is possible for the premisses of the resulting argument to be true and the conclusion false. There is much that is obscure and questionable about this surprising turn in Forbes's anti-realism. For example, how is the defined term 'valid for **L**' related to the notion of validity that he had been discussing earlier? The question I raise here is clearly related to Davidson's puzzle: how is relative truth—truth in a model—related to absolute truth?[14] Is this new notion supposed to replace the earlier one? Has he given up his proof-theoretical account of validity? It is not clear. And if he has given up his earlier account, is he not faced anew with the problem of attempting to explain (in this new setting) the naturalness and seeming correctness of possible worlds semantics? Perhaps the substitutional account is only supposed to supplement the other in some way. But if so, how does it supplement the earlier account? They can't both be right, can they? Forbes offers very few details of this new approach, sending the curious reader off to search for an unpublished doctoral dissertation by S. G. Williams, which is not even cited in his bibliography. If this account of validity is supposed to give his *principal anti-Modal Realist version* of modal logic, I find it curious to learn that he has left the reader with the task of searching for such crucial elements in an unpublished dissertation.

Doubts about Forbes's Explanations of how Worlds Semantics Works

I now turn to Forbes's attempts to explain how possible worlds semantics is related to the use of the modal operators. Recall that there were two elements in his explanation. Let us start our investigation with Forbes's attempt to show that we can obtain [PW] from [Box]. To do this, Forbes resorts to using a rule of inference, which he calls "Quine's Principle". This is a principle that allows one to reason from

$$\text{'}\Box\phi\text{' is true}$$

to

$$\Box(\text{'}\phi\text{' is true}).$$

[14] See Etchemendy 1990: 12–14 for a discussion of Davidson's puzzle, or see below, Chapter 5, Section 3.

Quine's Principle is based upon Quine's contention that '□' can best be understood as a predicate of sentences and not as a sentential operator. However, given that Forbes is decidedly unsympathetic to this view of Quine's, it is strange to find him appealing to Quine's Principle in this context (Forbes 1985: 51–3). After all, one cannot understand the box operator of Forbes's S5 quantificational modal logic as a predicate of sentences.[15] Furthermore, although Quine's Principle may seem to be plausible enough, from the viewpoint of one accepting the more traditional semantical explanation of the meaning of '□', it is not at all clear why anyone accepting Forbes's proof-theoretical account of the meaning of '□' should accept the principle. What is needed is a convincing justification of Quine's Principle that arises out of Forbes's proof-theoretic account of the meaning of the operators. Unfortunately, this is not supplied.

But there is another way that Forbes might reason to [PW] from [Box]. It might seem plausible that Forbes could make use of the principle expressed by the schema:

[T] □('ϕ' is true iff ϕ)

(where ϕ is to be replaced by any sentence of the language in question). Since [T] is a principle that is closely related to one made famous by Alfred Tarski and Donald Davidson,[16] one might suppose that it would be legitimate for Forbes to make use of this principle in deriving [PW] from [Box]. If so, Forbes could reason as follows: from [Box] and [T], he could infer □*P*. So, using □-E, he could infer *P*. Then, using [T] again, he could conclude '*P*' is true. Then, using □-I, he could infer

□('*P*' is true).

Using a natural language version of *Trans*, he could infer [PW]. Plausible as this line of reasoning may seem, it is not obvious that Forbes's proof-theoretic account of the use of '□' will provide any grounds for accepting [T].

I turn now to Forbes's explanation of how possible worlds semantics works in allowing us to infer the invalidity of a modal argument from the invalidity of the corresponding possible worlds argument. The explanation, you will recall, was based on the assumption that if the modal sentences are synonymous with their corresponding possible worlds sentences, this would guarantee that each modal sentence must behave, logically, in a modal argument in the same way its possible worlds counterpart behaves in the corresponding possible worlds argument. Now, initially, such reasoning may seem plausible. But doubts begin to arise when

[15] See, for example, Montague 1963.

[16] See Tarski 1956*a*: 275 for Tarski's classic statement of his famous "convention T".

one ponders just what it is that needs to be explained. If we merely treat possible worlds sentences as reverse translations of their modal counterparts, we do not get a useful and enlightening method of determining validity and invalidity of modal arguments. It is only when we treat possible worlds sentences as *standard quantificational sentences,* with all the logical features that quantificational sentences have, that we obtain Forbes's useful and enlightening method of determining validity and invalidity of modal arguments. The first example of the use of possible worlds semantics to determine the invalidity of a modal argument that was discussed earlier makes this point evident, since it was said that "our prior understanding of simple quantificational logic allows us to see that the latter inference is not valid". So what needs to be explained is *why it is legitimate and useful to treat possible worlds sentences in the standard quantificational way*, when, according to Forbes, these sentences are mere reverse translations of their modal counterparts. It is evident that we do not obtain the required explanation by asserting that possible worlds sentences are synonymous with their modal counterparts. Forbes's reverse translation thesis has only replaced one mystery with another.

Now it may be thought that Forbes has explained this mystery by proving the technical result I mentioned earlier. Recall that Forbes claimed that it could be proved that if X is a set of modal sentences and σ is a modal sentence, then there is a proof of σ from X in a natural deduction system with $\Box$-E and $\Box$-I if, and only if, there is a first-order proof of $Trans(\sigma)$ from $Trans(X)$. Forbes evidently thought that what this result would show is that "even though world quantifiers . . . do not mean what quantifiers usually mean in first-order languages, the difference is not sufficient to prevent us from employing our familiarity with first-order logic in assessing modal arguments" (Forbes 1985: 86). Let us examine the consequences of Forbes's technical claim. It is being maintained that:

[1] $X \vdash_{S5} \sigma$ iff $Trans(X) \vdash_{1st} Trans(\sigma)$.

We know that the version of quantificational S5 that Forbes employs in his book is complete.[17] We also know that first-order logic is complete. Thus we have:

[2] $X \vdash_{S5} \sigma$ iff $Trans(X) \models_{1st} Trans(\sigma)$.
[3] $Trans(X) \vdash_{1st} Trans(\sigma)$ iff $Trans(X) \models_{1st} Trans(\sigma)$.

It follows from [1], [2], and [3] that

[4] $X \models_{S5} \sigma$ iff $Trans(X) \models_{1st} Trans(\sigma)$.

[17] The deductive system of Kit Fine is proved to be complete (Fine 1978).

So, in particular, we have:

[5] For any sentence δ of quantificational modal logic,
$\vDash_{S5} \delta$ iff $\vDash_{1st}$ *Trans*(δ).

That is, δ is a valid sentence of quantificational S5 modal logic iff *Trans* (δ) is a valid sentence of first-order logic. But that is absurd! Consider, in particular, the sentence '$(\exists x)E^1x$'. This sentence gets translated into the possible worlds language *W* as:

[6] $(\exists w)(\exists x)(E^2xw \;\&\; E^2xw)$

where 'E^2xw' is interpreted, in the possible worlds semantics for *W*, as the existence relation: *x exists in w*.[18] Clearly, [6] is not a valid sentence of first-order logic; but its reverse translation, '$\Diamond(\exists x)E^1x$', is a valid sentence of quantificational S5 modal logic.[19]

What has gone wrong? Forbes has informed me that what he should have claimed, instead of [1], is:

[7] $X \vdash_{S5} \sigma$ iff $Trans(X) \vdash_{\mathbf{T}} Trans(\sigma)$

where **T** is a first-order theory whose axioms would capture the definition of an interpretation of the modal language.[20] Thus, **T** would contain an axiom

[8] $(x)(\exists w)E^2xw$

which would capture the clause in the definition of interpretation that asserts that every possible object *x* exists in at least one possible world (see the clause in the definition of 'interpretation of **M**' dealing with the function **d**). With such an axiom, [6] could be proved in **T**.[21]

But this way of salvaging the technical claim makes all the more evid-

[18] As an examination of [6] above will suggest, strictly speaking, what Forbes translated modal sentences into is a two-sorted first-order language. Forbes considers this a "minor complication" (Forbes 1985: 86 n. 18) and refers the reader to Enderton's well-known textbook for further details.

[19] This was shown in Chapter 1.

[20] In a letter dated 27 June 1990.

[21] How complex would **T** be? First, it should be noted that the reference to first-order logic in the technical claim being made must be a reference to first-order free logic. One reason for thinking so is: the sentence '$Faw^* \rightarrow (\exists x)Fxw^*$' is a valid sentence of standard first-order logic; and it gets reverse translated by the rules in the appendix into '$\mathcal{A}Fa \rightarrow \mathcal{A}(\exists x)Fx$', which is not a valid sentence of Forbes's S5 Quantificational Logic. Given the free logical framework, it would seem that among **T**'s axioms would be sentences of the form '$(\exists w)E\alpha w$' for each individual constant α, since given the semantics of Forbes's system, each individual constant must denote a possible object (see Forbes 1985: 44). In short, it would appear that **T** is not finitely axiomatizable.

ent a weakness in Forbes's reverse translation explanation of why it is legitimate and useful to employ possible worlds semantics in determining the validity and invalidity of modal arguments. For it makes evident the necessity to make use of the semantical definitions of possible worlds model theory in order to give the sort of justification that Forbes wants to give for assessing modal arguments in the possible worlds manner. But now we want to know why, in order to justify the method of assessing the validity of modal arguments that Forbes wishes to use, we need to treat the possible worlds sentences that are counterparts of modal sentences as if they possessed the logical structure derived from possible worlds semantics, and why we have to appeal to axioms that express semantical features generated by the supposedly literally false possible worlds model theory.

To clarify the point being made, let us consider how someone, who takes the possible worlds translations of modal sentences to be the *analysans*, and not the *analysandum* as Forbes maintains, could explain the technical fact upon which Forbes basis his explanation, namely:

$$X \vdash_{S5} \sigma \text{ iff } \mathit{Trans}(X) \vdash_{\mathbf{T}} \mathit{Trans}(\sigma).$$

Such a person could note that the rules of S5 modal logic were devised to be sound and complete. Similarly, the rules of first-order logic were devised to be sound and complete. So to explain the above, one would only need to explain

$$X \vDash_{S5} \sigma \text{ iff } \mathit{Trans}(X) \vDash_{\mathbf{T}} \mathit{Trans}(\sigma).$$

But this can be explained by the fact that for any sentence of the modal language ϕ, *Trans*(ϕ) just expresses in the worlds language the meaning that ϕ has been given by the semantics of S5 modal logic, and one needs the theory **T** on the left side of the biconditional because translating the sentences of the modal language does not capture the stipulation that the **D** of every interpretation must be non-empty.

But what happens if one adopts the reverse translation doctrine, according to which the direction of analysis is reversed? Then the above explanation can no longer be given. The point here can be seen clearly by reconsidering Russell's Theory of Descriptions. Russell wished to explain why the sentence

[D] The present King of France is bald

is not meaningless, but is instead false. His explanation was that the apparent grammatical form [D], its "surface grammar", is misleading. [D] should be analysed as having the meaning of the existential sentence

[E] $(\exists x)(y)((y$ is presently King of France iff $y = x)$ & x is bald).

Then [D] is really just an existential sentence and hence will not be meaningless if there is no one who is presently King of France; instead, it will just be false.

But what happens if one takes the position that [E] is not really an existential sentence at all and that [E] just has the meaning of [D]. In that case, one can no longer give the above explanation of why [D] is not meaningless. We find ourselves right back at the position Russell found himself at the start.

It is now time to investigate the reverse translation thesis itself, since there is much in this doctrine that is suspicious.

Doubts about the Reverse Translation Doctrine

What grounds have we been given for accepting the reverse translation thesis? Very little, if any, grounds have been supplied by Forbes. We are told that

[9] $(\exists w)P(w)$

does not mean what possible worlds semanticists take it to mean: it does not literally assert the existence of a possible world at which P holds. Rather, it means simply

[10] $\Diamond P$.

Now it is noteworthy that Forbes presents no empirical linguistic evidence to support this semantical thesis. The view seems to be put forward simply to avoid ontological commitments to worlds. Yet, evidence is surely called for, since the thesis is counterintuitive. Possible worlds semantics involves much more than formal sentences: it consists of a theory containing, among other things, a set theory, principles of mathematical logic, and a theory of interpretations, of truth (with respect to an interpretation), and of validity. Given the sort of semantical theory that underlies possible worlds sentences, I find it implausible to insist that [9] does not assert what it appears to. I also find it implausible to claim that the possible worlds sentences that do not translate into any modal sentence by $Trans^{-1}$, such as '$(w)(w = w)$', are meaningless or are uninterpreted stipulations which enable us to establish facts of interest about the interpreted sentences. After all, '$(w)(w = w)$' is treated in the underlying semantic theory as meaningful and true under every interpretation of the language.

Forbes supplies us with rules for "reverse translating" certain types of possible worlds sentences into sentences of **M.** But the possible worlds sentences he considers in this connection are, themselves, taken to be sentences of a formal quantificational language. It was pointed out earlier that Forbes's reverse translation thesis applies also to the worlds sentences of natural languages. Of course, when the possible worlds sentences that are to be reverse translated are taken to be English sentences, it is even more counterintuitive to maintain the thesis Forbes has put forward. For in that case it must be maintained that possible worlds English sentences do not mean what they seem to mean to competent English speakers and are, in fact, mere reverse translations of their corresponding modal sentences. Think of the counterintuitiveness, in this setting, of Forbes's thesis that such English sentences as 'Every possible world is identical to itself' are meaningless or are "uninterpreted stipulations which enable us to establish facts of interest about the interpreted sentences"! Most English speakers would not classify such sentences as uninterpreted stipulations. And if all such sentences are uninterpreted stipulations, as Forbes contends, then what are we to say about some of the possible worlds sentences that Forbes himself asserts in the course of defending various philosophical positions. I have in mind such sentences as

> if u and v are worlds which at any time have some existent object in common, then u and v have some initial segment of their courses of history in common.[22]

This is a sentence about worlds that cannot be reverse translated into Forbes's modal quantification language by analogues of the kind of rules of translation Forbes provides. So is it, too, to be considered an uninterpreted stipulation which enables us to establish facts of interest about the interpreted sentences? If so, how is it that Forbes gives every indication that he believes what it seems to say and, indeed, bases his defence of his principle of grounding upon what it evidently says?

The difficulties grow when we consider Forbes's treatment of transworld identity. Consider Forbes's principle [PSI], which you will recall said:

> For all worlds u and v, if x is an organism in world u with exactly the propagules $z_1, z_2, \ldots, z_k$, and y is an organism at world v with exactly the propagules $t_1, t_2, \ldots, t_j$, then x is the same organism as y iff (i) $\{z_n\} = \{t_m\}$; and (ii) the sort of x at u is the same as the sort of y at v.

There is no way of reverse translating this principle into Forbes's modal language. Yet it is hard to see how Forbes could regard this worlds statement as a meaningless or uninterpreted stipulation, given that his biological essentialism rests upon this principle.

At the end of the chapter on the necessity of origin, Forbes writes: "The theses we are advancing depend only on the correctness of the doctrine that facts about transworld identity and non-identity must be intrinsically grounded" (Forbes 1985: 159). Notice that he speaks of "facts about transworld identity". But how could there be facts about transworld identity if there are no worlds at all? Indeed, what sense can we make of

[22] Forbes 1985: 151.

transworld identity, given Forbes's anti-realist position? It would seem that Forbes simply forgot about his anti-realism when he turned his attention to the problem of transworld identity and the necessity of origin.

I mentioned earlier that, in response to some objections raised by Hazen, Forbes expands his modal language to include subscripted modal operators. Melia has objected that these operators "were introduced solely through the evaluation clauses in the model theory, and our understanding of them is gained entirely from their model-theoretic clauses".[23] Melia also notes that Christopher Peacocke, who originally introduced such subscripted operators and from whom Forbes's operators were borrowed, explained the meaning of the operators in terms of possible worlds (Melia 1992*a*: 45). Again, this points to the implausibility of the thesis that each world sentence is a mere reverse translation of its corresponding modal sentence. If the direction of analysis is as Forbes would have us believe, one would think that he would explain the meaning of the world sentence in terms of the modal sentence with the indexed operators, not the other way round.

Forbes's Solution to a Paradox

I turn now to a consideration of Forbes's analysis of a version of Chisholm's Four Worlds Paradox, because we find confirmation of my claim that Forbes tends to forget his anti-realist doctrines in dealing with various problems of modality.

Consider the following possibility: $<w_1, w_2, w_3, \ldots, w_n>$ is a sequence of n worlds, $<c_1, c_2, c_3, \ldots, c_n>$ is a sequence of clocks such that c_i is in w_i ($i = 1, 2, 3, \ldots, n$) and each clock in the sequence differs from the one immediately preceding it in the sequence by the fact that the later clock has exactly one part that is not in the earlier one; and the sequence is such that c_n is made up of entirely different parts from c_1. The paradox arises when we apply an intuitively acceptable principle:

> Necessarily, any clock could have originated from a slightly different collection of parts from any one collection from which it could have originated.

One can then conclude that each clock in the sequence is identical to the one that precedes it in the sequence, so that, by the transitivity of identity, we can conclude that $c_1 = c_n$. That seems to be absurd, since c_1 has no parts in common with c_n. For Forbes, such a possibility runs counter to his principle of grounding, since we would have a bare identity that is not grounded.

[23] Melia 1992*a*. It should be noted that this argument was made in the context of attacking Forbes's Modalism—not the reverse translation thesis.

Forbes classifies this paradox as a sorites paradox—a paradox of vagueness. A typical paradox of this sort is the paradox of the heap. The following seems to be an obviously true principle:

> [P] For every x and every y, if x is a heap of pebbles and y is what remains after removing just one pebble from x, then y is a heap of pebbles.

Suppose that one has a heap of pebbles, say **a**. Then by [P], removing one pebble from **a** will result in a heap of pebbles, say **b**. Then removing one pebble from **b** will result in a heap of pebbles, say **c**. But continuing this line of reasoning will result in a heap of pebbles consisting of just one pebble—an absurdity.

Forbes proposes to deal with the modal paradox about the clocks in the way he attacks standard sorites paradoxes. His diagnosis of such paradoxes is that they arise from the application of classical two valued logic to situations that require a logic for inexact concepts. Let us examine just how he analyses the paradox of the heap.

Basing his analysis on J. A. Goguen's logic for inexact concepts, Forbes maintains that objects fall under the inexact concepts of the paradox only to a degree.[24] For this reason, in Forbes's fuzzy logical semantics of the sentential logic, an interpretation or model assigns to each sentential letter a member of the closed unit interval [0, 1]. The degree of truth of a molecular sentence is then calculated according to the following rules:

$$\text{Deg}[\phi \vee \psi] = \max\{\text{Deg}[\phi], \text{Deg}[\psi]\}$$
$$\text{Deg}[\phi \,\&\, \psi] = \min\{\text{Deg}[\phi], \text{Deg}[\psi]\}$$
$$\text{Deg}[\text{-}\phi] = 1 - \text{Deg}[\phi].$$

If one took $\text{Deg}[\phi \rightarrow \psi]$ to be that of $\text{Deg}[-\phi \vee \psi]$, since $(\phi \rightarrow \psi)$ is equivalent to $(\text{-}\phi \vee \psi)$ in classical logic, then we would have:

$$\text{Deg}[\phi \rightarrow \psi] = \text{Deg}[-\phi \vee \psi] = \max\{1 - \text{Deg}[\phi], \text{Deg}[\psi]\}$$

However, Forbes defines it this way:

$$\text{Deg}[\phi \rightarrow \psi] = 1 - (\text{Deg}[\phi] - \text{Deg}[\psi]) \text{ if } \text{Deg}[\phi] > \text{Deg}[\psi] = 1 \text{ otherwise.}$$ [25]

Finally, the degrees of truth of universal sentences can be calculated as follows:

$$\text{Deg}[(\alpha)\theta] = \text{glb}\{\text{Deg}[\theta\alpha/\beta] : \beta \in \text{the domain of the quantifier}\}.$$

[24] Goguen 1969. Goguen, a Berkeley Ph.D., approaches the sorites situation from the perspective of fuzzy logic, under the influence of his thesis supervisor Lofti Zadeh.

[25] Actually, this difference will not affect my assessment of his solution to the sorites.

With the semantics of inexact concepts set, Forbes turns to an analysis of the sorites paradoxes. There seem to be only two alternatives that are open to us: either some of the premisses of the paradox are not true, or some inferences are not acceptable. In the case of the paradox of the heap, there seems to be only one premiss to consider, namely [P]. Now [P] can be considered a long conjunction, each conjunct of which is a conditional. Suppose that **a** consists of 15,000 pebbles piled together in a mound, and **b** is what results from just removing one pebble from **a**. Then both the antecedent and the consequent of the conditional

> If **a** is a heap, and **b** is what results from just removing one pebble from **a**, then **b** is a heap

will have degrees of truth close to 1. Given the way degrees of truth are calculated, Forbes holds that such premisses as [P] are "almost wholly true" and hence are premisses that it "would be absurd to deny" (Forbes 1985: 174).

Forbes's reason for rejecting such premisses as [P] seems to parallel that of R. M. Sainsbury, who has written concerning this very question:

> [T]he conditional premises, though very nearly true, are not quite true. Hence we need not be fully committed to them. Hence, the paradoxical argument does not commit us to the paradoxical conclusion. However, this would not be a good account. Even if we are not fully committed to the premises, they are very nearly true. The degree theorist has to explain how premises that are all very nearly true can lead to a conclusion that is wholly false. (Sainsbury 1988: 42)

As these two philosophers analyse the paradox, then, the solution must be found elsewhere. Both Forbes and Sainsbury follow Goguen in concluding that it is the rule of inference *modus ponens* that is the source of the absurdity.[26] Notice that '**b** is a heap' can be obtained by *modus ponens* from the two premisses:

> (1) **a** is a heap;
> (2) If **a** is a heap, and **b** is what results from just removing one pebble from **a**, then **b** is a heap.

Now the degree of truth of the premisses, as well as that of the conclusion, will be high; but the degree of truth of the conclusion, it is argued, will be slightly less than the degree of truth of the premisses. This leads Sainsbury to diagnose the situation as one in which "the application of *modus ponens* can lead to a "leakage" of truth. The leakage may be small for each application, but can be large if the number of applications is large, as is the case of the paradoxical argument" (Sainsbury 1988: 43).

[26] See Goguen 1969: 330.

Although this diagnosis of the paradox is attractive, I feel I should mention some questionable points in the analysis. First, it is by no means clear just what the numbers that supposedly represent degrees of truth really measure. We have, in this case, nothing that corresponds to the Bayesian's explanations and definitions correlating degrees of belief with betting behaviour. There is, I am afraid, much that is unclear at the very foundations of this approach. Second, I am not convinced that the method of calculating degrees of truth Forbes uses is always reasonable. Take the case of railway car *A*, which is filled with 4,000 apples. Every apple in *A* is perfect, except for one that is so bruised that the degree of truth of the proposition that this apple is perfect is ½. By the rules Forbes is using, the proposition

Every apple in *A* is perfect

has degree of truth ½. Now consider railway car *B*, which is also filled with 4,000 apples. Suppose, however, that every single apple in *B* is bruised to the same degree as is the one apple in *A*. It then will be the case that the proposition

Every apple in *B* is perfect

will have the same degree of truth as the former proposition, namely ½. But this seems wildly at odds with our intuition that the former proposition is much closer to the truth than the latter. If you were the retailer who purchased the apples in car *A*, you would not quarrel with the claim that every apple in *A* is perfect: one imperfect apple out such a large totality does not detract significantly from the truth of the statement. On the other hand, if you were the retailer who purchased the apples in car *B*, you would find the claim that every apple in *B* is perfect to be outrageous. The first claim would be regarded as very close to the truth; but not the second.

Another feature of the analysis that seems dubious is the tacit assumption that all the conditionals implicit in the main premiss of the paradoxical argument, i.e. the conditionals in [P], have a degree of truth close to 1. Certainly, if a collection **a** of pebbles piled into a mound is very large, then it is reasonable to attribute to the conditional

> If **a** is a heap, and **b** is what results from just removing one pebble from **a**, then **b** is a heap

a degree of truth extremely close to 1. What is not so clear to me is that, if we are dealing with a very small collection of pebbles, say one that consists of just three pebbles, such a conditional should also be given a degree of truth very close to 1. Suppose that the three pebbles happen to look like Fig. 6. Removing just one pebble from this arrangement will result in a collection that looks like Fig. 7. Now it would not be unreasonable to at-

FIGURE 6

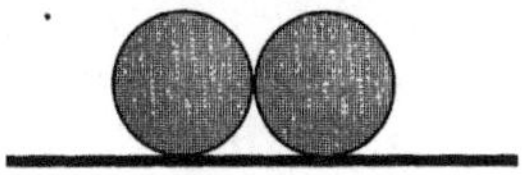

FIGURE 7

tribute to the statement 'The three-pebble collection is a heap' a low degree of truth, say 0.25, and to attribute to the statement 'The two-pebble collection is a heap' a much lower degree of truth, say 0.001. The conditional would then receive the degree of truth 0.76, which is not very close to 1. Thus, it is not so obvious to me that principle [P] should be given the sort of assessment that both Forbes and Sainsbury have given it, namely one that assigns to the principle a degree of truth very close to 1. And although something like the solution offered by Forbes and Sainsbury may be on the right track, I feel that much more work needs to be done on working out the wrinkles of the analysis before one could have much confidence in it.

Returning to Forbes's analysis of the modal paradox, the principal idea is to treat it in the way the sorites paradoxes were analysed above. The details of Forbes's solution to the paradox need not concern us here. The salient features are these: the reasoning is formalized in a possible worlds *counterpart theoretic* language. The sentences of his counterpart theoretic language are partially ordered according to the "degrees of truth" that they have. Degrees of truth are needed because, it is argued, the counterpart relation admits of degrees: a clock in one world is similar to a clock in another world only to some degree. Of course, clock *A* in world *u* may be very similar to clock *B* in world *v* with respect to size, say, but very dissimilar with respect to colour. To account for these different respects, the degrees of truth are given by *n*-tuples of real numbers.

But in analysing the counterpart relation in this way, Forbes oversimplifies the situation. This can be seen in his crossworld condition asserting that "objects [such as clocks] which are similar across worlds in design (and anything else deemed to be relevant) and which have most of their parts in common, are counterparts" (Forbes 1985: 187). This condition greatly simplifies the situation by treating the parts of clocks as if they can occur in different possible worlds. Thus, he uses a possible worlds framework in which there is transworld identity of parts, even though there is no transworld identity of the manufactured objects, made up of these parts: the manufactured objects have counterparts. Such a framework needs to be investigated, since it is not clear if such a framework is philosophically coherent.

Within the counterpart framework he developed earlier in his book, he should have taken the parts of a clock to have counterparts in other worlds; and hence there should be a fuzzy counterpart relation for the parts too—or more accurately, there should be different counterpart relations for different sorts of parts. Thus, the respects in which one mainspring is similar to another mainspring would differ considerably from the respects in which one alarm clock bell would be similar to another. Now, how all of this could be worked into a coherent theory is not at all clear.

I suspect that Forbes would claim that these are mere details and not something to worry about, since the main idea is clear. But the point is: a more accurate coherent theory of this sort would no doubt have to have a much more complicated way of recording degrees of truth than is indicated by Forbes. Perhaps counterpart theoretic sentences of this sort would have to have degrees of truth measured by finitely descending trees of real numbers.

Near the end of this chapter, Forbes reminds us that "we do not take the meta-language sentences of counterpart theory to have the literal meaning which they apparently have; that is, their true meaning is fixed by reverse-translation into the object language" (Forbes 1985: 189). The trouble is, in this setting, it is very implausible to maintain Forbes's reverse translation doctrine that the worlds sentence is the *analysandum* and the corresponding modal sentence is the *analysans*. Take the modal sentence, 'This clock could have had a different mainspring.' According to Forbes's analysis, it would make sense to attribute to this sentence the degree of truth <0.5, 0.8, 0.2, . . ., 0.9>. But if, as was noted above, we adopted a more accurate account of this case, we would have to attribute to the sentence much more mathematical structure, say a finitely descending tree. Now such an attribution has been explained in terms of counterpart theoretic possible worlds sentences. Here again, we would have a counterpart theoretic *analysis* of the modal sentence, not the other way round. So it would seem that the modal sentence would have to be the *analysandum*, and the worlds sentence would have to be the *analysans*: I would not know how to understand the kinds of attributions of degrees of truth to modal sentences that Forbes's solution to the paradox requires, if the counterpart theoretic explanation were abandoned. What this suggests is that the reverse translation doctrine and the proof-theoretic account of meaning are difficult for Forbes to maintain. It is not surprising that, as soon as he leaves the metaphysical arena in which his anti-realism is espoused, he reverts to writing about, and thinking in terms of, possible worlds.

Some Concluding Comments

What lessons can we draw from Forbes's valiant, but ultimately unsatisfying, attempt to develop an anti-realist account of the semantics of modal logic? One striking feature of the approach is the complete absence of any attempt to carve out a role for the mathematical structures of standard possible worlds semantics: it is as if modal logicians only concerned themselves with the possible worlds translations of modal sentences, so that one could dispense with structures once one made the translations. This absence of any mention of structures, in Forbes's account of the relationship between modal sentences and possible worlds sentences, is surely in need of explanation. Why does not Forbes offer at least a word of explanation?

These features of Forbes's anti-realism seem to be related to another noteworthy peculiarity of the account: almost all of Forbes's account is focused on the task of explaining how possible worlds sentences can be used to assess the validity and invalidity of modal arguments. But possible worlds semantics is used for much more than assessing validity and invalidity, just as the model theory of first-order logic is concerned with much more than just the evaluation of the validity of arguments.[27] In the case of first-order logic, for example, the Homomorphism Theorem is applied to provide information about definability within structures, elementarily equivalent structures, and decidability.[28]

The above discussion of Forbes's anti-realism, coupled with the earlier objections to Forbes's reverse translation thesis, strongly suggests that we should look for an account of possible worlds semantics that analyses the relation between sentences of **M** and possible worlds sentences in a more complex way than is acknowledged by Forbes. We should expect such an account to assign an important logical role to the mathematical structures of the semantics of modal logic. And we should expect the account to explain much more than just how possible worlds model theory can be used to assess the validity and invalidity of modal arguments. How such an account can be developed from an anti-realist perspective will be explored in Chapters 6 and 7.

[27] See, for example, van Benthem 1983 for discussions of some important results obtained through a study of possible worlds structures. See also, in this connection, Goldblatt 1993: ch. 1.

[28] See, for example, Enderton 1972: 91–7.

3. Modal Fictionalism: Exposition

One radical anti-realist analysis of modality that has been receiving a lot of attention in the literature recently is *modal fictionalism*—a doctrine put forward by Gideon Rosen (Rosen 1990). Rosen's idea is to obtain the fruits of Lewis's labours without paying the price of taking on his objectionable ontology of possible worlds. The central idea is to say, roughly speaking, that a modal sentence ϕ is true iff the Lewisean translation of ϕ, say ϕ^*, would be true, were Lewis's theory of worlds true. (The Lewisean translation of 'It is possible that donkeys talk' is 'There is a world in which donkeys talk'.) To be more precise, the modal fictionalist sketches a piece of fiction (which he calls *PW*) based on the theory of worlds given in Lewis 1986. It is maintained that *PW* enters into the truth conditions of modal statements in the following way: if ϕ is a modal statement and ϕ^* is the Lewisian translation of ϕ, then

ϕ is true iff, according to *PW*, ϕ^*

where the phrase 'according to *PW*' is the modal fictionalist's main primitive. Just as one could say that, according to the Disney version of *Snow White,* the wicked stepmother dies in a fall, one can say that, according to *PW*, there is a world in which there are tailless kangaroos. Rosen offers the following paraphrases of the right side of the above biconditional to give the reader some indication of what the primitive means:

If *PW* were true, ϕ^* would be true;
If we suppose *PW*, ϕ^* would follow;
It would be impossible for *PW* to be true without ϕ^* being true.

From these paraphrases, one can see that ϕ^* *being true according to PW* essentially comes down to ϕ^* being some sort of consequence of *PW*. Obviously, the primitive is a *modal* primitive. Rosen allows that he has not put forward a reduction of modality to the non-modal; and he states that the prospects of analysing away the modal element from his primitive are not bright. The reason is, more must be true according to *PW* than is explicitly stated. So "the problem is to specify the *consequence relation* which determines this larger class in a non-modal way" (Rosen 1990: 347). This, he says, he cannot do.

Another point that needs to be made regarding the above biconditional is that the 'iff' must be so understood that, in the case of a modal sentence ϕ which is such that neither ϕ^* nor $-\phi^*$ is a consequence of *PW*, Rosen declares that ϕ is to "lack a truth value" (Rosen 1990: 343). As a result, Rosen declares that the connectives of the modal language are not truth-functional. To see why Rosen is led to this position, take some modal sentence ϕ which is such that neither ϕ nor $-\phi$ is true according to *PW*. Since $(\phi \vee -\phi)$

must be true according to *PW* (all tautologies being taken by *PW* to be true), the connectives cannot be truth-functional.

Finally, I should note that there are indications in Rosen's writings that he does not wholeheartedly espouse modal fictionalism. And I have learned from him personally that he does not consider himself a modal fictionalist. Thus, we should take him to be merely exploring what he takes to be a promising position. However, for purposes of simplicity of exposition and evaluation, in what follows, I shall write as if Rosen is the modal fictionalist he is investigating.

What is the Fictionalist's Fiction?

The fictionalist's fiction, *PW*, consists of an encyclopaedia and the following seven postulates:

1. Reality consists in a plurality of worlds.
2. One of these is what we ordinarily call *the* universe; the largest connected spatiotemporal system of which we are parts.
3. The others are things of roughly the same kind: they are systems of objects connected by a network of relations like the spatiotemporal distances that connect the objects of our universe.
4. Each universe is isolated from the others (they are not spatiotemporally related).
5. The totality of universes is closed under a *principle of recombination.* This principle is only given roughly by Rosen as follows: For any collection of objects from any number of universes, there is a single universe containing any number of duplicates of each, provided there is a space-time large enough to hold them. (We are referred to Lewis 1986: 87–90 for a more precise statement of the principle.)
6. There are no arbitrary limits on the plenitude of universes.
7. Our universe is not special. That is, there is nothing remarkable about it from the point of view of the system of universes.[29]

This set of postulates requires some explanation. Postulate 2 simply asserts that what I have been calling "the actual world" is one of the worlds in the "plurality of worlds" mentioned in postulate 1. Postulate 5 does almost all the work of generating possibilities—for example, it is used, with the encyclopaedia described below, to allow us to conclude that it is possible that there are tailless kangaroos. Postulate 7 is vague in so far as it is

[29] Rosen suggests that the above postulates may need to be supplemented with an additional postulate that tells us that the universe of postulate 2 is actual, whereas the other worlds are possible worlds. We are told that this additional postulate would be brought in to make *PW*, the realist's view, incompatible with Actualism. I shall make no further mention of this supplemental postulate, since Rosen asserts that it "will play no role in his account" (Rosen 1990: 334).

not specified what it is to be "special". Would our universe be "special" if all possible objects were combinations of duplicates of the objects in our universe? Certainly, there are grounds for thinking that, in that case, our universe would be very "special". On the other hand, there could be an enormously large number of such universes, so one could make a strong case for concluding that our universe would not be "special" after all. I surmise that Rosen intends this postulate to be understood in such a way that it rules out all possible objects being duplicates of the objects in our universe, for he writes: "If our universe contained an instance of every natural property, it would be a remarkably rich world, in violation of [Postulate 7]" (Rosen 1990: 333 n. 13).

The encyclopaedia consists of a list of the "non-modal truths about the intrinsic character of this universe"—where a sentence is a non-modal truth iff it contains no modal vocabulary and entails neither the existence nor the non-existence of anything not in our universe (Rosen 1990: 335). (It is worth noting here that Rosen's characterization of even the encyclopaedia is modal, in so far as it makes use of the notion of "entailment".) Rosen says very little about this encyclopaedia, leaving the reader to imagine how this list of truths is to be specified. It might be thought that it will be something like the *Britannica*, which is supposed to provide the reader with all the reliable knowledge society possesses at the time of publication. But the fictionalist's encyclopaedia will have to give us much more information than any ordinary one would give, since we are told that the encyclopaedia specifies that our universe contains objects which are rather like ordinary kangaroos but for the lack of a tail (at least in the form of undetached kangaroo parts). This is supposed to allow us to infer, by the principle of recombination, that there are other universes containing freestanding tailless kangaroos. Presumably, the encyclopaedia will also have to specify that our universe contains objects which are rather like ordinary kangaroos but for the lack of ears or eyes or a pouch or . . . But what about specific kangaroos? We want it to be the case that Johnny's pet kangaroo "Roo-pert" could have been tailless, don't we? And how could we infer, using the principle of recombination and the encyclopaedia, that Roo-pert could have been tailless, unless specific things are mentioned in the encyclopaedia? It is now beginning to look as if the encyclopaedia would have to list all the actual objects that exist. What about objects that do not now exist but that have existed or will exist? Presumably, the encyclopaedia would have to list all those objects too—otherwise *PW* would not capture all the possibilities. Clearly, this encyclopaedia could never be completed by mere mortals.

Another point regarding this encyclopaedia that needs to be cleared up is the question of mathematical truths. Mathematical truths are generally held to be necessary truths, so it would seem that *PW* must somehow give us all the mathematical truths. But it is hard to see how it can do that.[30] There does not seem to be any way that the principle of recombination can give us the mathematical truths from just a description of the non-mathematical universe, so that leaves only the encyclopaedia to provide us with all the mathematical truths. Does it just list the members of the (non-recursive) set of mathematical truths expressible in English? Rosen does not disclose how he thinks such a feat of listing could be accomplished, so it is not clear to me that this proposed encyclopaedia of *PW* is constructible, even in principle.

Central Claims made by the Fictionalist

At the end of his article, Rosen sketches an alternative form of fictionalism, which he calls "timid fictionalism". Timid fictionalism claims to offer only a theory linking modal facts with facts about *PW* (Rosen 1990: 354). It does not claim to shed any light on the nature of modal facts or upon the meaning of the modal operators. It is a form of fictionalism that is not espoused by Rosen. He notes that the view "raises as many questions as it answers" (Rosen 1990: 354).

The form of modal fictionalism that Rosen is inclined to defend is the view I have been discussing above. It is a view that is supposed to deliver *a reduction of the modal operators to just one primitive notion* (Rosen 1990: 345 and 347): Rosen tells us that he has been attempting to give us an "analysis" of modal language and a *theory of possibility*, "that is, an account of the truth conditions for modal statements" (Rosen 1990: 354). Thus, Rosen writes:

> It should be utterly uncontroversial that *according to the realist's conception of the plurality of universes*, there is a universe with a blue swan in it. And this is as it should be, because it is equally uncontroversial that there might have been blue swans. And for the fictionalist, these come to the same thing.[31]

That the fictionalist has supplied an *analysis* of the modal operators is an important feature of Rosen's position; it is a key ingredient of one argument he gives in support of modal fictionalism. Why is it that the principles that guide our imagination correctly describe the realm of possible worlds—a realm from which we humans are completely cut off?

[30] It might be thought that mathematical truths are to be generated somehow by the primitive consequence relation postulated by the fictionalist. Thus, there is a suggestion (in Rosen 1990: 348 n. 27) that all analytic truths are to be given by the consequence relation. However, it seems utterly implausible to suppose that there is any such consequence relation that would give one every mathematical truth. Consider, for example, the recent proof of Fermat's Last Theorem: imagine it being claimed that, according to *PW*, Fermat's Last Theorem is true.

[31] Rosen 1990: 336 (the italics are Rosen's, the underline is mine).

From the Modal Realist's perspective, this question cries out for an answer, suggests Rosen, whereas this problem does not arise for the modal fictionalist:

> If the realist is right in suggesting that we are guided in the imaginative construction of possibilities by principles like the postulates of *PW*, then when we engage in imaginative experiments, *the least we discover is what is true according to PW*. But for the fictionalist, that is enough. The modal facts just are facts of this kind. (Rosen 1990: 340, my underline)

Later in the article, Rosen takes up the objection that "fictionalism may be thought to strain credulity" by identifying Humphrey's regret that he could have won the election with Humphrey's regret regarding the content of an arcane story. Rosen responds: "For the fictionalist, the fact about the story just is the modal fact; and a lucid fictionalist must have the same attitude towards both" (Rosen 1990: 351).

Rosen attributes a number of other wonderful features to his favoured account of the modal operators.

> [The fictionalist] has offered a powerful reduction of a wide variety of modal notions to one—a streamlining of ideology—with no cost in ontology. His account is adequate to common sense modal opinion; it makes good sense of the source of modal knowledge. (Rosen 1990: 347)

Rosen goes on to claim that, "in central cases, *the realist and the fictionalist will concur in their theoretically informed modal judgements*. That is, whenever the realist believes that his paraphrase of a modal claim is true, the fictionalist should have the same view of *his* paraphrase" (Rosen 1990: 337).

4. Modal Fictionalism: Problems and Doubts

The Brock–Rosen Problem

The fictionalist's analysis has been subjected to a number of objections in the literature. One that Rosen found especially troublesome is raised by Stuart Brock (Brock 1993). Let θ be the statement that there are at least two worlds. Then

$$\Box\theta \text{ iff according to } PW, (\Box\theta)^*.$$

Now $(\Box\theta)^*$ is analysed by Brock to be the statement that, at every world *w*, there are at least two worlds. But it can be seen that, according to *PW*, at every world *w*, there are at least two worlds. Hence, the fictionalist is committed to $\Box\theta$. But one of the fundamental laws of modal logic is

$$[\text{N}]\ \Box\phi \rightarrow \phi.$$

Hence, the fictionalist seems to be forced to accept the truth of θ, that is, the proposition that there are at least two worlds—a conclusion that would completely undermine the motivation for being a fictionalist!

In Rosen 1993, Rosen discusses what is essentially the same objection to fictionalism. As Rosen views the situation, the fictionalist has only three options: (1) reject [N]; (2) resist the reasoning that $\Box\theta$ must be accepted by the fictionalist; or (3) give up his Actualism, that is, the belief that there is only one world. At the time of writing this article, Rosen evidently could find no satisfactory way out of this dilemma for the fictionalist, leaving the fictionalist with the daunting challenge of explaining why his position is not committed to the above absurdity.[32]

Harold Noonan, however, came to the rescue (Noonan 1994). Unlike Brock and Rosen, who merely sketch the above objection, Noonan turns to Lewis's original paper on counterpart theory in an attempt to determine precisely how one is to translate modal sentences into world sentences. What, in particular, is the sentence $(\Box\theta)^*$? To apply Lewis's translation rules,[33] we need to express θ as a sentence of first-order logic. θ can be taken to be the sentence

$$[1]\ (\exists u)(\exists v)(Wu\ \&\ Wv\ \&\ u \neq v).$$

Then, according to Lewis's translation scheme, $(\Box\theta)^*$ is the sentence

$$[2]\ (w)(Ww \rightarrow (\exists u)(\exists v)(Iuw\ \&\ Wu\ \&\ Ivw\ \&\ Wv\ \&\ u \neq v)).$$

Now it is easy to see that, according to *PW*, [2] is false. From postulate 4 of *PW*, each world is isolated from all the others; and hence, two distinct worlds cannot be *in* any world w. It follows that, contrary to what the Brock–Rosen objection had assumed, $\Box\theta$ is false. Thus, by using the translation scheme of Lewis 1979*a*, Noonan argues, this apparent dilemma can be obviated.

In Rosen 1995, Noonan's proposal is taken by Rosen to be an acceptable solution to the difficulty. It is not that he thinks that Noonan's analysis saves the *original* fictionalist position of Rosen 1990. However, he does think that

> Noonan is right to point out that a *modified* fictionalism which proceeds exclusively in terms of Lewis's 1968 translation scheme is not vulnerable to the objections. The fictionalist should therefore accept Noonan's proposal as a friendly amendment. (Rosen 1995: 68)

In what follows, I shall be discussing this modified version of fictionalism.

[32] Rosen does argue that Armstrong's version of fictionalism is equally open to the above objection. See Rosen 1993: sect. 4.

[33] These will be given in Chapter 9, Section 4, below.

Consider the sentence θ again. With the Noonan amendment, not only is it the case that the fictionalist will declare □θ to be false, he will also conclude that –◇θ is true. However, that there are other universes that are spatiotemporally disconnected from our universe seems to be a definite conceptual possibility, so the fictionalist's analysis has a counterintuitive implication. As we shall see, Rosen is aware of the fact that his revised fictionalist position has the above unattractive consequence. In Rosen 1995 he makes a suggestion for revising *PW*, so as to block this consequence. This suggestion is made in the context of replying to another objection.

Hale's Objection to Fictionalism

We saw earlier that the fictionalist is committed to the view that θ is necessarily false. Since *PW* asserts θ, *PW* is itself necessarily false. Take any modal sentence ϕ. The fictionalist maintains that

ϕ is true iff, according to *PW*, ϕ*.

But if *PW* is necessarily false, then, it is suggested by Bob Hale (Hale 1995), the right side of the biconditional must be true regardless of what ϕ* is. The reason given by Hale is: a subjunctive conditional with a necessarily false antecedent is true; and Rosen offered the following gloss (to explain the fictionalist's prefix 'according to *PW*, ϕ*'): *if PW were true, then* ϕ* *would be true*. Needless to say, this objection, if unanswered, would reduce the fictionalist's position to absurdity.

Rosen's response to this difficulty is to reject Hale's interpretation of the fictionalist's principal modal primitive, namely, the consequence relation that Rosen expressed with the words 'according to *PW* . . .' Rosen denies that, if *PW* is necessarily false, then for any sentence ϕ, we have: according to *PW*, ϕ. The fictionalist's primitive consequence relation, it is claimed, just does not behave that way (Rosen 1995: 69–70). Now I certainly do not wish to deny the fictionalist his right to explicate his central primitive in any way he wishes; and as Ben Hansen has emphasized to me, fictional operators of the sort Rosen is using as his central primitive do seem to work as Rosen is claiming. So I shall accept Rosen's response.

Rosen's Alternative Way of Responding to Hale

Another way of obviating Hale's objection is explored by Rosen. The idea is to reject the Lewis manner of characterizing worlds in terms of spatiotemporal relations. Instead, one would use a different characterization in the fictionalist's fiction. The new fiction, *PW**, would introduce another primitive, 'worldmate', in terms of which worlds would be defined. A world, in this revised version, would be a maximal system of worldmates (Rosen 1995: 71). In that case, the relation of worldmate would be a purely fictional relation, in the way that 'kryptonite' is a purely fictional substance. This would allow the fictionalist to argue that such troublesome sentences as θ are not necessarily false, as he had first maintained; rather, they would simply lack a truth value in the way that 'Kryptonite is deadly for Superman' is. In this way, the fictionalist could avoid having to insist that the possible worlds fiction is necessarily false. And there would be no need to reject the standard account of counterfactuals.

Let us use the term 'island universe' for the old sense of 'world'. In other words, *an island universe is a maximal system of spatiotemporally related objects*, whereas a world is a maximal system of worldmates. Rosen has argued that one advantage of the new translation scheme would be that the fictionalist would no longer be saddled with the doctrine that it is not broadly logically possible that more than one island universe exists, the suggestion being that we can conclude from *PW** that there are some worlds containing several island universes. Rosen suggests that, "given the fictionalist's new translation scheme, he can happily agree that there might have been many [such universes]" (Rosen 1995: 72). Thus, Rosen concludes, the question of how many island universes there are is, with this new scheme, a contingent matter.

Again, let θ be the sentence that there are at least two island universes. Now with the new translation scheme, we can no longer conclude $\Box-\theta$, as was the case with the old translation scheme. But is it the case that we can conclude $\Diamond\theta$, as Rosen suggests? I don't see how. An examination of *PW** will show that the only principle we have for inferring the existence of other worlds with specific inhabitants is the principle of recombination. But if, as seems quite possible, the actual world contains no island universes, then how can the principle of recombination imply the existence of other worlds containing many island universes?[34] It would seem, then, that neither $(\Diamond\theta)*$ nor $(-\Diamond\theta)*$ is a consequence of *PW**. In that case, neither $\Diamond\theta$ nor $-\Diamond\theta$ has a truth value. But such a result would be at odds with the strong intuition we have that $\Diamond\theta$ is true. In short, we evidently still have the kind of conflict with our modal intuitions that Rosen sought to avoid.

I shall not say anything more about this alternative version of modal fictionalism for two reasons: (1) Rosen is not very enthusiastic about it, suggesting at the end of his paper that the defects of this proposal would have to be explored; and (2) most of my objections (to follow) will apply to both versions.

Rosen's Arcane Example

Let us take up the question of whether the fictionalist and the modal realist agree on which modal statements are true. Rosen allows that, "in some

[34] If the principle did have such implications, then would not Lewis have seen this and drawn the conclusion that, even within his chosen framework, such island universes are possible?

rather marginal cases", the fictionalist disagrees with the realist. Let κ be some cardinal number greater than the number of space-time points in our universe; and let τ be the statement:

> There might have been κ non-overlapping physical objects.

Then τ^* will be the statement:

> There is a universe containing κ non-overlapping physical objects.

Lewis's modal realist maintains that τ^* has a definite truth value. Rosen determines, on the other hand, that *PW* is silent on this point, and so concludes that τ^* has no truth value (Rosen 1990: 342). We thus have a disagreement: the realist holds that τ has a definite truth value, whereas the fictionalist maintains that τ has no truth value. Still, the disagreement is about a rather arcane matter, and it can argued that they agree about all modal statements "that strike us as obviously true or obviously false" (Rosen 1990: 341). I suppose that Rosen could also argue that the disagreement over $\Diamond\theta$ is an arcane matter.

Cartesian Egos

But does the fictionalist's truth conditions for modal sentences yield the same modal truths as does the realist's, at least for the case of all modal truths that are not "marginal"? It will be agreed by most philosophers that there are in fact no spirits or Cartesian egos in existence; so it can be agreed that such things will not appear in the encyclopaedia of *PW*. But most philosophers believe that such things could exist—that is, that the existence of such things is *logically possible* (in the broad sense of the term).[35] And Lewis explicitly allows for such a possibility (Lewis 1986: 73). But Rosen's fictionalist is not in a position to allow for the possibility of Cartesian egos, since there is no way of deducing, from the postulates of *PW* (specifically, the principle of recombination) and the encyclopaedia, that there is world containing Cartesian egos.

Let us investigate how disagreements of this sort can arise. The realist maintains that, for every modal statement ϕ, ϕ iff ϕ^*; whereas the fictionalist holds that ϕ iff, according to *PW*, ϕ^*. Comparing the two views, it can be seen that, regarding world statements, the realist is concerned with what is the case; whereas the fictionalist is only concerned with what *PW* says to be the case. When *PW* is silent about the truth or falsity of some ϕ^*, the fictionalist is forced to conclude that the modal statement ϕ is not true. The realist, however, can maintain that ϕ has a truth value—we just don't know what its truth value is. (One is reminded here of the disagree-

[35] Bigelow and Pargetter argue (Bigelow and Pargetter 1987: 106) that Lewis should not propound a theory of modality that implies that there could not be Cartesian egos.

ment between Intuitionists and Platonists in the philosophy of mathematics.) The disagreement on this matter is logically substantial. (The differences between Intuitionistic logic and classical logic are not trivial.) It suggests that the modal fictionalist should have a significantly different modal logic. Thus, I find it surprising that, having noted that the fictionalist's connectives are not classical, Rosen does not go on to explore, or at least discuss, the question of just what sort of modal logic would be appropriate for the fictionalist's analyses.

Does the Fictionalist's Analysis Agree with our Modal Intuitions?

The particular story the fictionalist has picked to give the truth conditions of modal statements is not supposed to be arbitrary. Yet it is hard to see what about that particular story is special. Why pick that story? It might be argued that what makes that story special just is the fact that the set of modal truths we get from this choice is intuitively the right set. But that position is hard to reconcile with the fictionalist's claim that an analysis of modal facts and the very meaning of the modal operators has been given. Furthermore, there are grounds for being sceptical of the claim that the set of modal truths generated by this analysis is the intuitively right one. As we have seen, the set of modal truths generated by the fictionalist's analysis does not match the set determined by the modal intuitions of many philosophers. But there are more mundane examples that raise further doubts about the fictionalist's claims.

Consider the modal sentence 'I could have had eggs for breakfast this morning'. If this is our ϕ, will it be true, according to the fictionalist's account? Notice that, although ϕ does not involve any recondite notions, ϕ^* will involve the notion of *counterpart*. Notice also that *PW* makes no mention of counterparts. Recall, furthermore, that Lewis never gives sufficient conditions for something to be a counterpart. But this raises a special problem for the fictionalist. Since the fictionalist regards worlds as fictions, he must hold that there is no genuine counterpart relation to obtain or to not obtain among the individuals of different worlds; the counterpart relation also turns out to be a mere fiction. Given this situation, it would seem that the Lewisian translation of 'I could have had eggs for breakfast this morning' is a purely fictional statement having no truth value. Thus, it is hard to see how ϕ^* could be a consequence of *PW*. How could we possibly deduce ϕ from *PW*? It would seem then that it is not the case that, according to *PW*, ϕ^*. Similarly, it is not the case that, according to *PW*, $(-\phi)^*$. Evidently, ϕ is neither true nor false—which would be clearly absurd. By extrapolation, it can be seen that there are countless modal sentences of this sort that we believe to be true, but that, for the above reason, will lack a truth value according to the fictionalist's analysis.

Since Rosen is constructing the fiction *PW*, why can't he simply lay down some sufficient conditions for something to be a counterpart of me so that we could infer from *PW* that there are worlds containing counterparts of me? I suspect that it is not so easy to come up with plausible and reasonably statable sufficient conditions; otherwise, Lewis would have given us some clear examples of such conditions. But even if this could be done, there is the additional problem of trying to determine if the time at which a counterpart of me has eggs for breakfast is the temporal counterpart of "this morning". Evidently, we will need sufficient conditions for temporal counterparts. And even if these can be supplied, what about other common objects in the universe? Won't there be countless modal truths involving other objects to be considered? To capture all the intuitive possibilities, it would seem that Rosen would have to give, for every concrete thing x that has, does, or will exist, sufficient conditions for something to be a counterpart of x—an impossible task, as I see it.

The Implausibility of the Fictionalist's Analysis

At the beginning of Rosen's article on fictionalism, it is suggested that his fictionalism would be able to use the fruits of the modal realist's possible worlds analysis of modality, without incurring the onerous ontological commitments of that theory. But it now begins to look as if the fictionalist's analysis of modality does not conserve all the nice features of the realist's analysis after all. It yields, for example, a very different kind of modal logic—one which has not even been explored, so far as I can tell—and it does not yield the same set of modal truths as does the realist's analysis. Furthermore, we have been given nothing like an anti-realist *account of possible worlds semantics*. And as John Divers has argued (Divers 1995), there are grounds for thinking that the fictionalist cannot deliver a "genuine possible worlds semantics for ordinary modal discourse". The fictionalist's theory of possibility no longer seems to have the wonderful advantages of "theft over honest toil" that was advertised for it. At this point, the reader may begin to wonder whether there are any good reasons for accepting the fictionalist's account.

Is the account at all intuitively attractive? Initially, the analysis will strike most philosophers, I believe, as being extremely implausible. Why should a fictional story—especially one that is obviously false and made up at some particular time—*determine* what is and what is not possible (and make true sentences of the forms $\Diamond\phi$ and $\Box\phi$)? Who or what gave this particular story its privileged position? To a person who is sure that spirits are possible beings, the fact that an appropriate world sentence is not a consequence of *PW* will seem hardly a conclusive reason for conceding that spirits are not possible after all.

Suppose that, upon being told by the fictionalist that Cartesian egos are not possible because the appropriate world sentence is not a consequence of *PW*, the Cartesian responds that, in that case, we should *change the fictional story* so that the appropriate world sentence ends up being a consequence of the revised version. "What is so special about that piece of fiction?" it might reasonably be asked. There is something bizarre about letting some philosopher's fictional story logically determine what is and what is not possible. From the intuitive point of view, the difference between the possible and impossible does not logically depend upon what is said in a piece of fiction.

Of course, the fictionalist can simply dig in his heels and respond as he did in the case of the Kripke example: "What is so special about the story," he can say, "is just that it enters into the very meaning of the modal operators; the fact that Cartesian egos are impossible just is the fact that, according to *PW*, there is no world in which Cartesian egos exist." But such a response simply makes apparent the implausibility of the fictionalist's analysis. I believe practically all English speakers will find the fictionalist's analysis of the modal operators to be completely counterintuitive, especially when it is noted that English speakers used the terms 'possible' and 'necessary', and made many judgements involving these concepts, long before this bit of fiction was created. How could such speakers have been using modal terms with the "analytic" truth conditions that Rosen's analysis postulates?[36]

The fictionalist is concerned, in Rosen 1990, solely with broadly logical necessity (or necessity in the "broadly logical sense"). What about causal necessity? Or technological necessity? If the fictionalist's analysis of broadly logical necessity were correct, then it would seem that these related notions should also be analysed in terms of possible worlds stories. In Lewis 1979*a*: sect. 5, Lewis indicates how he would give a possible worlds account of such "relative modalities". According to Lewis's analysis, a sentence ϕ is causally necessary (relative to our world) if it would be true in all worlds that are compatible with the laws of nature of our world. In other words, the Lewisian translation of '$\Box$(no physical object travels faster than the speed of light)' is given by: for every world *w* compatible with the laws of nature of our world, no object in *w* travels faster than the

[36] If it is admitted that, at those times, the modal operators could not be legitimately analysed in terms of the story *PW*, then I would want to know how can they *now* be so analysed? I assume that the fictionalist does not want to maintain that modal operators took on new meanings when the story was created.

speed of light. Thus, to give the truth conditions of causal necessity, the fictionalist would probably specify, in addition to a new Lewisian translation scheme, a new story, *PW**, and a new encyclopaedia—where the new encyclopaedia would list every law of nature of the actual world. He could specify, *mutatis mutandis*, similar truth conditions for technological necessity, epistemic necessity, and other related notions as well. The epicycles required to fill out the fictionalist's analyses, in my judgement, renders this theory of modality even more implausible.

'Possible' and 'necessary' are terms in a family of tightly interrelated terms. Another term in this family is 'could'. Does the fictionalist maintain that this modal word is also to be analysed in terms of *PW*? It would seem so, since 'it is possible that' can be defined in terms of 'could'. For example, what is expressed by 'it is possible that donkeys talk' comes to essentially the same thing as what is expressed by 'the world could have been such that donkey talked'. Others in this family of interrelated terms are 'can', 'might', 'may', and 'must'. Imagine trying to teach a child what these words mean, on the assumption that the fictionalist's analysis is right. How would one go about it? Five-year-olds use modal terms like 'could' and 'can'. Can you imagine yourself trying to explain the postulates of *PW* and *PW** to a five-year-old? What would you say about "connected spatiotemporal systems"? The principle of recombination? Other worlds? And how would you explain the encyclopaedia? As I noted earlier, even Rosen did not try to explain how he was to get all mathematical truths into it.

The Connection between Rosen's Truth Conditions and Necessity

One would think that the truth conditions of necessarily true sentences are linked to what 'necessarily true' means, so that if ϕ is necessarily true, then it should be the case that (1) ϕ is true, and (2) ϕ would have been true, no matter how the world might have been. Thus, if the fictionalist's analysis really gave the truth conditions of all modal sentences, then, at the very least, one should be able to show that when the fictionalist's truth conditions of $\Box\phi$ are satisfied, ϕ is true and ϕ would have been true no matter how the world might have been. At the very least, we should be able to prove, from $\Box\phi$, that ϕ. Of course, the principle $\Box\phi \rightarrow \phi$ is one of the fundamental principles of modal logic (it is principle [N] above), so one would think that this principle would be easily provable, given any adequate analysis of '$\Box$'. It is striking that Rosen offers no proof of [N] and just takes it for granted that it will come out valid. Had he attempted such a proof, I suspect that he would have come to see that the analysis of modality he proposed is untenable. For a study of the fictionalist's proposed truth conditions would convince most modal logicians that it is very doubtful that [N] is provable. In fact, the principle seems to be clearly invalid. The antecedent tells us that, according to *PW*, $(\Box\phi)^*$. Why should that enable us to conclude that ϕ is true? After all, *PW* is acknowledged to be a piece of fiction. Furthermore, much of the story is explicitly declared to be false by Rosen. Indeed, some of the things *PW* declares to be the case, such as "There are many worlds", is determined to be necessarily false by the analysis.[37] So how can it be legitimate to infer the truth of ϕ from what this wildly implausible story says about $(\Box\phi)^*$? For this reason and for the others set forth in this section, I find the fictionalist's analysis of modality decidedly unpromising. In the following chapters, I shall develop a completely different anti-realist approach to the semantics of modal logic.

[37] Well, it might be suggested that one can infer the truth of ϕ from *PW* because *PW* is necessarily false. But in that case, one could also infer the truth of $-\phi$ from *PW*.

5

The Semantics of Classical Predicate Logic

One striking feature of both Forbes's and Rosen's anti-realism was the absence of any analysis of the role that possible worlds *interpretations* or *structures* play in modal logic. To avoid following in their footsteps, I shall begin my alternative anti-realist account by investigating the nature of the interpretations of classical sentential and (first-order) predicate logic. One reason for this investigation is this: the possible worlds model theory of modal logic is very closely related to the model theory of classical first-order logic,[1] so much of the philosophical investigations of this chapter will be highly relevant to what I will be discussing in subsequent chapters. Furthermore, this chapter has an important didactic function to perform. The basic anti-Realist approach I shall adopt in subsequent chapters is much easier to understand when it is applied in the thoroughly familiar and understood context of classical logic. The investigations of this chapter are intended to prepare the reader for the complications to follow, by providing a sort of simplified example worked out in familiar surroundings. Among the things I shall be discussing in this chapter will be John Etchemendy's analyses of standard model theory as presented in his controversial Etchemendy 1990.[2] Etchemendy's analyses presuppose his useful distinction between representational semantics and interpretational semantics—a distinction that is especially helpful in grasping the significance of various sorts of interpretations I shall be discussing in the next chapter.

It may be claimed that Plantinga has provided us with a satisfactory understanding of the role of possible worlds structures in modal logic from,

[1] So much so that some mathematical logicians tend to think of modal logic as just a variation on first-order logic.

[2] According to Etchemendy, there is a widespread belief that model theory provides us with an accurate analysis of the intuitive, preanalytic notions of *logical consequence* and *logical truth*, at least when the sentences to which these notions are to be applied are those of certain formalized languages. To undermine this belief is the principal goal of Etchemendy's book. Vann McGee has continued this attack in McGee 1992 and in a paper delivered at the Pacific Division meeting of the APA, March 1993. I raise a number of objections to the above works in Chihara 1998. For other objections to Etchemendy's criticisms of contemporary model theory, see Sher 1996, Ray 1996, and Bach 1995.

at least, the perspective of his Actualistic Modal Realism. Such a claim may be especially convincing in view of Jager's formalization of an Actualistic system of S5 quantificational modal logic that is thought to capture Plantinga's philosophical ideas. But has Plantinga truly explained how possible worlds structures function? True, he has given us an intended interpretation for Jager's Actualistic versions of S5 modal logic. But does that make it clear why we need the various set theoretical structures for our metatheoretical studies of modal logic? Plantinga's intended interpretation comes to this: for the intended structure **I** = <**W**, **D**, **w***, **d**, **r**, **e**>, **W** is the set of all maximal possible states of affairs, **D** is the set of all essences, **w*** is that maximal possible state of affairs that obtains, **r** and **e** are the appropriate language-specific reference functions. Recall that, in response to the various paradoxes that infected his system, Plantinga came to deny that there is a set of all maximal possible states of affairs. Indeed, it would seem that he would also have to deny that there is a set of all essences. Given this revised position, how are the various *set theoretical structures*, which we study in the model theory of modal logic, related to the intended interpretation? And why can we use *being true in all set theoretical structures* as the criterion for being a logical law of modal logic?

1. Interpretational and Representational Semantics

Interpretational Semantics

In Etchemendy 1990, Etchemendy characterizes the standard model theory of predicate logic—what he calls "the Tarskian view"—as an *interpretational view*, according to which the different models are "meant to canvass semantically well-behaved reinterpretations of the atomic sentences of the languages" (Etchemendy 1990: 58–9). The different "well-behaved reinterpretations" he speaks of here can be regarded as the different assignments of *meanings* or *senses* that could be supplied to the "variable terms" (parameters) of the logical language. The idea is this: in interpretational semantics, "each model provides a possible interpretation of certain expressions [in first-order logic, they will be the non-logical constants and quantifiers] appearing in the language". But what Etchemendy means by 'possible interpretation' here is something like 'possible meaning'. So, one of these possible interpretations might take the sentential letter 'P' to mean 'Snow is white'; and another might take it to mean 'Grass is green'. When such an interpretation or "reinterpretation" is given to the language, the sentences are supposed to be like the declarative sentences of a natural language: they are thought to express statements that are true or false. A sentence of the language can then be said to be "true in a model" if "what it would have said about the world on the suggested interpretation is, in fact, the case". 'True in a model' thus means: 'true, given that the parameters have the meanings assigned to them by the model.' The valid sentences of the language—the sentences true in all models—are held to be those sentences that are true "regardless of how we interpret a subset of their component expressions" (Etchemendy 1990: 60).

Representational Semantics

This interpretational view is contrasted, by Etchemendy, with the *representational view*, according to which the different models are meant to represent "different possible configurations of the world" (Etchemendy 1990: 58). In representational semantics, one is dealing with a language whose sentences are already meaningful: these sentences already express statements that are either true or false (Etchemendy 1990: 59). To provide a representational account of the first-order models of such a language, we first assume that all the expressions of the language have "their ordinary interpretation". Then the assignments made in specifying a model allow us to construct different representations of how the world could have been. For example, if sentential letter 'P' of the logical language has the meaning 'Snow is white', one model may represent the world as being such that 'P' is true (and hence that snow is white), whereas another may represent the world as being such that 'P' is false (and hence that snow is not white). The fundamental difference between these two views can be seen, he suggests, by pondering the quite different counterfactual implications that the model theory is supposed to have according to these different positions:

> Thus, from the *representational* perspective our semantic theory supports the claim that the sentence 'Snow is white or snow is not white' would have been true even if snow had not been white; that contingency is, after all, depicted by various of our models. However, from the *interpretational* perspective no claim is made about what would have happened to the truth value of our sentence had snow not been white. Rather, our theory supports the quite different claim that this sentence would still be true even if the component expression 'Snow is white' were somehow reinterpreted, perhaps given the interpretation presently enjoyed by the English sentence 'George Washington had a beard.' (Etchemendy 1990: 59)

How these two views differ can be further clarified by contrasting the different criteria they use to determine truth in a model. In the case of interpretational semantics, we must ask if the sentence in question would be true if the non-logical constants and the universe of discourse were to be interpreted (that is, given meaning) in the way specified by the model, *given that the world is as it in fact is.* In the case of representational semantics, we should ask if the sentence in question would have been true *had the world been as it is depicted by the model.*

How does the interpretational semanticist regard the different universes of discourse of the various interpretational models of his logical language? To answer this question, we should take note of a general characterization of the difference between representational and interpretational semantics: the former keeps the interpretation of the language fixed and considers, by means of the models, different ways the world could have been; whereas the latter keeps the world fixed and considers, by means of the models, different interpretations of the language. According to the interpretational characterization, the different universes of discourse that come with the various models are merely different subsets of the set of all things that actually exist (exist in the actual world). One of Etchemendy's principal objections to the interpretational view is aimed at this actual-world feature of the view.[3]

As I said above, Etchemendy maintains that classical first-order model theory is an interpretational view. How he arrived at such a view of first-order model theory is not clear. It is possible that he first concluded that the model theory is not representational, and then inferred from this (and various things that Tarski said in Tarski 1956*b*) that the model theory had to be interpretational. One reason for maintaining that *the model theory of first-order logic is not representational* is obvious: according to the above characterization, the sentences of first-order logic, in representational semantics, are supposed to express statements (independently of the different models), so that it would make sense to ask whether this or that sentence of the language would have been true had the world been as it is represented as being by the various models. The sentences of classical first-order logic, however, do not express statements at all. As the non-logical constants and quantifiers of a typical first-order language are treated as parameters, it makes no sense to say of the uninterpreted sentences of such languages that it would have been true had the world been thus and so. These sentences are only true or false *under an interpretation* (or in a structure). In the absence of such an interpretation, the non-logical constants do not have any meaning (in the usual sense of 'meaning'), and hence the sentences of the language do not express statements.

Given that our contemporary model theory of first-order logic is not representational, can we conclude that this model theory must then be interpretational? Clearly not. We have not been given any reasons for thinking that these two semantical views exhaust all the possibilities. It is conceivable that this model theory is neither representational nor interpretational. Certainly, there are grounds for viewing with suspicion the claim that contemporary model theory is an interpretational semantics.

Central to the semantics of first-order logic is a function which, in

[3] See Etchemendy 1990: ch. 8. For criticisms of this objection, see Chihara 1998.

effect, tells us what the individual constants denote, what extensions the predicates have, and what the variables of quantification range over. It is not unusual to find logicians describing such a function as "assigning meaning to the parameters".[4] But this kind of "assignment" does not precisely fit the kind of assignment that Etchemendy's *interpretational* semantics is supposed to make. Given that some interpretation **S** has "assigned" to the sentential letter 'P' the truth value 'true', we cannot determine from that what 'P' means or what statement 'P' expresses: 'P' could have the meaning of 'Snow is white' or 'Clinton is President'. Or suppose that **S** has given the predicate 'R^2' an extension that is some set of ordered pairs. We still do not know what, if any, *meaning* it has—at least of the sort Etchemendy talks about in connection with interpretational semantics. One could argue that what an interpretation does is to abstract from what a genuine assignment of meaning would do, retaining just enough from such an assignment to provide the logician with the extensional elements required to define truth, relative to these elements. Thus, it is something of a stretch to claim that a structural assignment "gives meaning to the predicate".

NL Interpretations

There is, however, a kind of "interpretation" or "model" of logical languages that seems to fit very nicely the kind of model that Etchemendy talks about in connection with interpretational semantics. These are the "interpretations" that *philosophically trained* logicians are apt to consider when "translations" of the logical language into some natural language are seriously contemplated. These "interpretations" do more than mathematical structures do: they not only assign the relevant sort of sets and objects to the parameters of the logical language in question, they also supply meanings or senses to the parameters. For example, in Mates 1972, "interpretations" of this sort assign to each individual constants the sense of some English name or definite description; and they provide each predicate of the language with the sense or meaning of an *English predicate*, where English predicates are obtained from English declarative sentences by replacing occurrences of names or definite descriptions with occurrences of circled numerals.[5] I shall call interpretations of this sort *natural*

[4] See, for example, Enderton 1972: 79.

[5] Actually, Mates requires that the occurrences be "direct occurrences", in order to avoid having the circled numerals occur within the scope of terms of psychological attitude or modal operators, but it is not necessary, for our purposes, to go into these complications. See Mates 1972: 77 for Mates's introduction to English predicates, which he tells us he borrowed from Quine (see, for example, Quine 1959: 131–4). However, it should be noted that Quine used the expression 'predicate' to refer to what I am here calling an 'English predicate'. I follow Mates in calling these linguistic items 'English predicates' in order to distin-

language interpretations or *NL interpretations* for short.[6] When a logical language is given an NL interpretation, the sentences of the language can be regarded as expressing statements that are true or false. For example, if the interpretation I assigns to 'R^2' the sense of the English predicate '① is married to ②', and to 'a' and 'b' the senses of 'The President of the United States' and 'Hillary Rodham Clinton' respectively, the sentence 'R^2ba' expresses the statement 'Hillary Rodham Clinton is married to the President of the United States'. I shall say that a logical language equipped with the meanings or senses provided by an NL interpretation is an *NL interpreted language*.

Suppose that we defined validity and consequence for a logical language not in terms of structures, but rather in terms of NL interpretations. For example, we might say that a sentence of the language is valid iff it comes out true (of the actual world) no matter what NL interpretation is given to the language. And a sentence ϕ of the language is a logical consequence of a set Γ of sentences of the language iff every NL interpretation under which all the members of Γ come out true is an NL interpretation under which ϕ comes out true. Then we would be adopting the interpretational approach that Etchemendy talks about. But, in fact, none of the highly respected logic texts I know of define validity and consequence in terms of such NL interpretations.[7]

Summary Comments on Interpretational and Representational Semantics

Returning, now, to Etchemendy's position that the standard model theory of first-order logic is an interpretational rather than a representational semantical theory, it can be seen that not only has he not justified this counterintuitive position, there are grounds for thinking that the position is untenable.[8] At this point, then, it would seem that standard first-order

guish them from the predicates of the formal language. I also follow Mates in treating English predicates as devices used in giving a kind of natural language interpretation to formal languages. See Mates 1972: 77–86 for a discussion of English predicates which more accurately gives the point of view of this work.

[6] Quine discusses such NL interpretations in Quine 1959: sect. 18.

[7] In Quine 1959, Quine eschews defining logical truth (or valid sentence) as truth under all NL interpretations because, were we to do so, whether a sentence is valid or not would be made to depend on "how rich a vocabulary we assume" (p. 95). However, in Quine 1970, he takes the position that truth under all NL interpretations would coincide with truth under all models, so long as the object language is rich enough to express elementary number theory (p. 53). He justifies his position by citing a number of logical theorems (especially one by Hilbert and Bernays). But Van McGee has written to me that the consequence relation defined in this way in terms of NL interpretations would not coincide, extensionally, with the standard model-theoretically defined consequence relation.

[8] It is possible that he came to his position because he believed that Tarski developed an interpretational account of model theory. But even if he is right about Tarski—a controversial interpretation (see Ray 1996, and also Sher 1991: 137–9)—this is no reason for

model theory is neither interpretational nor representational. In the next section, an examination of standard sentential logic will provide us with further evidence of the falsity of Etchemendy's thesis that first-order logic is interpretational. We can grant Etchemendy, however, that representational semantics is a "perfectly legitimate" approach, and that it bears no relation at all to the interpretational view, despite some superficial similarities (Etchemendy 1990: 10).

2. The Semantics of the Sentential Calculus

To explore these matters more deeply, let us confine our attention for now to the Sentential Calculus. First, let us define a language **SC** to serve as the language of our Sentential Calculus. The non-logical constants of **SC** will be the upper case letters of the English alphabet, with or without numerical subscripts. These nonlogical constants will be called *sentential letters*. The set of logical constants of **SC** will consist of the connectives and parentheses of **M**. Every sentential letter is an atomic sentence of **SC**. To obtain the sentences of **SC**, use just those five syntactic rules of well-formedness of formulas of **M** that concern the five connectives.

The key notion of the semantics of **SC** is the notion of *interpretation* (or structure). Now an interpretation of **SC** will just be a function that assigns to each and every sentential letter a truth value. 'Truth under an interpretation' is first defined for the atomic sentences in the obvious way: A sentence θ is true under **I** iff **I** assigns the truth value 'true' to θ. Then 'truth under an interpretation' is defined for all the sentences of **SC** by using the five clauses for the five connectives that occur in the definition of 'truth at a world under an interpretation' for **M.** Sound and complete rules of derivation can be devised for **SC** that are analogous to the ones provided for **M**, and the usual metalogical definitions can be given for this system.[9]

The question that we need to investigate now is: what kind of semantic theory do we obtain as we develop the usual logical theory of the Sentential Calculus? Is it a representative semantics? Is it interpretational? Or neither? I suggest that we explore these questions by examining the use of the following truth-table to determine the validity of a simple sentence of **SC**:

thinking that *contemporary model theory* is an interpretational view. It should be mentioned that Sher also questions Etchemendy's premiss that contemporary model theory is either representational or interpretational (see Sher 1996). For a detailed rejection of Etchemendy's thesis that first-order model theory is interpretational, see Bach 1995.

[9] See Mates 1972: ch. 6.

P	Q	$P \to (Q \to P)$
T	T	T
T	F	T
F	T	T
F	F	T

According to Etchemendy's characterization of representational semantics, if the above truth-table were being used in the development of a representational theory, the different rows of the table would represent different ways that the world could have been. Evidently, we are to view the sentential letters as having definite meaning, so that had the world been as is indicated by the first row, then 'P' would have been true and 'Q' would have been true.

As an analysis of standard model theory of the Sentential Calculus, this does not seem to be quite right. As I have defined 'interpretation', the sentential letters need not be given any definite meaning by an interpretation, in virtue of which they can be said to be true or false of the actual world. Furthermore, had the sentential letters been given a definite meaning, so that they could be said to be true or false, then the question would arise as to whether all the rows of the standard truth-tables should be allowed. Thus, if 'P' had the meaning of the sentence 'Bill Clinton is married' and if 'Q' had the meaning of the sentence 'Bill Clinton is a bachelor', then there should be no row in the truth-table in which these sentential letters have the same truth value; for those sentences could not both have been true (or both false) no matter how the world might have been. It seems clear that classical model theory is not representational.

But is it interpretational? If so, we are to view the rows of the truth-table as corresponding to various "reinterpretations" of the sentential letters. Thus, the first row might be taken to correspond to interpreting 'P' as having the meaning of '2 + 2 = 4' and 'Q' as having the meaning of 'Bill Clinton is married'; the second row might be taken to correspond to interpreting 'P' as having the meaning of '2 + 1 = 3' and 'Q' as having the meaning of 'Bill Clinton is a bachelor'.

This account does not seem quite right either. Many of the contemporary textbooks in mathematical logic I have examined do not require that the sentential letters be so interpreted or "reinterpreted".[10] Indeed, had the authors of these books truly intended to be presenting an interpretational semantic theory, as Etchemendy suggests, they would have been guilty of having left out a number of crucial elements of the account. One could rightly accuse these authors of incompetence in logic. Furthermore,

[10] See, for example, Enderton 1972.

it would be reasonable to expect that highly respected mathematical logicians, who use these textbooks in their teaching, would have pointed out to their students the errors of omission committed by these authors. One would think that such gross errors would be well known to graduate students of mathematical logic. But they aren't. This fact strongly suggests that there is a way of developing the model theory of first-order logic which does not involve adopting the interpretational point of view.

Evidently the model theory of the Sentential Calculus is neither representational nor interpretational. This gives rise to the problem of characterizing the goals of classical model theory. If we treat the interpretations of **SC** as purely mathematical entities, *how can the classical model theory of **SC** shed any light on truth and logical inference*? What does it have to do with *logic* in the traditional sense of that term? These questions are related to the following problem.

3. Davidson's Problem

Etchemendy describes a puzzle which he attributes to Donald Davidson. The notion of *truth under an interpretation* is a relativized notion: it is truth *relativized to an interpretation.* On the other hand, the notion of truth that we have all been using since childhood is not so relativized: it is an absolute notion. So the question arises: how is the relativized notion related to the absolute notion? In particular, how can model theory, which deals with the relativized notion, shed any light on our ordinary absolute notion? Etchemendy puts it this way: before the relativized notion can be judged to have implications about our ordinary notion of truth, we need to see how the former is related to the latter. This is Davidson's Problem.[11]

Consider Davidson's Problem in connection with the language **SC**. Interpretations of **SC** are just mathematical functions. Sentences of **SC** are true or false under interpretations. So what, if anything, does the truth or falsity of such a sentence *under an interpretation* tells us about genuine truth?

4. A Connecting Theorem

There is a kind of interpretation of formal logical languages that philosophical logicians are apt to make use of in their theorizing which should be distinguished from the kind of extensional mathematical interpretation

[11] Etchemendy 1990: 13.

I have been discussing in the previous section. Consider a kind of interpretation of **SC** which specifies what truth value each sentential letter is to have, not by assigning a truth value to the sentential letter directly, but rather by assigning to it the meaning of some particular declarative sentence, the idea being that the sentential letter will have the truth value that the declarative sentence has. So we can distinguish two sorts of interpretations of **SC**. On the one hand, there is the kind of interpretation which is simply a mathematical function that assigns truth values to the sentential letters directly. On the other hand, there is the kind of interpretation which assigns truth values to the sentential letters only indirectly, by allowing us to understand the sentential letters as declarative sentences. Notice that when **SC** is given the second sort of interpretation, the sentences of the language can be said to be true or false (in the absolute or non-relative sense). Thus, we also have the relative notion of truth—truth under an interpretation—which applies to the sentences of **SC** when the language is given the first sort of interpretation; and the absolute notion of truth, which can be applied to the sentences of **SC** when the language is given the second sort of interpretation. The question arises: can we relate these two notions of truth so as to obtain some kind of answer to Davidson's Problem?

I first give some definitions. A *natural language proto-interpretation* of **SC** (or NL proto-interpretation of **SC**) will assign to each sentential letter the meaning of some declarative sentence of English. Furthermore, each connective will be given its standard truth functional meaning. Let ϕ be a sentential letter of **SC**, and suppose that **SC** has been given an NL proto-interpretation $\mathcal{I}$. Then ϕ will be assigned the meaning of some declarative sentence by $\mathcal{I}$. I will use the notation '$[\phi/\mathcal{I}]$' to refer to ϕ with the meaning it has been assigned by $\mathcal{I}$. For example, if 'P' has been assigned the meaning of 'Bill Clinton is a bachelor' by $\mathcal{I}$, then $[P/\mathcal{I}]$ will be the sentential letter 'P', understood to have the meaning of 'Bill Clinton is a bachelor'. Thus, it would be correct to say that $[P/\mathcal{I}]$ is false, since Bill Clinton is married to Hillary.

I shall say that an NL proto-interpretation $\mathcal{I}$ *conforms to interpretation* **I** iff the following condition is satisfied:

For every sentential letter ϕ, ϕ is true under **I** iff $[\phi/\mathcal{I}]$ is true.

For example, let **I** be the interpretation of **SC** that assigns true to every sentential letter, and let $\mathcal{I}$ be the NL proto-interpretation of **SC** that gives every sentential letter the meaning of '2 + 2 = 4'. Then, $\mathcal{I}$ is an NL proto-interpretation conforming to **I**. If there is an interpretation of **SC** to which an NL proto-interpretation $\mathcal{I}$ conforms, then $\mathcal{I}$ will be said to be an *NL interpretation of* **SC**. Thus, every NL interpretation of **SC** is an NL proto-interpretation of **SC**, but not every NL proto-interpretation of **SC** is an

NL interpretation of **SC**. For an NL proto-interpretation to be an NL interpretation, there must be some interpretation of **SC** to which it conforms.

Intuitively speaking, an NL proto-interpretation gives meaning to all the sentences of the logical language in question, but it may not determine (generate or yield) an appropriate logical structure for the language. When an NL proto-interpretation does determine an appropriate logical structure, it can be classified as an NL interpretation of the language.

Thus, suppose that $\mathcal{J}$ is an NL proto-interpretation of **SC** which assigns to 'P' the meaning of the sentence 'Snow White weighed exactly seven pounds at birth'. It can be seen that $\mathcal{J}$ is not an NL interpretation of **SC**. For there is no interpretation of **SC** to which $\mathcal{J}$ conforms. The reason is: the sentence 'Snow White weighed exactly seven pounds at birth' has no truth value, so [P/$\mathcal{J}$] is neither true nor false. Generally speaking, an NL proto-interpretation assigns meaning to all the sentences of the language, but that may not be sufficient to determine truth values for each and every sentential letter. So there can be NL proto-interpretations that are not NL interpretations.

I shall now extend my notation '[ϕ/$\mathcal{I}$]' so that ϕ can be any sentence of **SC**. Now a sentence of **SC** may contain occurrences of various logical connectives, in addition to sentential letters; so we need to specify that the logical connectives are to be understood in the classical truth functional way specified by the classical truth tables. Then we can say that '[ϕ/$\mathcal{I}$]' is to refer the sentence ϕ, when each sentential letter occurring in ϕ is to be understood to have the meaning assigned to it by $\mathcal{I}$, and the logical connectives occurring in ϕ are understood in the classical truth functional way. The following theorem can now be proved:

> Theorem: For every interpretation **I** of **SC**, every NL interpretation $\mathcal{I}$ conforming to **I**, and every sentence ϕ of **SC**, ϕ is true under **I** iff [ϕ/$\mathcal{I}$] is true.

The proof of this principle is so straightforward and trivial that it is somewhat misleading to label the above a 'theorem'. I do so here only in anticipation of what will be proven in the areas of modal sentential logic and modal quantificational logic: the analogous principles in these areas, as will become apparent in the following chapters, cannot be proven so easily.

The above "theorem" brings out a strong connection between the relative notion of true under an interpretation and the absolute notion of truth. To that extent, we get at least a partial answer to Davidson's Problem. We also obtain the beginning of an answer to our previous question regarding the semantics of the sentential logic: if interpretations of **SC** are merely mathematical functions, how can the classical model theory of **SC** shed any light on truth and logical inference? The following corollaries show how we can obtain answers to this question.

> Corollary [a]: For every NL interpretation $\mathcal{I}$ conforming to an interpretation **I**, for every set of sentences of **SC** Γ; and for every sentence ϕ of **SC**, if $\Gamma \models \phi$, then, if every member θ of Γ is such that $[\theta/\mathcal{I}]$ is true, then $[\phi/\mathcal{I}]$ is true.

> Corollary [b]: For every NL interpretation $\mathcal{I}$ conforming to an interpretation **I**, for every set of sentences of **SC** Γ; and for every sentence ϕ of **SC**, if $\Gamma \vdash \phi$, then, if every member θ of Γ is such that $[\theta/\mathcal{I}]$ is true, then $[\phi/\mathcal{I}]$ is true.

In short, given that the language has been given an NL interpretation, then from $\Gamma \models \phi$ or $\Gamma \vdash \phi$ one can infer that if the members of Γ are all true, so also is ϕ. Thus, one can see how the inference rules of the Sentential Logic allow one to infer the truth of a sentence from the truth of a set of sentences—and this by virtue of the "logical form" of the sentences in question.[12] And one can see why the theorems of the Sentential Calculus (sentences of **SC** that can be derived from the empty set) can be regarded as *principles of logic*: every such theorem θ is *valid* (in the technical sense of being true under every interpretation), and hence, for every NL interpretation $\mathcal{I}$ conforming to an interpretation **I**, $[\theta/\mathcal{I}]$ must be true. Also, since the truth of $[\theta/\mathcal{I}]$ depends (in some rough sense) only upon the meaning of the logical constants (connectives and parentheses), one can see a sense in which the sentence is true in virtue of the "logical form" of θ. Furthermore, since the truth of $[\theta/\mathcal{I}]$ can be inferred without appeal to experiment or observation, we have some rough grounds for classifying such sentences as "a priori". Thus, it can be argued that such valid sentences have many of the features one might expect to find in principles of logic.

Notice that to view the use of logic in the above way, there is no need to regard the interpretations, structures, or models of **SC** as anything more than straightforward mathematical functions. In other words, we do not have to suppose that there is some favoured model of **SC** that is the "intended interpretation" or the "canonical model". This suggests that we should look for an analogous way of viewing modal logic and the semantics of modal logic, in order to obtain an anti-Realist understanding of how modal logic is applied.

[12] The idea of "logical form" goes all the way back to Aristotle. See Mates 1972: ch. 1, sect. 5.

5. The Semantics of First-Order Logic

Before turning to an investigation of the semantics of modal logic, however, it should be noted that everything done above for the case of Sentential Logic can also be carried out for the case of Classical Quantificational Logic. Briefly, the language **M** can be cut down to a first-order language, by eliminating the modal operators and all the definitions pertaining to the operators. More specifically, the syntax of such a language can be obtained from **M** by simply eliminating the two modal operators from the list of logical constants and then eliminating all the syntactical rules and definitions (or parts of rules and definitions) that concern the modal operators. To obtain the semantics of this language, which I shall call '**FL**', simply require the first component of every interpretation **I**, namely **W**, to be a unit set. Since there is only one possible world, namely **w***, truth will always be evaluated at that one world, the domain of the quantifiers will always be **d**(**w***) = **D**, where **D** is non-empty; and the referent of every individual constant will be in **d**(**w***). Thus, an interpretation can be reduced to a simple ordered triple <**d**, **r**, **e**>, where **e** will now be a function taking an *n*-ary predicate of **FL** to a subset of the nth Cartesian product of **d**. This will give us, in effect, the semantics of first-order logic, where truth is always truth at one world. The free logical rules of derivation can then be dropped in favour of the standard rules of first-order logic.[13] Again, the interpretations of **FL**—the first-order structures—can be taken to be purely mathematical entities, which can be distinguished from the NL interpretations of **FL**.

NL Proto-Interpretations of ***FL***

These are specified by interpreting the connectives to have their standard truth functional sense. In addition, the meanings of the quantifiers, individual constants, and predicates are given by:

(1) selecting an English predicate of degree one—henceforth $\mathcal{P}$(**d**)—to delimit the domain of the quantifiers to just those things that satisfy $\mathcal{P}$(**d**), so that the existential quantifier '(∃α)' can be understood to express the idea that there is at least one *F*, say α, which is such that . . . (where the preceding occurrence of '*F*' is to have the sense of $\mathcal{P}$(**d**));

(2) assigning to each individual constant the sense of a name or rigidly denoting definite description of an object that satisfies the English predicate that gives sense to the quantifiers;

and

[13] See Mates 1972: ch. 7.

(3) designating the meaning of each predicate θ to be that of an assigned English predicate of degree n (where n is the degree of θ).

In the following, '$[\alpha/\mathcal{I}]$' will denote the individual constant α with the sense given to it by $\mathcal{I}$; similarly, '$[\theta/\mathcal{I}]$' will denote the predicate θ with the meaning given to it by $\mathcal{I}$. Thus, $[\alpha/\mathcal{I}]$ is just α, as it is interpreted by $\mathcal{I}$; and $[\theta/\mathcal{I}]$ is just θ, as it is interpreted by $\mathcal{I}$.

*NL Proto-Interpretations of **FL** that Conform to First-Order Structures*

Let $\mathbf{I} = \langle \mathbf{d}, \mathbf{r}, \mathbf{e} \rangle$ be a first-order interpretation and let $\mathcal{I}$ be an NL proto-interpretation of **FL**. Then $\mathcal{I}$ will be an NL proto-interpretation conforming to **I** iff:

[1] There is a bijection f from **d** to the extension of $\mathcal{I}(\mathbf{d})$ such that

(1) for every predicate θ of any degree n, and for every n-tuple of members of **d**, $\langle d_1, d_2, \ldots, d_n \rangle$:

$\langle d_1, d_2, \ldots, d_n \rangle \in \mathbf{e}(\theta)$ iff
$\langle \mathrm{f}(d_1), \mathrm{f}(d_2), \ldots, \mathrm{f}(d_n) \rangle \in$ the extension of $[\theta/\mathcal{I}]$;

and

(2) for every individual constant α, $\mathrm{f}(\mathbf{r}(\alpha))$ = the denotation of $[\alpha/\mathcal{I}]$.

Intuitively speaking, when an NL proto-interpretation determines (generates or yields) a first-order structure, it will be said to conform to the first-order structure and then classified as an NL interpretation.

A theorem analogous to the earlier theorem can be proved. Analogues of the above two corollaries can also be proved; and the above explanatory account of the application of logic can be given for the case of first-order logic. These details can obviously be developed in a straightforward fashion and they are left as an exercise for the industrious reader.

The above development of the semantics of first-order logic allows us to give an account of the interpretations, structures, or models of first-order logic that contrasts sharply with Etchemendy's account. Each of these structures can be regarded as providing a bare abstract mathematical representation[14] of a possible universe of discourse (or domain) and the referents of each of the non-logical constants of the language, showing, in particular, how the extensions of the predicates are related (structurally) to each other and to the domain. Such mathematical representations do not fit Etchemendy's characterization of the models of either interpretational or representational semantics, since there is no

[14] The idea of a mathematical representation will be explained in detail in the following chapter.

requirement here that the different structures must depict either different possible configurations of the world that an already interpreted language is "talking about" (as in the case of representational semantics) or different reinterpretations of the non-logical constants of the language, while the actual world being "talked about" is kept fixed (as in the case of interpretational semantics). The first-order structures provide abstract representations of different possible domains and extensions of predicates, without settling the question of whether it is the meanings of the parameters that are being held fixed, while the different possible situations are up for consideration, or it is the world that is being held fixed, while the different assignments of meanings to the parameters are up for consideration. We thus find confirmation of the conclusion we arrived at earlier: the semantics of first-order logic is neither representational nor interpretational.

What the above theorem allows us to conclude is that, in the case of a valid sentence ϕ, no matter how the world might be, and no matter what meaning we give to the non-logical constants, that is, no matter what NL interpretation $\mathcal{I}$ we may choose, so long as $\mathcal{I}$ conforms to some first-order structure, $[\phi/\mathcal{I}]$ must be true. Furthermore, no matter how the world might have been, and no matter what meaning we might have given to the non-logical constants, so long as $\mathcal{I}$ conformed to some first-order structure, $[\phi/\mathcal{I}]$ would have been true.

There is one other reason for suspecting that the semantics of mathematical logic is not interpretational. As I noted earlier, the model theory of quantificational model logic is very similar to the model theory of first-order logic. Now Etchemendy has emphasized that interpretational semantics "equates the logical truth of a sentence with the ordinary truth of another sentence, one that makes a nonmodal, nonepistemic, nonsemantical claim about the world, *about the world as it actually happens to be*" (Etchemendy 1990: 99). But if that is the case, then how can the model theoretic semantics of modal logic be interpretational? And if the model theoretic semantics of modal logic is not interpretational, does this not suggest that the model theoretic semantics of first-order logic is not interpretational either?

6

Modality without Worlds I

The Semantics of Modal Sentential Logic

The reader may wish to review, at this point, the argument for Modal Realism that was given in the Introduction. There, it was argued that philosophers who eschew including possible worlds in their ontology are thus precluded from making use of possible worlds semantics in their assessments of the validity and invalidity of modal arguments, and in their explanations and analyses of modal reasoning. As we have seen in previous chapters, for many Modal Realists, this is too high a price to pay for ontological economy.

In response to this argument, I shall put forward an account of modal logic that can be used both in the assessments of the validity and invalidity of modal arguments and also in the explanations and analyses of modal reasoning. This account, however, will not explain the modal operators in terms of possible worlds at all, but rather in terms of *how the world could have been.* I shall give this account in stages, considering for simplicity just the Modal Sentential Calculus.

1. An S5 Modal Sentential Calculus

The language of this modal logic will be called '**MSC**'. **MSC** can be obtained from **M** by simply eliminating all the quantifiers and replacing all the atomic and general sentences of **M** with sentential letters. To be more precise, I shall give the syntax and semantics of **MSC** below.

The Syntax of **MSC**

Primitive Symbols

The only non-logical constants of **MSC** are the upper-case letters 'A' through 'Z', with or without numerical subscripts. The logical constants are the five connectives and the two modal operators of **M**.

Formulas

Each sentential letter is an atomic formula of **MSC**, and nothing else is an atomic formula of the language. Dropping the clauses having to do with the quantifiers, the definition of 'formula of **MSC**' is the same as that of 'formula of M'.

The Semantics of **MSC**

S5 Interpretations

An S5 interpretation is an ordered triple <**W**, **w***, **v**>, where

> **W** is a non-empty set;
> **w*** is an element of **W**;
> **v** is a function whose domain is the Cartesian product S × **W** (S being the set of sentential letters) and whose co-domain is the set {0, 1}.

Truth at a World under an Interpretation

If ϕ is a sentential letter and $w \in$ **W**, then ϕ is true at w under an S5 interpretation **I** = <**W**, **w***, **v**> iff **v**(ϕ, w) = 0.

If ϕ is not a sentential letter, then use clauses (b)–(h) of the definition of 'true at w under **I**' in Chapter 1.

Comments

(1) According to the "canonical conception" discussed in earlier chapters, **v**(θ, w) would be considered to be the valuation function that tells one what truth value the sentential letter θ has in world w (taking '0' to denote *true* and '1' to denote *false*). However, according to the "official" anti-Realist position to be taken in this chapter, *S5 interpretations* (*S5 structures*) *are to be taken to be purely mathematical objects*—indeed objects of pure set theory. Thus, in defining what an S5 interpretation is, we can specify that: **W** is to be restricted to non-empty sets of pure sets (sets without urelements); sentences and sentential letters are to be taken to be the Gödel numbers of these expressions of the language; and functions, relations, and numbers are to be the appropriate sort of pure sets.

(2) It is helpful, at this point, to reconsider the two conceptions of model theoretic semantics that Etchemendy discussed in his book: the representational and the interpretational. Let us regard these two conceptions as being competing paradigms, which other conceptions more or less approximate. Later, I shall bring out various features of S5 interpretations that will allow us to regard them as rough approximations to the models of Etchemendy's representational semantics. To facilitate bringing this about, I shall first introduce another sort of semantical system.

2. C-Sentences

The aim of this section is to remove some of the mystery that surrounds the use of S5 structures. For this reason, I shall introduce certain sentences, sequences of which will function in place of S5 structures. These sentences—the C-sentences—should be regarded as mere heuristic (and didactic) devices brought in to aid our understanding of the way S5 structures function For this reason, there is no significant theoretical loss in simplifying certain features of the logical language or of the semantics of the language in discussing these devices.

In order to simplify the reasoning, let us suppose that we cut down the stock of sentential letters of our modal language to some finite number, say

$$P_1, P_2, \ldots, P_n$$

and suppose that $<\mathbf{W}, \mathbf{w^*}, \mathbf{v}>$ is an S5 interpretation of this modal language. Let us suppose further, for the sake of simplicity, that **W** is a countably infinite set whose members are $w_1, w_2, \ldots, w_j, \ldots$. Let $\mathbf{C}(w_i)$ be a sentence constructible from the following template:

The world could have been such that
'P_1' was $\mathbf{t}(P_1, w_i)$
'P_2' was $\mathbf{t}(P_2, w_i)$
.
.
.
.
'P_n' was $\mathbf{t}(P_n, w_i)$

where each expression '$\mathbf{t}(P_k, w_i)$' in the above is to be replaced by either the word 'true' or the word 'false', depending on the value $\mathbf{v}(P_k, w_i)$. The replacement is to be 'true' if $\mathbf{v}(P_k, w_i) = 0$; otherwise the replacement is to be 'false'. For example, suppose that the only sentential letters in our language are 'P_1' and 'P_2'. One $\mathbf{C}(w)$ sentence then will be:

The world could have been such that
'P_1' was true
'P_2' was false.

This displayed sentence literally tells us how the world could have been in so far as the truth values of the sentential letters of our language are of concern. The basic idea, then, is to form a kind of semantics for the sentential modal language that uses, in place of possible worlds, these *statements of how the world could have been.*

According to this interpretation, the sequence of sentences

$\mathbf{C}(w_1), \mathbf{C}(w_2), \ldots, \mathbf{C}(w_n), \ldots$

will be generated by an S5 interpretation <**W**, **w***, **v**>, and truth will be defined in terms of the sequence (and hence only indirectly in terms of <**W**, **w***, **v**>). I first define:

Truth at a world according to a sequence. For any sentence θ of **MSC**, any **C**-sequence Γ generated by the S5 interpretation **I** = <**W**, **w***, **v**>, and for any w ∈ **W**:

[i] If θ is a sentential letter, then θ is true at *w* according to the sequence iff **C**(*w*) contains an occurrence of the phrase that consists of θ with single quotes around it followed by 'was true'.
[ii] If θ is the sentence –ψ, then θ is true at *w* according to the sequence iff ψ is not true at *w* according to the sequence.
[iii] If θ is the sentence (ψ & τ), then θ is true at *w* according to the sequence iff both ψ and τ are true at *w* according to the sequence.
[iv] If θ is the sentence (ψ ∨ τ), then θ is true at *w* according to the sequence iff either ψ is true at *w* according to the sequence or τ is true at *w* according to the sequence.
[v] If θ is the sentence (ψ → τ), then θ is true at *w* according to the sequence iff either ψ is not true at *w* according to the sequence or τ is true at *w* according to the sequence.
[vi] If θ is the sentence (ψ ↔ τ), then θ is true at *w* according to the sequence iff either both ψ and τ are true at *w* according to the sequence or both ψ and τ are not true at *w* according to the sequence.
[vii] If θ is the sentence ◇ψ, then θ is true at *w* according to the sequence iff some member *u* of **W** is such that ψ is true at *u* according to the sequence.
[viii] If θ is the sentence □ψ, then θ is true at *w* according to the sequence iff every member *u* of **W** is such that ψ is true at *u* according to the sequence.
[ix] A sentence θ is true according to the sequence iff θ is true at **w*** according to the sequence.

We now have the notion of *true according to a sequence*. *Validity*, *consequence*, and (semantic) *consistency* can then be defined in essentially the standard way. Then it can be shown that the usual S5 system of derivation rules for the modal sentential calculus are sound and complete. In other words, we can get what is essentially the usual S5 sentential logic without bringing in such abstract entities as S5 structures. Thus, there will be little temptation to suppose that possible worlds are being presupposed at this stage in the development.

3. NL Proto-Interpretations that Conform to C-Sequences

In order to use this modal logic to analyse modal reasoning and evaluate modal principles, we shall want to interpret the sentences of the modal language in such a way that the interpreted sentences turn out to be *true* or *false* in a straightforward untechnical sense of these words. Thus, we shall want to interpret the language in such a way that its sentences express thoughts directly, as do the declarative sentences of a natural language. This will be done via NL proto-interpretations of **MSC** and, more specifically, in terms of NL proto-interpretations of **MSC** that conform to a **C**-sequence. For this reason, the notion of NL proto-interpretation that was developed in the previous chapter will be extended to apply to **MSC**.

> [Def] An *NL proto-interpretation* of **MSC** will assign to each sentential letter the meaning of some declarative sentence of English. Each connective will be given its standard truth functional meaning and, in addition, the modal operators will be understood not in terms of possible worlds, but rather by the conditions:
>
> > [T-d] A sentence '$\Diamond\phi$' is true just in case the world could have been such that ϕ was true.
>
> and
>
> > [T-b] A sentence '$\Box\phi$' is true just in case it is not the case that the world could have been such that ϕ was not true.

Continuing the development that is analogous to what was done in the previous chapter, when **MSC** has been given an NL proto-interpretation, I shall use the expression '$[\phi/\mathcal{I}]$' to refer to the sentence ϕ interpreted according to $\mathcal{I}$. Also, since any specific member of a **C**-sequence, say **C**(u), can be understood as representing how the world could have been (in so far as what are of concern are the truth values of the sentential letters) *when the sentential letters are interpreted as having the meanings assigned to them by* $\mathcal{I}$, I shall use the notation '$\mathbf{C}_{\mathcal{I}}(u)$' to indicate that I am taking **C**(u) in this more specific way.

To illustrate these ideas, let us again suppose, for purposes of simplicity, that there are only two sentential letters in **MSC**, 'P_1' and 'P_2'. Let **I** = <**W**, **w***, **v**> be an S5 interpretation of **MSC**, and let **S** be the two element **C**-sequence generated by **I**:

The world could have been such that
'P_1' was true
'P_2' was false.

> The world could have been such that
> 'P$_1$' was false
> 'P$_2$' was true.

Let $\mathbf{S}(w_1) = \mathbf{S}(\mathbf{w}^*)$ be the former and $\mathbf{S}(w_2)$ the latter. Suppose an NL proto-interpretation $\mathcal{J}$ interprets 'P$_1$' to have the meaning of the English sentence 'At no time, did a human being exist who had three heads' and it interprets 'P$_2$' to have the meaning of 'At some time, a human being existed who had three heads'. Then, $[P_1/\mathcal{J}]$ is true iff the world is such that, at no time, did a human being exist who has three heads. Thus, $\mathbf{S}_{\mathcal{J}}(w_1)$ asserts that the world could have been such that 'At no time, did a human being exist who had three heads' was true, and 'At some time, a human being existed who had three heads' was false. I believe that we can reasonably assume that what $\mathbf{S}_{\mathcal{J}}(w_1)$ asserts could have been the case not only could have been the case, but is in fact the case. Similarly, $\mathbf{S}_{\mathcal{J}}(w_2)$ asserts that the world could have been such that 'At no time, did a human being exist who had three heads' was false, and 'At some time, a human being existed who had three heads' was true. And it would seem that the world could have been as $\mathbf{S}_{\mathcal{J}}(w_2)$ asserts that it could have been.

NL Proto-Interpretations that Conform to C-Sequences

As before, an NL proto-interpretation of **MSC** must satisfy additional conditions to be an NL proto-interpretation that conforms to a **C**-sequence.

> [Def] An NL proto-interpretation $\mathcal{J}$ is an *NL proto-interpretation that conforms to* **C**-*sequence* **C** iff the following conditions are satisfied:
>
> [i] For every w_k, the world could have been as $\mathbf{C}_{\mathcal{J}}(w_k)$ says that it could.
> [ii] It is not the case that the world could have been such that none of the $\mathbf{C}(w_k)$ says that it could have been that way.
> [iii] For every sentential letter ϕ, ϕ is true according to **C** iff $[\phi/\mathcal{J}]$ is true.

Returning to the above **S** example, it can be seen that $[P_1 \vee P_2/\mathcal{J}]$ will be the disjunction '$P_1 \vee P_2$' interpreted to express the statement:

> Either, at no time did a human being exist who had three eyes, or, at some time, a human being existed who had three eyes.

Since '$P_1 \vee P_2$' is true, according to **S**, at both w_1 and w_2, '$\Box(P_1 \vee P_2)$' is true, according to **S**. Also, $[P_1 \vee P_2/\mathcal{J}]$ would have been true, had the world been as described by either $\mathbf{S}(w_1)$ or $\mathbf{S}(w_2)$' and hence $[P_1 \vee P_2/\mathcal{J}]$ would have been true, regardless of how the world might have been; that is, it is necessarily true.

The question naturally arises: will it always be the case that, for any sentence ϕ of **MSC**, any **C**-sequence Γ, and any NL proto-interpretation $\mathcal{J}$ of **MSC** that conforms to Γ, $\Box\phi$ is true according to Γ iff $[\phi/\mathcal{J}]$ is necessarily true? Later, this question will be answered.

An Objection

The above question is related to an objection that may occur to the reader. How can $[P_1 \vee P_2/\mathcal{J}]$ be necessarily true, since the world could have been such that no sentence existed? If it is admitted that the world could have been such that $[P_1 \vee P_2/\mathcal{J}]$ did not exist, then how can it be maintained that, no matter how the world might have been, $[P_1 \vee P_2/\mathcal{J}]$ would still have been true?

Notice that this objection does not arise because of any specific features of my anti-Realist semantics. It could just as well be raised against the various semantical systems developed by Modal Realists. Still, it comes up so often that it deserves an answer. Here is mine.

To say that $[P_1 \vee P_2/\mathcal{J}]$ is necessarily true is not to say that, in every possible world *w*, the sentence $[P_1 \vee P_2/\mathcal{J}]$ exists and is true in *w*; rather, to continue using the realist's language, it is to say that what the sentence expresses (in this world) would hold in every possible world. Etchemendy's representational semantics provides us with a hint as to how that could be. Recall that, in representational semantics, the language is held fixed, and the models represent the different ways the world could have been. A sentence is necessarily true if it is true in all the models. We do the same here: the language is to be held fixed; and we see if the world could have been such that the sentence in question, with the meaning it actually has, said correctly what was the case. Thus, the sentence $[P_1 \vee P_2/\mathcal{J}]$ is necessarily true if it is not the case that the world could have been such that what $[P_1 \vee P_2/\mathcal{J}]$ says to be the case (remember, here, that we are keeping the language fixed) was not the case. Thus, $[P_1 \vee P_2/\mathcal{J}]$ is true iff it is the case that either at no time did a human being exist who had three eyes, or at some time a human being existed who had three eyes. Now regardless of how the world might have been, and regardless of whether the sentence $[P_1 \vee P_2/\mathcal{J}]$ would or would not have existed, it still would have been the case that either at no time did a human being exist who had three eyes, or at some time a human being existed who had three eyes. So the sentence $[P_1 \vee P_2/\mathcal{J}]$ is necessarily true.

4. Connecting Theorems

We are now in a position to answer Davidson's Problem as it covers the relative notion of truth we have been discussing in this chapter. The answer will be given by way of a theorem that is analogous to the "theorem" of the previous chapter.

Let $\mathcal{I}$ be an NL proto-interpretation conforming to some **C**-sequence Γ generated by some S5 interpretation **I** = <**W**, **w***, **v**>, and let ϕ be some sentence of the modal sentential language. Then

> Theorem [1]. A sentence ϕ is true at w_i according to a sequence **C** iff [ϕ/$\mathcal{I}$] would have been true, had the world been as $\mathbf{C}_{\mathcal{I}}(w_i)$ says that it could have been.

Since this theorem plays a crucial role in the logical developments to follow, I shall provide below a detailed proof of it. However, readers who find lengthy proofs tedious or boring can skip the following with the assurance that they will not miss the most important elements of the overall account.

Proof. This will be proved by induction on the *index* of the formulas of the modal language.[1] Thus, we assume that the above holds for all sentences ϕ of index less than *k*. I shall prove that it holds for all sentences of index *k*.

Suppose that θ is a sentence of index *k*. Now if $k = 0$, then θ is a sentential letter. Then, from the definition of 'true at *w* according to **C**',

> θ is true at w_i according to **C** iff $\mathbf{C}(w_i)$ contains the phrase that consists of θ with quotes around it followed by 'would have been true'.

But if $\mathbf{C}(w_i)$ contains the phrase consisting of θ with quotes around it followed by 'would have been true', then [θ/$\mathcal{I}$] would have been true, had the world been as $\mathbf{C}_{\mathcal{I}}(w_i)$ says that it could have been. And if [θ/$\mathcal{I}$] would have been true, had the world been as $\mathbf{C}_{\mathcal{I}}(w_i)$ says that it could have been, then $\mathbf{C}(w_i)$ must contain the phrase consisting of θ with quotes around it followed by 'would have been true'. So we can infer:

> ϕ is true at w_i according to **C** iff [ϕ/$\mathcal{I}$] would have been true, had the world been as $\mathbf{C}_{\mathcal{I}}(w_i)$ says that it could have been.

Suppose that $k > 0$. Then θ must be a sentence of one of the following forms:

(1) $\theta = -\psi$
(2) $\theta = (\psi \,\&\, \chi)$
(3) $\theta = (\varphi \vee \chi)$
(4) $\theta = (\varphi \rightarrow \chi)$
(5) $\theta = (\varphi \leftrightarrow \chi)$
(6) $\theta = \Diamond\psi$
(7) $\theta = \Box\psi$.

[1] The definition of 'index' is given on p. 209 below.

I shall consider each of these possibilities in the above order.

Case (1). Suppose that $\theta = -\psi$. Then ψ is of index k–1. Hence, by the inductive hypothesis, we have:

> ψ is true at w_i according to **C** iff $[\psi/\mathcal{J}]$ would have been true, had the world been as $\mathbf{C}_{\mathcal{J}}(w_i)$ says that it could have been.

Hence,

> $-\psi$ is true at w_i according to **C** iff $[\psi/\mathcal{J}]$ would not have been true, had the world been as $\mathbf{C}_{\mathcal{J}}(w_i)$ says that it could have been.

Hence,

> $-\psi$ is true at w_i according to **C** iff $[-\psi/\mathcal{J}]$ would have been true, had the world been as $\mathbf{C}_{\mathcal{J}}(w_i)$ says that it could have been.

Hence,

> θ is true at w_i according to **C** iff $[\theta/\mathcal{J}]$ would have been true, had the world been as $\mathbf{C}_{\mathcal{J}}(w_i)$ says that it could have been.

Case (2). $\theta = (\psi \,\&\, \chi)$. Both ψ and χ must be of index less than k. Hence, by the inductive hypothesis, we have:

> ψ is true at w_i according to **C** iff $[\psi/\mathcal{J}]$ would have been true, had the world been as $\mathbf{C}_{\mathcal{J}}(w_i)$ says that it could have been,

and

> χ is true at w_i according to **C** iff $[\chi/\mathcal{J}]$ would have been true, had the world been as $\mathbf{C}_{\mathcal{J}}(w_i)$ says that it could have been.

Hence,

> both ψ and χ are true at w_i according to **C** iff both $[\psi/\mathcal{J}]$ and $[\chi/\mathcal{J}]$ would have been true, had the world been as $\mathbf{C}_{\mathcal{J}}(w_i)$ says that it could have been.

Hence,

> $(\psi \,\&\, \chi)$ is true at w_i according to **C** iff $[(\psi \,\&\, \chi)/\mathcal{J}]$ would have been true, had the world been as $\mathbf{C}_{\mathcal{J}}(w_i)$ says that it could have been.

Hence,

> θ is true at w_i according to **C** iff $[\theta/\mathcal{J}]$ would have been true, had the world been as $\mathbf{C}_{\mathcal{J}}(w_i)$ says that it could have been.

Cases (3)–(5) are essentially the same as case (2), so I shall not bother with these cases. I turn now to

Case (6). $\theta = \Diamond\psi$. From the definition of 'true at w according to a sequence', we have:

> [A] θ is true at w_i according to **C** iff there is some $u \in \mathbf{W}$ which is such that ψ is true at u according to **C**.

ψ is of index k–1. Hence, by the inductive hypothesis, we have:

> [B] ψ is true at u according to **C** iff $[\psi/\mathcal{I}]$ would have been true, had the world been as $\mathbf{C}_{\mathcal{I}}(u)$ says that it could have been.

From [A] and [B], we can infer

> [C] θ is true at w_i according to **C** iff there is some $u \in \mathbf{W}$ which is such that $[\psi/\mathcal{I}]$ would have been true, had the world been as $\mathbf{C}_{\mathcal{I}}(u)$ says that it could have been.

The next step consists in showing that the right side of the biconditional [C] holds iff:

> [D] The world could have been such that $[\psi/\mathcal{I}]$ was true.

I shall show this in two steps. First assume that the right side of [C] holds. Then $\mathbf{C}_{\mathcal{I}}(u)$ describes how the world could have been such that $[\psi/\mathcal{I}]$ was true. Furthermore, since $\mathcal{I}$ is an NL proto-interpretation conforming to **C**, the world could have been as $\mathbf{C}_{\mathcal{I}}(u)$ says that it could have been. It follows that the world could have been (as $\mathbf{C}_{\mathcal{I}}(u)$ says that it could have been) such that $[\phi/\mathcal{I}]$ was true. That is, [D] follows.

Now assume that [D] holds. Then since $\mathcal{I}$ is an NL proto-interpretation conforming to **C**, it is not the case that the world could have been such that there is no $\mathbf{C}_{\mathcal{I}}(w)$ that describes how it could have been. It follows that there is some $w \in \mathbf{W}$ which is such that $\mathbf{C}_{\mathcal{I}}(w)$ describes how the world could have been such that $[\psi/\mathcal{I}]$ was true. That is, there is some $w \in \mathbf{W}$ which is such that $[\psi/\mathcal{I}]$ would have been true, had the world been as $\mathbf{C}_{\mathcal{I}}(w)$ says that it could have been. In short, the right side of [C] follows.

Replacing the right side of [C] with [D], we obtain:

> [E] θ is true at w_j according to **C** iff the world could have been such that $[\psi/\mathcal{I}]$ was true.

But, according to truth-condition [T-d] given in Section 3, the right side of the biconditional [E] holds iff $[\Diamond\psi/\mathcal{I}]$ is true. So we can conclude:

> [F] θ is true at w_i according to **C** iff $[\Diamond\psi/\mathcal{I}]$ is true.

Now if something is possible, in the sense of broadly logical (or conceptual) possibility, then it would have been possible no matter how the world might have been. So if $[\Diamond\psi/\mathcal{I}]$ is true, then no matter how the world might have been, [$\Diamond\psi/\mathscr{I}$] would still have been true. Similarly, if [$\Diamond\psi/\mathscr{I}$] would have been true, had the world been as $\mathbf{C}_{\mathscr{I}}(w_i)$ says that it could have been, then [$\Diamond\psi/\mathscr{I}$] is true. So we have:

> [G] [$\Diamond\psi/\mathscr{I}$] is true iff [$\Diamond\psi/\mathscr{I}$] would have been true, had the world been as $\mathbf{C}_{\mathscr{I}}(w_i)$ says that it could have been.

So, from [F] and [G], we can infer:

> [H] θ is true at w_i according to **C** iff [$\theta/\mathscr{I}$] would have been true, had the world been as $\mathbf{C}_{\mathscr{I}}(w_i)$ says that it could have been.

That completes the reasoning for case (6). It can be seen that case (7) will be essentially the same as the above *mutatis mutandis*.

An Objection Considered; the Author's Aims Stated

Notice that some of the steps in the above reasoning involve modal reasoning. For example, in justifying [G], I make use of the modal principle that if something is possible, then it would have been possible no matter how the world might have been—a principle that seems to differ little from the S5 principle '$\Diamond\phi \rightarrow \Box\Diamond\phi$'. This may strike some readers as involving a vicious circle of some sort. Am I not attempting to justify the principles of S5 modal logic? And how can I do that if I assume some of the very principles I am attempting to justify?

I certainly did not claim, nor did I want to suggest, that I could carry out the above proof without the use of modal reasoning. The series of metatheoretic theses advanced above is not aimed at providing an ultimate justification of the principles of S5 modal reasoning. I am attempting to give an anti-Modal Realistic account of possible worlds semantics. I want to explain how possible worlds semantics can be used even though there aren't any possible worlds. In other words, my goal is to account for the use of possible worlds semantics in modal logic, in a way that does not presuppose the existence of possible worlds.

It should now be obvious that I am not a foundationalist, attempting to justify all our modal principles from a certain set of self-evident or a priori truths. Nor am I attempting to provide a semantical foundation for modal logic from within some austere non-modal framework, such as set theory. Instead, I make free use of such modal notions as is expressed by 'the world could have been such that', and I have no compunctions about appealing to subjunctive conditionals. Thus, I do not attempt to provide analyses of modal notions in the way David Lewis does.[2] In this respect,

[2] Lewis suggests in a number of places that he regards his own account of modality as providing an explication of modal primitives in terms of only nonmodal terms (see, for example, Lewis 1986: 165). However, William Lycan and Stewart Shapiro express doubts

my way of proceeding is closer to Plantinga's, who also makes frequent use of subjunctive conditionals in giving his account of modal logic.

To those who believe that there is some sort of circularity involved in the above reasoning, I suggest that they consider the usual proofs of the soundness of the natural deduction rules of the sentential calculus. One proves the soundness of, for example, the rule *modus ponens* by means of metatheoretic reasoning that presupposes a form of *modus ponens*. That is, one assumes *modus ponens* in the metatheory being used to prove the soundness of the object theory version of *modus ponens*. Does that mean that the proof of soundness is viciously circular and accomplishes nothing?

First of all, there is a difference between the object-language version of *modus ponens* and the metatheoretic version. After all, there is a significant difference between the object-language (an artificial, logical language) and the natural metalanguage.[3] Second, it is clear that our confidence in the soundness of a natural deduction system of logic is enhanced by a rigorous proof of soundness, even if one uses in the metatheory some forms of the object theory principles whose soundness is being proved. (Those who believe that proofs of soundness accomplish nothing should take note of the various systems of logic that have turned out to be unsound!) Third, a proof of soundness is aimed primarily at showing something about how the system of (natural deduction) rules functions *as a whole*. This goes beyond the soundness of any particular rule (such as *modus ponens*).

that Lewis's account of modality does succeed in analysing modality in nonmodal terms, writing: "One might think that, in particular, Lewis uses no modal primitives, since his treatment of worlds as flesh-and-blood avoids the sorts of abstraction that actualists resort to, but not so: Lewis's primitive is the notion of "world" itself. "World" for him means "*possible* world," since he admits no impossibilia, and he no more provides an account of how it is that some sets of sentences correspond to "worlds" and some do not than we do of how it is that some are satisfiable and some are not" (Lycan 1986: 358).

Now it might be objected that, strictly speaking, 'world' is not a primitive of Lewis's account, since he can be taken to have defined 'world' in terms of mereological sums and objects that are spatio-temporally related. But the point Lycan and Shapiro are making does not depend upon taking 'world' to be a primitive. They would argue that Lewis is assuming that mereological sums admit no impossibilia. Lycan and Shapiro also express doubts that anyone will ever succeed in giving an adequate explication of all our modal notions in entirely nonmodal terms. More recently, Scott Shalkowski has argued, in a fashion that shows some agreement with Lycan and Shapiro, that Lewis's reduction is defective. The question of whether Lewis succeeds in "reducing" our modal notions to the nonmodal will be discussed in detail in Chapter 8.

[3] All of this brings to mind the dispute between Poincaré and Weyl, on the one side, and the Hilbert school, on the other, regarding the question of whether there is a circularity in Hilbert's justification of mathematical induction. See van Heijenoort's introduction to Weyl's contribution in van Heijenoort 1967: 480–2.

Corollaries

Continuing the development of the theory, from Theorem [1] and Principle [C], we can infer:

> Corollary [a]. If **C** is a **C**-sequence generated by interpretation <**W**, **w***, **v**>, then sentence φ is true at w_k under <**W**, **w***, **v**> iff [φ/$\mathcal{I}$] would have been true, had the world in fact been as $\mathbf{C}_{\mathcal{I}}(w_k)$ says that it could have been.

And,

> Corollary [b]. If the world *is* in fact as $\mathbf{C}_{\mathcal{I}}(\mathbf{w}^*)$ says that it could have been, then [φ/$\mathcal{I}$] is true iff φ is true under <**W**, **w***, **v**>.

Discussion

The above account not only provides us with a preview of how my anti-Realist account of the semantics of the S5 Modal Sentential Calculus will be given, it also facilitates giving the anti-Realist account itself (as will become obvious in what follows). The crucial element to this account will be a theorem that does for the S5 Modal Sentential Calculus what the "connecting" theorem of the previous chapter did for the Sentential Calculus and what Theorem [1] does for the **C**-sequence system.

5. The Anti-Realist Account

The goal of this section is to show is that S5 structures function in the standard semantical system for **MSC** in essentially the way the sequences of **C**-sentences discussed above function in the semantical system of the previous sections. As a preliminary step in achieving this goal, I note:

> Principle [C]. A sentence of **MSC** is true according to a **C**-sequence iff it is true under the S5 interpretation that is the generator of the **C**-sequence.

That Principle [C] holds can be proved in a straightforward manner by mathematical induction on the index of the formulas of **MSC**, where 'index' is defined below as:

> [Def] For any formula φ of **MSC**, the *index* of φ is the number of occurrences of connectives and/or operators in φ.

It is evident, from an examination of the definitions of 'true according to a **C**-sequence' and 'true under an interpretation' that, if the principle holds for all sentences of index 0, then it will hold for all sentences of higher index. I give below the proof that the principle holds for all sentences of index 0.

Proof. Suppose that ϕ is a sentence of **MSC** of index 0. Then ϕ is a sentential letter; and ϕ is true according to Γ iff ϕ is true at **w*** according to Γ. But that is so iff the phrase consisting of ϕ with quotes around it followed by 'was true' occurs in **C(w*)**. And that is so iff **v**(ϕ, **w***) = 0. And that is so iff ϕ is true under **I**.

I now turn to what I have been calling Davidson's Problem. We have, on the one hand, the relative notion of truth under an S5 interpretation, and we have, on the other hand, the absolute notion of truth. So much of what follows will be directed at proving a theorem, similar to the one proved in the previous chapter, that links the two notions. To facilitate explaining the key ideas of this anti-Realist account, it will be useful to develop a way of regarding certain elements of an S5 structure. As the first step in such a development, let us investigate the role of a certain function.

V-Functions

Consider now the close structural relationship that obtains between the **C**-sentences and the **v**-function. Notice that, for fixed w, **v**(ψ, w) gives the truth value that **C**(w) ascribes to the sentential letter ψ. An examination of the template for constructing the **C**-sentences makes it obvious that **v**(ψ, w), regarded as a function of the sentential letters, provides all the information about the truth values of the sentential letters that is provided by **C**(w), and hence can be taken to represent what **C**(w) represents.

To clarify the principles at work in the above reasoning, an illustration will be helpful. A switching network consists of wires capable of carrying electrical current connected to switches each of which can be in one of two possible states: open or closed.[4] Current will flow through a switch only if the switch is closed. Let *E* be a switching network with three switches *A*, *B*, and *C*, wired according to Figure 8.

Suppose that *B* and *C* have been so constructed that when one is open the other must be closed, but that *A* is independent of the others in the sense that it may be open or closed regardless of the states of the other two. Then there is a total of four *possible states the network could be in, in so far as the states of the switches are concerned:*

[1] *A* open, *B* open, and *C* closed;
[2] *A* open, *B* closed. and *C* open;

[4] This example is obviously related to the use of the sentential calculus in the design of switching networks. See Lightstone 1965: ch. 1, sect. 1.7, for a discussion of this use of the sentential calculus.

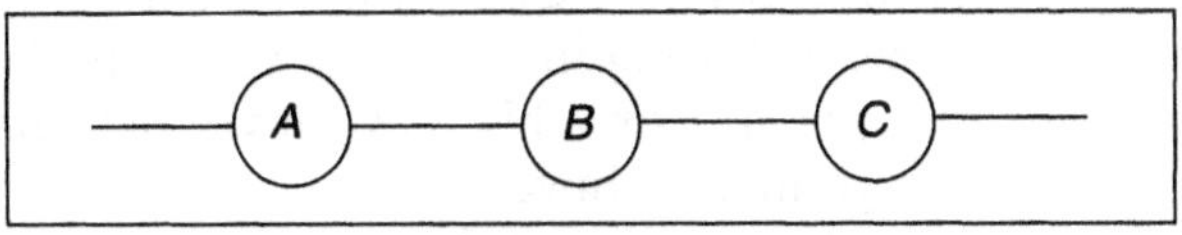

FIGURE 8

[3] *A* closed, *B* open, and *C* closed;
[4] *A* closed, *B* closed, and *C* open.

Let us use '1' to represent the state of being closed and '0' to represent the state of being open.

Table 1

States	*Switches* *A*	*B*	*C*
1	0	0	1
2	0	1	0
3	1	0	1
4	1	1	0

Then Table 1 will represent the totality of possible states described above. It is noteworthy that Table 1 represents the possibilities in question in a much more economical fashion than is accomplished in Lewis's possible worlds semantical analyses. There, you will recall, possibilities are represented by full-blooded counterparts in possible worlds. For example, the possibility represented by the second row of Table 1 might be represented by some real network of switches existing in some possible world. We can suppose that this network is wired according to Fig. 1 in such a way that the switch in the middle is closed and the other two are open. Obviously, such a real network will put forth much more information than is contained in Table 1. Thus, the real wire through which the current flows will have a specific diameter and length, it will be made of a material that has such-and-such a resistance, and it will have been manufactured at a specific time and by a specific factory. The switches will themselves have definite dimensions, and be constructed of certain materials, etc. For many purposes, all this extra information is irrelevant and may actually hinder the utilization of the relevant information by cluttering up the mind with useless information.

Suppose, now, that a valuation function **f** has been defined so as to capture the information given by the above table. That is, for any letter α ($\alpha \in \{A, B, C\}$) and any numeral j ($j \in \{1, 2, 3, 4\}$), $\mathbf{f}(\alpha, j)$ is to be the numerical value found in column α and row j. It is a simple matter to establish that the network, being in series, will have current flow when its switches are describable as "being in state j" if and only if the product $\mathbf{f}(A,j) \times \mathbf{f}(B,j) \times \mathbf{f}(C,j)$ is 1. From this, one can see that current will not flow in the network diagrammed above.

In this example, I made use of a simple mathematical function to represent different *ways in which a switching network can function, in so far as current flow through the switches is of concern.* Similarly, I argued above that a certain mathematical function can be regarded as representing different *ways in which the world could have been, in so far as the truth values of the sentential letters are of concern.* Since, for fixed w, $\mathbf{C}(w)$ explicitly describes a way that the world could have been, in so far as the truth values of the sentential letters are of concern, $\mathbf{v}(\psi, w)$ can also be regarded as *representing a way that the world could have been, in so far as the truth values of the sentential letters are of concern*, much as, for fixed j, the function $\mathbf{f}(a, j)$ can be regarded as representing a way that the switching network described earlier could have functioned, in so far as current flow through the switches is of concern.

NL Proto-Interpretations that Conform to S5 Interpretations

We can now drop the restriction we made earlier that there are only finitely many sentential letters among the non-logical constants of the language. Also, **W** can be of any cardinal size whatever. As for the case of **SC**, an *NL proto-interpretation* of **MSC** will assign to each sentential letter the meaning of some declarative sentence of English, and every connective and operator will be interpreted as was specified above for the sequence case.

As for the previous case, I shall use the expression '$[\phi/\mathcal{I}]$' to refer to the sentence ϕ interpreted according to the NL proto-interpretation $\mathcal{I}$ that conforms to S5 interpretation **I**. Notice that if Γ is a **C**-sequence generated by S5 interpretation $\mathbf{I} = \langle \mathbf{W}, \mathbf{w^*}, \mathbf{v} \rangle$, and if $\mathcal{I}$ is an NL proto-interpretation conforming to Γ, then it follows that $\mathcal{I}$ is an NL proto-interpretation conforming to **I**. It can also be seen that, for any $w \in \mathbf{W}$, not only can $\mathbf{v}(\sigma, w)$ be regarded as a representation of how the world could have been, in so far as the truth values of the sentential letters are of concern, but when the language is given an NL proto-interpretation $\mathcal{I}$ that conforms to the S5 interpretation **I**, $\mathbf{v}(\sigma, w)$ can be regarded more specifically as representing how the world could have been, in so far as what are of concern are the truth values of the sentential letters, *when they are interpreted as expressing the statements assigned to them by* $\mathcal{I}$. To emphasize that we are taking $\mathbf{v}(\sigma, w)$ to be representing how the world could have been, in this more specific way, I shall use the notation '$\mathbf{v}_{\mathcal{I}}(\sigma, w)$'.

I can now give the definition of *NL proto-interpretation that conforms to an S5 interpretation of* **MSC**:

[Def] An NL proto-interpretation $\mathcal{J}$ is an *NL proto-interpretation that conforms to S5 interpretation* **I** = <**W**, **w***, **v**> of **MSC** iff the following conditions are satisfied:

[i] For every w, the world could have been as each $\mathbf{v}_{\mathcal{J}}(\sigma, w)$ says that it could.

[ii] It is not the case that the world could have been such that no $\mathbf{v}_{\mathcal{J}}(\sigma, w)$ represents that it could have been that way.

[iii] For every sentential letter ϕ, ϕ is true under **I** iff $[\phi/\mathcal{J}]$ is true.

As in the previous chapter, an NL proto-interpretation that conforms to an S5 interpretation of **MSC** will be said to be an NL interpretation.

The following theorem can be proved by induction in basically the way Theorem [1] was proved.

Theorem [2]. For every S5 interpretation **I** = <**W**, **w***, **v**> of **MSC**, for every NL proto-interpretation $\mathcal{J}$ conforming to **I**, and for every $w \in \mathbf{W}$, a sentence ϕ is true at w under **I** iff $[\phi/\mathcal{J}]$ would have been true, had the world in fact been as it is represented as being by $\mathbf{v}_{\mathcal{J}}(\phi, w)$.

Discussion

Although I made some effort, in the first part of this section, to explain how the function $\mathbf{v}(\psi, w)$ can be regarded, it needs to be emphasized that this way of regarding $\mathbf{v}_{\mathcal{J}}(\sigma, w)$ was given primarily for didactic and heuristic purposes and also to facilitate the expression of conditions, theorems, and proofs. For example, in the statement of Theorem [2], one finds the expression "had the world in fact been as it is represented as being by $\mathbf{v}_{\mathcal{J}}(\sigma, w)$". Taking $\mathbf{v}_{\mathcal{J}}(\sigma, w)$ to be simply a mathematical function, I could have expressed the above in the somewhat more cumbersome words:

. . . had the world in fact been such that, for every sentential letter θ, $[\theta/\mathcal{J}]$ would have been true iff $\mathbf{v}_{\mathcal{J}}(\theta, w) = 0$.

Similarly, my expression of the conditions that must be satisfied in order that an NL interpretation be an NL interpretation conforming to an S5 interpretation is a shortening of a condition involving quantification over the sentential letters of **MSC**. My way of regarding $\mathbf{v}_{\mathcal{J}}(\sigma, w)$ is meant to help the reader to follow the logical developments, but nothing essential in the development depends upon regarding the **v**-function in this way: in particular, nothing essential to the proof of Theorem [2] depends upon regarding the function in this way.

Some Consequences

I shall now give, without proof, some consequences of Theorem [2].

Corollary [a]: For every S5 interpretation **I** = <**W**, **w***, **v**> of **MSC**, and for every NL proto-interpretation $\mathcal{I}$ conforming to **I**, assuming that the world *is* as it is represented as being by $\mathbf{v}_{\mathcal{I}}(\sigma, \mathbf{w}^*)$, then ϕ is true under **I** iff $[\phi/\mathcal{I}]$ is true.

We can also conclude that:

Corollary [b]: For every S5 interpretation **I** = <**W**, **w***, **v**> of **MSC**, and for every NL proto-interpretation $\mathcal{I}$ conforming to **I**, there is a $w \in \mathbf{W}$ such that ϕ is true at w under **I** iff there is a $w \in \mathbf{W}$ such that $[\phi/\mathcal{I}]$ would have been true, had the world in fact been as it is represented as being by $\mathbf{v}_{\mathcal{I}}(\sigma, w)$.

From the conditions that must be satisfied by an NL proto-interpretation conforming to **I**, it follows that the right side of the above biconditional holds iff the world could have been such that $[\phi/\mathcal{I}]$ was true. And the latter is the case iff it is possible that $[\phi/\mathcal{I}]$ is true. The left side of the biconditional of Corollary [b] holds iff $\Diamond\phi$ is true under **I**. Hence, we obtain:

Corollary [c]: For every S5 interpretation **I** = <**W**, **w***, **v**> of **MSC**, and for every NL proto-interpretation $\mathcal{I}$ conforming to **I**, $\Diamond\phi$ is true under **I** iff it is possible that $[\phi/\mathcal{I}]$ is true.

It then follows that:

Corollary [d]: For every S5 interpretation **I** = <**W**, **w***, **v**> of **MSC**, and for every NL proto-interpretation $\mathcal{I}$ conforming to **I**, $\Box\phi$ is true under **I** iff it is necessary that $[\phi/\mathcal{I}]$ be true.

Clearly, then,

Corollary [e]: For every S5 interpretation **I** = <**W**, **w***, **v**> of **MSC**, and for every NL proto-interpretation $\mathcal{I}$ conforming to **I**, if ϕ is valid, then it is necessary that $[\phi/\mathcal{I}]$ be true.

In short, if ϕ is valid, then any NL instance of ϕ is necessarily true.

We can also see how a theorem that $\Gamma \models \phi$ can be used to draw conclusions about the validity of modal reasoning. For another corollary of Theorem [2] is:

Corollary [f]: For every S5 interpretation **I** = <**W**, **w***, **v**> of **MSC**, for every NL proto-interpretation $\mathcal{I}$ conforming to **I**, for every set of sentences Γ, and every sentence ϕ, if $\Gamma \models \phi$, then, if every member θ of Γ is such that $[\theta/\mathcal{I}]$ is true, so also is $[\phi/\mathcal{I}]$.

In short, given that $\Gamma \models \phi$ and that the language **MSC** is given an NL interpretation, then, if all the members of Γ are true, so is ϕ. Similarly,

Corollary [g]: For every S5 interpretation **I** = <**W**, **w***, **v**> of **MSC**, for every NL proto-interpretation $\mathcal{I}$ conforming to **I**, for every set of sentences Γ, and every sentence ϕ, if $\Gamma \vdash \phi$, then, if every member θ of Γ is such that $[\theta/\mathscr{I}]$ is true, so also is $[\phi/\mathscr{I}]$.

Thus, one can see how the inference rules of S5 Modal Sentential Logic allow one to infer the truth of a sentence from the truth of a set of sentences—and this by virtue of the "logical form" of the sentences in question. And one can see why the *theorems* of the S5 Sentential Calculus (that is, sentences of **MSC** that can be derived from the empty set) can be regarded as *principles of modal logic*: every such theorem θ is *valid* (in the technical sense of being true under every interpretation), and hence, for every NL proto-interpretation $\mathscr{I}$ conforming to an interpretation **I**, $[\theta/\mathscr{I}]$ must be true—indeed, true in virtue of the logical form of θ. Since the truth of $[\theta/\mathscr{I}]$ can be inferred without appeal to experiment or observation, we have grounds for classifying such sentences as "a priori". Thus, it can be argued that such valid sentences have many of the features one might expect to find in principles of logic.

Notice also that to view the use of logic in the above way, there is no need to regard the structures or models of **MSC** as being anything more than straightforward mathematical entities. In other words, we do not have to suppose that there is some favoured model of **SC** that is the "intended interpretation" or the "canonical model". Nor do we have to suppose, in our applications of this logic, that there is a set of all possible worlds.

Let us now see if we can extend this account to the case of S5 Modal Quantificational Logic.

7

Modality without Worlds II

The Semantics of Modal Quantificational Logic

Let us reconsider the question with which I began Chapter 3: what is the connection between the set theoretical structures studied in the model theory of modal logic and anything objective in virtue of which we can explain how a study of these structures can yield genuine insights into the principles and inferences of modal logic? My answer to this question will be given by means of a statement and proof of a theorem for S5 Modal Quantificational Logic that is analogous to the "connecting" theorem of Chapter 5 and Theorem [2] of Chapter 6. With such a theorem, I will be able to give the same sort of anti-Realist account of the semantics of this logic that was given earlier. As will become evident shortly, the logical material to be discussed in this chapter is considerably more complicated than the material discussed in the two preceding chapters. However, the basic ideas are the same.

1. Actualism and Serious Actualism

The Pollock–Plantinga Dispute

'Serious Actualism' was characterized (in Chapter 1) to be the doctrine that a thing cannot have a property or be in a relation at a world if it does not exist in that world. John Pollock has argued that this doctrine is either false or uninteresting. Suppose that *w* is a world in which Socrates does not exist (Pollock 1985: 126). Then, it is tempting to say that Socrates has the property of *not existing* at *w*, even though Socrates does not exist in *w*. Now Pollock anticipates the rejoinder that there is no such property as "the property of not existing": the phrase 'not existing', he anticipates it being said, simply does not express a genuine property. In that case, says Pollock, we have what amounts to a "partial stipulation" of how the term 'property' is to be used: "then we cannot object to it except on the grounds that it makes serious actualism trivial and uninteresting" (Pollock 1985: 129).

Pollock's strategy is to give Plantinga that use of the term 'property'. However, there is nothing to prevent him from introducing a more general term to cover both properties (in the above sense) and things like *not existing*. Thus, he defines a *condition* to be a function from objects to states of affairs and then defines what it is for an object to "satisfy a condition at a world":

> *x* satisfies condition *C* at world *w* iff *C*(*x*) obtains at *w*.

Since there are worlds at which Socrates does not exist, the state of affairs of *Socrates not existing* obtains at some worlds; and it can be argued that Socrates can satisfy a condition at a world in which he does not exist. Thus, although it is true that no object can have a property at a world in which it does not exist, an object can satisfy a condition at a world in which it does not exist. If Serious Actualism is the doctrine that no object can satisfy a condition at a world in which it does not exist, the doctrine is false. We can see why Pollock thinks that Serious Actualism is either false or uninteresting.

Plantinga, however, believes that he can state a true and non-trivial doctrine that can appropriately be called 'Serious Actualism'. He responds to Pollock's objection by giving Pollock his use of the expression "satisfy a condition at a world". He thus allows that an object can satisfy a condition at a world in which it does not exist. But he introduces a new expression 'satisfy a condition <u>in</u> a world' in such a way that he believes he can show that, for an object *x* to satisfy a condition *C* <u>in</u> some world *u*, *x* must satisfy *C* at *u* and, in addition, *x* must exist in *u*. If his demonstration were successful, then he could defend his Serious Actualism by claiming that no object could satisfy a condition <u>in</u> a world in which it does not exist. And he could charge his critics with confusing satisfaction at a world with satisfaction <u>in</u> a world (Plantinga 1985*a*: 321 and 347).

The definition he gives is:

> An object *x* satisfies a condition *C* <u>in</u> a world *w* iff, necessarily, had *w* obtained, it would have been the case that *x* satisfied *C*. (Plantinga 1985*a*: 321)

But what does the phrase 'it would have been the case that *x* satisfied *C*' mean? Not: it would have been the case that *x* satisfied *C* <u>at</u> *w*. Plantinga allows, it was just noted, that *x* could satisfy *C* at *w* even if *x* does not exist in *w*. Notice that, just as the notion of *an object satisfying a condition in a world* is a technical notion, so also is the notion of *an object existing <u>in</u> a world.* Recall (from Chapter 1) that the latter technical notion was cashed out, by Plantinga, in terms of existence (simpliciter) in the following way:

> Object *x* exists in world *w* iff *x* would have existed, had *w* obtained.

Thus, Plantinga gave an explication of *an object existing in a world* which is quite similar to his explication of the technical notion that concerns us here: in both cases, the indicative clause to the left of the 'iff' is analysed on the right in terms of a subjunctive. In his earlier explication, when Plantinga introduced talk of what would have been the case, had *w* obtained, the framework shifted from talk about possible objects to subjunctive talk about whether the object would have existed (simpliciter). Similarly, I suggest, the clause 'it would have been the case that *x* satisfied *w*', in Plantinga's explication of the notion of *an object satisfying a condition in a world,* should be understood as saying that *x* would have satisfied (simpliciter) the condition, dropping all talk of possible objects satisfying a condition at other possible worlds. Within this subjunctive framework, there are no possible objects. In short, just as *existing in a world* had been explained in terms of what would have existed (simpliciter), so also *satisfying a condition in a world* is explained in terms of what would have satisfied (simpliciter) a condition.

I shall now argue, on the basis of Actualism, that any object satisfying a condition <u>in</u> a world exists in that world. Suppose *x* satisfies *C* <u>in</u> *w*. That means: had *w* obtained, *x* would have satisfied *C*. But *x* could not have satisfied a condition *C* if *x* had not existed. The reason is, if *x* had not existed, *x* would have been a merely possible object; and according to Actualism, necessarily, there are no merely possible objects. Hence, had *w* obtained, *x* would have existed. But that means *x* exists in *w*. We can conclude: If *x* satisfies a condition <u>in</u> a world, then *x* exists in that world. We can see why Plantinga holds that Serious Actualism follows from Actualism.[1]

The Actualistic Position to be Taken

Let us now take stock and see what, if any lessons, we should carry away from this dispute. Like Plantinga, I wish to make sense of the semantics of modal logic in terms of modal sentences involving subjunctive counterfactual clauses, without making reference to merely possible beings. For example, I will make assertions with the grammatical structure of:

> The world could have been such that there was a person who first was married to Elizabeth Taylor and then was married to Debbie Reynolds, but who was such that, had his mother not stopped to pick up a hitchhiker one Christmas Eve, he would never have been born.

Such assertions do not require referring to merely possible beings. Of course, there is no person who first was married to Elizabeth Taylor and then was married to Debbie Reynolds. But like Plantinga, I see no good

[1] Plantinga 1985*a*: 321. Compare the above argument with the one put forward by Michael Bergmann in order to refute several authors who have denied that Actualism implies the truth of Serious Actualism (Bergmann 1996).

reason to postulate a merely possible person who first was married to Elizabeth Taylor and then was married to Debbie Reynolds. A thing that does not exist, but is a possible being, has no place in my philosophical account of modal logic. The Actualistic perspective will be noteworthy, later in this chapter, when the semantics of a formal system of modal logic is set up. Since there will be no assumption of the existence of any merely possible objects, I shall not allow assignments of merely possible objects to individual constants; nor shall I make any reference to merely possible objects in presenting the semantical system.

I also agree that the Serious Actualistic position Plantinga arrives at is reasonable. The Serious Actualistic view motivates the principal feature of the formal modal language $\mathbf{M_1}$ (to be presented in the next section) that distinguishes it from $\mathbf{M}$: the (primitive) predicates of $\mathbf{M_1}$ will be regarded as expressing "genuine properties", the possession of which requires existence. Thus, the **e**-function will always assign to an n-ary predicate at world w a subset of $\mathbf{d}(w)^n$. The Serious Actualistic system I shall develop should be contrasted with Jager's Actualistic version, discussed earlier, that was based upon Plantinga's views, and the very different Actualistic S5 system of McMichael's (McMichael 1983*a*), according to which the Brouwerian axiom, $\phi \rightarrow \Box\Diamond\phi$, is not valid.

Gupta's Argument for Possibilism

But before turning to the formal language, I should like to consider one sort of objection that might be raised to the position I have outlined above. Consider Allan Gibbard's example of two lumps of clay, a and b, modelled in such a way that just by pressing the two lumps together, one will obtain a statue (a would form the bottom half of a statue of a man; b would form the top half).[2] Of these two lumps of clay, Gupta has written:

> Clearly, we can name the piece of clay that *would* result if a and b were to get stuck to each other. Let us call this possible piece of clay c. 'c' denotes a possible object. Further, c can bear relations to actual objects. c and a, respectively, bear the relation "heavier than". (Gupta 1980: 20 n. 15)

According to Gupta, then, there could be (and are) merely possible objects. Obviously, he is no Actualist. He also rejects Serious Actualism, since he maintains that non-existent objects can bear primitive relations ("properties") to actual objects. About these matters he is quite explicit: "We can have names for merely possible objects, and possible objects bear interesting relations to actual objects which can be expressed by predicates" (Gupta 1980: 20).

Since there is something very persuasive about the above example,

[2] This type of example is discussed throughout Gibbard 1975.

something should be said in reply. I believe that essentially all the really strong intuitions that lie behind Gupta's Possibilist argument can be captured by the Actualist without requiring the acceptance of the paradoxical doctrine that one can name merely possible objects. An Actualist can certainly allow that had she pressed the two lumps a and b together, there would have been a statue c which would have been heavier than either of the two parts. And she can allow that she would have named the statue c had she formed a statue out of a and b, and that c would have been placed in the cabinet above the piano. But acceptance of these, and countless other, related subjunctives in no way commits one to the acceptance of Possibilism. The letter 'c' in the first of the two subjunctives is no more a *name* of a merely possible object than is the variable 'x' in the sentence 'There could have been a statue x on that table'. I, of course, have no objections to using the letter 'c' in such ways, but I see no reason to conclude from such uses that there are merely possible objects or that it is possible to name such non-existent things. What are eschewed by the Actualist, then, are not subjunctives of the above sort, but the sort of ontological leaps taken by Gupta.

2. The Serious Actualistic Compossibility System

I shall present two new systems of modal quantificational logic, which differ in small ways from the standard one previously discussed.

The Serious Actualistic System

The S5 modal quantificational language $\mathbf{M_1}$ is syntactically identical to $\mathbf{M}$. $\mathbf{M_1}$ differs from $\mathbf{M}$ primarily because the semantics of the new language reflects the position of Serious Actualism. In particular, the semantics of the new language will require that, for every n-ary predicate ϕ, every S5 interpretation $\mathbf{I} = \langle \mathbf{W}, \mathbf{D}, \mathbf{w^*}, \mathbf{d}, \mathbf{r}, \mathbf{e} \rangle$, and every $w \in \mathbf{W}$, $\mathbf{e}(\phi, w)$ be restricted to subsets of $\mathbf{d}(w)^n$. The idea is that the primitive predicates of the language will only be satisfied at a world by things that exist in that world. It follows, of course, that everything that doesn't exist in that world will automatically satisfy $-\phi$ at that world.

The only other feature of the new language which distinguishes it from $\mathbf{M}$ is to be found in the definition of 'true at w under an S5 interpretation'. The individual constants of $\mathbf{M}$ perform two functions that are quite distinct. On the one hand, individual constants function like names: they simply denote something in $\mathbf{D}$. But individual constants also function in a way similar to the way variables do. This can be seen from the definition of 'true at w under an S5 interpretation'. When θ is the general sentence $(\exists\alpha)\phi$, the truth of θ at w under $\mathbf{I}$ depends upon whether the instantial sentence $\phi\alpha/\beta$—where β is the first individual constant that does not occur in ϕ—is true under at least one β-variant of **I** that assigns to β some object from **d**(w). Roughly speaking, we let β take all possible values from **d**(w) to see if one of these values results in $\phi\alpha/\beta$ being true at w under **I**.

In what follows, it will be useful to divide the class of individual constants into two subclasses: one will only include those that perform this "variable" function and the other will only contain those that perform only the "naming" function. Thus, the definition of 'true at w under **I**', for the case in which θ is a general sentence, will require the instantial sentence $\phi\alpha/\beta$ to be such that the instantial constant β will be the first individual constant *with an odd subscript* that does not occur in ϕ. (Henceforth, an individual constant with an odd subscript will be said to be an 'odd-subscripted individual constant'.) In short, from now on, we shall be dealing with a formal language in which only odd-subscripted individual constants will perform the "variable" function described above. All the other individual constants will only perform the "naming" function. Since unsubscripted individual constants will be classified as even-subscripted individual constants, we can say that all even-subscripted individual constants function only as names, whereas odd-subscripted individual constants function as names and also perform the "variable" function. Since this is a trivial difference, essentially the only significant respect in which $\mathbf{M}_1$ differs from **M** is in the definition of 'S5 interpretation'.

This change in the definition of 'true under **I**' will allow me to treat a sentence of $\mathbf{M}_1$, in which odd-subscripted individual constants occur, as a kind of open-sentence for which a relation of *satisfaction at a world under an interpretation* will be defined. It should be noted, however, that this relation of satisfaction is not being specified in order to define 'true at a world under an interpretation', as one might reasonably expect. The basic semantical notion of truth at a world under an interpretation is defined for $\mathbf{M}_1$ in essentially the way the notion was defined for **M**, and the fundamental semantical features of $\mathbf{M}_1$ are no different from those of **M**.

Here, the logically acute reader may well wonder why *satisfaction at a world under an interpretation* is being defined at all, given that it is not to be used in the definition of *true at a world under an interpretation*. To appreciate the rationale for this definition of satisfaction, one needs to keep in mind what I aim to accomplish in this chapter. I am setting out to establish links between the relational notion of *truth under an S5 interpretation* and the monadic notion of *truth* (of NL interpreted sentences) in virtue of which a theorem analogous to the "connecting theorem" of Chapter 5 and Theorem [2] of Chapter 6 can be proved. Now NL proto-interpretations will not assign meaning or sense to odd-subscripted individual constants and will, instead, treat such constants as variables. Thus, the NL interpreted sentences of $\mathbf{M}_1$, in which odd-subscripted individual constants occur, turn out to be open-sentences that are satisfied by sequences of the appropriate sort. The notion of *satisfaction under an interpretation* being defined here will correspond to this notion of satisfaction; and this correspondence will enable me to link the relational notion of *truth under an S5 interpretation* to the monadic notion of *truth* (of NL interpreted sentences).

Regarding the derivational system for this new language, notice that every S5 interpretation of $\mathbf{M_1}$ is an S5 interpretation of **M**. Thus, it is easy to see that the rules of derivation that are sound for **M** will also be sound for $\mathbf{M_1}$. However, since there are S5 interpretations of **M** that are not S5 interpretations of $\mathbf{M_1}$, it is clear that the set of rules of derivation that are complete for **M** will not be complete for $\mathbf{M_1}$. The additional (serious) actualistic requirement that a sextuple <**W**, **D**, **w***, **d**, **r**, **e**> must satisfy to be an interpretation of $\mathbf{M_1}$ clearly indicates that the previous rules of derivation will have to be supplemented. This can be done by adding the Falsehood Principle which was discussed in Chapter 1. More specifically, we can add the following rule:

> *F* (Falsehood). One may enter on a line any sentence of the following type:
>
> $\Box(x_1)\Box(x_2)\ldots\Box(x_n)\Box(\theta^n x_1\, x_2 \ldots x_n \rightarrow (E^1 x_1\ \&\ E^1 x_2\ \&\ \ldots\ \&\ E^1 x_n)),$
>
> where θ_n is a predicate of degree n. As premiss numbers of this new line, take the empty set.

The Completeness of the Derivational System

I shall be concerned here only with weak completeness. I shall prove that, for any sentence θ of $\mathbf{M_1}$, if θ is valid, then θ is derivable from the empty set of premisses. Suppose that θ is a valid sentence of $\mathbf{M_1}$. Let $\psi_1, \psi_2, \ldots, \psi_k$ be all the predicates that occur in θ, and let $\psi_{i\#}$ be obtained from the Falsehood Principle by taking the schematic predicate in the statement of the principle to be ψ_i ($i = 1, 2, \ldots, k$). Now the sentence

$$(\psi_{1\#}\ \&\ \psi_{2\#}\ \&\ \ldots\ \&\ \psi_{k\#}) \rightarrow \theta$$

must be true under all interpretations of **M**, since, on the one hand, the consequent must be true under all interpretations of $\mathbf{M_1}$, and on the other hand, any interpretation of **M** under which θ is false must be one under which the antecedent is false. Since the derivational system of **M** is complete, the conditional must be derivable in **M** from the empty set of premisses. Hence, the consequent must be derivable, from the empty set of premisses, in the derivational system for $\mathbf{M_1}$.

The Serious Actualistic Compossibility System

The second new S5 modal quantificational language to be taken up is, like $\mathbf{M_1}$, syntactically identical to $\mathbf{M}$, and differs semantically from $\mathbf{M_1}$ in only one small respect. To see what lies behind the small change to be made in the semantics of the formal language, let us consider a possible "natural language" interpretation of the language, according to which the ranges of the quantifiers are determined by the following English predicate θ:

> ① is a piece of furniture *originally* constructed out of *m* and nothing else but glue, so that, after the construction, nothing of *m* remained

(where '*m*' rigidly denotes a particular slab of oak).[3] Now it would seem that, even though the world could have been such that there was a chair that satisfied θ, and the world could have been such that there was a table that satisfied θ, it is not the case that the world could have been such that there was both a chair that satisfied θ and also a table that satisfied θ. In other words, using the Possibilist's terminology, the possible chair and the possible table would not be compossible. Now the semantics of this new language rules out such a situation. In other words, the following intuitive condition must be satisfied by all interpretations:

> [The Principle of Compossibility] If the world could have been such that there was something x_1, and the world could have been such that there was something $x_2, \ldots,$ and the world could have been such that there was something x_n, then the world could have been such that $x_1, x_2, \ldots, x_n$ existed.

The idea underlying this condition is to rule out interpretations that violate a kind of *principle of recombination* that David Lewis has described as, roughly speaking, "the principle that anything can coexist with anything else".[4] Formally, then, every S5 interpretation $\mathbf{I} = \langle \mathbf{W}, \mathbf{D}, \mathbf{w^*}, \mathbf{d}, \mathbf{r}, \mathbf{e} \rangle$ of the new language, $\mathbf{M^*}$, must satisfy the following condition:

> For every finite ordinal n, if $b_1, b_2, \ldots, b_n \in \mathbf{D}$, then there is a $u \in \mathbf{W}$ such that $b_1, b_2, \ldots, b_n \in \mathbf{d}(u)$.

The derivational system for the new language differs from the one for $\mathbf{M_1}$ only by the addition of one new rule:

> *Cp* (Compossibility) One may enter on a line any sentence of the following type:

[3] In order that a piece of furniture be "originally" constructed out of *m*, it must be made directly from the slab of oak and not from some other piece of furniture which was constructed out of *m*.

[4] Lewis 1986: 88. I discuss this principle and Lewis's acceptance of a related principle in Chapter 8.

$$\Box(x_1)\Box(x_2)\ldots\Box(x_n)\Diamond(E^1x_1 \,\&\, E^1x_2 \,\&\ldots\&\, E^1x_n)$$

As premiss numbers of this new line, take the empty set.

It is clear that the resulting derivational system is sound. Is it complete? As before, I shall concern myself only with weak completeness.

The Completeness of the Derivational System

Let χ_n be the sentence:

$$\Box(x_1)\Box(x_2)\ldots\Box(x_n)\Diamond(E^1x_1 \,\&\, E^1x_2 \,\&\ldots\&\, E^1x_n)$$

and let θ be a valid sentence of **M*** that is not a valid sentence of $\mathbf{M_1}$. Suppose that **I** is any S5 interpretation of $\mathbf{M_1}$. Now if $\mathbf{I} \not\models \theta$ then it must be because, for some finite ordinal k, $\sigma_1, \sigma_2, \ldots, \sigma_k \in \mathbf{D}$, but there is no $u \in \mathbf{W}$ such that $\sigma_1, \sigma_2, \ldots, \sigma_k \in \mathbf{d}(u)$. It follows that $\mathbf{I} \not\models \chi_k$. Thus, $\mathbf{I} \models (\chi_k \rightarrow \theta)$. Now if $\mathbf{I} \models \theta$, then $\mathbf{I} \models (\chi_k \rightarrow \theta)$. So in either case, $(\chi_k \rightarrow \theta)$ must be true under **I** and hence a valid sentence of $\mathbf{M_1}$. It follows that it must be derivable in the derivational system of $\mathbf{M_1}$ from the empty set. We can conclude that θ is derivable in the derivational system of **M*** from the empty set.

Henceforth, in this chapter, I shall be working with the language **M***.

3. Satisfaction Under an S5 Interpretation

In order to state the connecting theorem, it is useful to divide the class of sentences of **M*** into two subclasses: an *open*-sentence contains one or more occurrences of odd-subscripted individual constants; a *closed*-sentence contains no occurrences of odd-subscripted individual constants. A relation of *satisfaction at a world under an interpretation* that obtains between sentences of **M*** and denumerable sequences of objects will be defined. To facilitate giving this definition, several new notions will be introduced.

Ordering the Odd-subscripted Individual Constants. The odd-subscripted individual constants are to be ordered

$$a_1, b_1, c_1, \ldots, t_1, a_3, b_3, c_3, \ldots, t_3, a_5, b_5, c_5, \ldots$$

so that each odd-subscripted individual constant α will have a unique place in this ordering. Let $P(\alpha)$ be the ordinal place of α in this ordering. For example, $P(a_1) = 1$; $P(a_3) = 21$; and $P(t_3) = 40$.

Denumerable Sequences. Suppose that $\mathbf{I} = \langle \mathbf{W}, \mathbf{D}, \mathbf{w^*}, \mathbf{d}, \mathbf{r}, \mathbf{e} \rangle$ is an S5 interpretation of **M***. A denumerable sequence of elements of **D** will be called a **D**-*sequence*. (It should be noted that such sequences may contain repetitions.) Any **D**-sequence which is such that б_k occurs in place $P(\beta_k)$, $k = 1, 2, \ldots, n$, will be called a **D**-$[\beta_1/\text{б}_1, \beta_2/\text{б}_2, \ldots, \beta_n/\text{б}_n]$-sequence. Thus, suppose that John Searle occurs in the 21st place in a **D**-sequence. Then that sequence is a **D**-$[a_3/$John Searle$]$-sequence.

Variants of S5 Interpretations. The idea of a $\beta_1\beta_2 \ldots \beta_n$-variant of an S5 interpretation is obtained by an obvious generalization from the idea of a β-variant of an S5 interpretation. Thus, if **I** and **J** are S5 interpretations and $\beta_1, \beta_2, \ldots, \beta_n$ are odd-subscripted individual constants, then **J** is a $\beta_1\beta_2 \ldots \beta_n$-variant of **I** iff **J** is exactly like **I** with the possible exception of what is assigned to the constants $\beta_1, \beta_2, \ldots, \beta_n$. ($\mathbf{J}(\beta_k)$ may differ from $\mathbf{I}(\beta_k)$, $k = 1, 2, \ldots, n$.)

Satisfaction at a World Under an Interpretation. Suppose that **I** = <**W**, **D**, **w***, **d**, **r**, **e**> is an S5 interpretation of **M***, that $w \in \mathbf{W}$, and that ϕ is a sentence in which the distinct odd-subscripted individual constants $\beta_1, \beta_2, \ldots, \beta_n$ occur, and in which no other odd-subscripted individual constants occur. Then, for all objects $\text{б}_1, \text{б}_2, \ldots, \text{б}_n$ that are elements of **D**, ϕ is *satisfied at w under* **I** by σ, a **D**-$[\beta_1/\text{б}_1, \beta_2/\text{б}_2, \ldots, \beta_n/\text{б}_n]$-sequence, iff ϕ is true at w under the $\beta_1\beta_2 \ldots \beta_n$-variant of **I**, say **J**, which is such that $\mathbf{J}(\beta_k) = \text{б}_k$, $k = 1, 2, \ldots, n$.

Let us consider a simple example. Let **I** = <**W**, **D**, **w***, **d**, **r**, **e**> be an S5 interpretation of **M*** in which $w \in \mathbf{W}$ and $\mathbf{d}(w)$ is the set of humans actually alive today. And let $\mathbf{e}(F^1, w)$ be the set of philosophers actually alive today. Then 'F^1b_1' is satisfied at w under **I** by any **D**-$[b_1/$John Searle$]$-sequence. Of course, this example does not conform to the "official view" I have adopted that requires all S5 interpretations to be set theoretical structures involving only pure sets. The above example is only performing a didactic function.

For the case in which ϕ is a closed sentence, ϕ is satisfied at w under **I** by every **D**-sequence iff ϕ is true at w under **I**; ϕ is satisfied at w under **I** by no **D**-sequence iff ϕ is false at w under **I**.[5]

Note. If a **D**-$[\beta_1/\text{б}_1, \beta_2/\text{б}_2, \ldots, \beta_n/\text{б}_n]$-sequence satisfies a sentence ϕ at w under interpretation **I**, then every **D**-$[\beta_1/\text{б}_1, \beta_2/\text{б}_2, \ldots, \beta_n/\text{б}_n]$-sequence satisfies ϕ at w under **I**.

Davidson's Problem Again

The official position I have adopted requires that every S5 interpretation be a purely mathematical structure whose elements are pure sets. Hence,

[5] My definition of satisfaction is loosely based upon the ideas in Tarski 1956*a*. See Mendelson 1987: 47–8, and Quine 1970: 37–41 for more recent treatments of satisfaction to which my account more closely corresponds.

every **D**-sequence that satisfies a sentence at *w* under **I** will have as elements only pure sets. The questions that now arise are these. (1) What do these pure sets have to do with modality? (2) How is satisfaction under an interpretation related to satisfaction? The second of these questions is obviously just a satisfaction version of Davidson's Problem. Following in the footsteps of the previous chapters, we need to bring in NL proto-interpretations.

4. NL Interpretations of M*

I shall be concerned, in this section, with developing "natural language" interpretations of **M*** that perform for modal quantificational logic what the NL proto-interpretations conforming to an interpretation of **MSC** did for the modal sentential logic. The idea is to give meaning to the symbols of the formal language so as to yield sentences that directly express statements that can be true or false. The class of NL proto-interpretations conforming to an interpretation to be defined in this section will be a subset of the class of what I shall call 'NL proto-interpretations'. These proto-interpretations give meaning to the primitive symbols of the formal language: connectives, modal operators, predicates, individual constants, and quantifiers of the formal language.

When **M*** is given an NL proto-interpretation, the connectives and the modal operators are understood in the way described in the previous chapter. However, instead of sentential letters, it is predicates, individual constants, and quantifiers to which proto-interpretations give meaning. Some preliminary comments on how this will be done will be made in order to simplify the definitions that follow.

1. *Predicates*

I shall make use of English predicates, which I discussed in Chapter 5. As I noted earlier, English predicates are much like open-sentences except for the fact that circled numerals occur in English predicates where variables occur in open-sentences. Every English predicate will have a degree: the circled numeral of highest value occurring in the predicate gives the degree of the predicate. If an English predicate has degree n, then each numeral of a positive integer less than n must occur circled in the predicate. For example, an English predicate of degree 3 will contain one or more occurrences of the numeral '3', circled, and also occurrences of the numerals '2' and '1', each circled. Thus,

① is a brother of ②

is an English predicate of degree 2.

What sorts of things satisfy English predicates? If θ is an English predicate of degree n, then anything that satisfies θ will be an ordered n-tuple. For example, the ordered pair <2, 7> satisfies the English predicate

① is a natural number less than ②,

but the ordered pair <7, 2> does not. To see if an ordered pair satisfies the above English predicate, take '①' as a name of the first element of the pair and take '②' as a name of the second; then see if the resulting sentence is true. It should be kept in mind (in view of the previous section) that if some n-tuple $<б_1, б_2, б_3, \ldots, б_n>$ satisfies the English predicate θ, then the objects $б_1, б_2, б_3, \ldots, б_n$ must all exist. An n-tuple of non-existent objects (to speak somewhat misleadingly) cannot satisfy an English predicate.

An NL proto-interpretation, then, will correlate, with each predicate θ of **M***, an English predicate of degree n (where n is the degree of θ): the English predicate will give the meaning of its correlated predicate.

2. *Quantifiers*

What does the existential quantifier mean? Quine has maintained that '$(\exists x)Fx$' just means: there is an entity (or object) x such that x is F. Although, I am willing to grant that in certain contexts, talk of entities or objects makes sense, I do have doubts that translating the existential quantifier in the above way is satisfactory. It is questionable that, without a great deal of supplementary stage setting, translating the quantifiers in that way yields sentences that are true or false. I do not wish to go into a long digression on this point here, since I have discussed it elsewhere.[6] I shall require of proto-interpretations that they fix the domain of the quantifiers with an English predicate of degree one, and in this way give a definite sense to the quantifiers. Thus, if the correlated English predicate is '① is an animal', then the general sentence '$(\exists x)Fx$' will mean: there is at least one animal x such that x is F.[7]

3. *Individual Constants*

An NL proto-interpretation will correlate, with each *even-subscripted individual constant*, a name or definite description (in English) of an actual thing that satisfies the English predicate that gives sense to the quantifiers. The individual constant can then be regarded as having the sense of the name or the definite description (taken to rigidly designate the actual

[6] See Chihara 1990: 69–71. Anil Gupta hypothesizes that whenever "there is variable binding (or its analogue) in the syntax of a language, there is a common noun (or a proper name) binding the variable (or its analogue)" (Gupta 1980: 9). See also Putnam 1991: ch. 7 for other arguments supporting the position I take here.

[7] If Gupta is right (see preceding footnote), the English predicate in question would have to be one that contains a common noun.

thing). Thus, if the predicate 'R^2' is interpreted to function as the English predicate '① is married to ②', and the individual constants '*a*' and '*b*' are assigned the senses of the names 'Bill Clinton' and 'Hillary Rodham Clinton', then the sentence 'R^2ab' will mean: 'Bill Clinton is married to Hillary Rodham Clinton'.

How are *odd-subscripted individual constants* to be interpreted? Since these can function in a way similar to the way variables do, they will not be regarded as having the sense of a name or definite description. Instead, sentences of **M*** that contain occurrences of odd-subscripted individual constants—that is, what I have called "open sentences"—will be regarded as what Mates calls "sentence forms" to be satisfied, if at all, by denumerable sequences.[8]

Satisfaction of Interpreted Sentences

To explicate the satisfaction relation for interpreted open-sentences, I will again need to characterize the kind of sequences which can enter into this relation. In this section, let θ be an English predicate of degree 1. Then a θ-sequence will be a denumerable sequence of objects each of which satisfies θ. (In much of what follows, θ will be the English predicate that fixes the domain of the relevant quantifiers, so the elements of the θ-sequence will just be the things in the range of the quantifiers.) As before, such a sequence may contain repetitions. Any denumerable sequence which is such that $б_k$ occurs in place $P(\beta_k)$, $k = 1, 2, \ldots, n$, will be called a $[\beta_1/б_1, \beta_2/б_2, \ldots, \beta_n/б_n]$-sequence. A θ-sequence that is also a $[\beta_1/б_1, \beta_2/б_2, \ldots, \beta_n/б_n]$-sequence will be called a 'θ-$[\beta_1/б_1, \beta_2/б_2, \ldots, \beta_n/б_n]$-sequence'.

Suppose that **M*** is suitably interpreted so that every sentence of the language in which no odd-subscripted individual constant occurs will be meaningful and either true or false. Suppose further that ϕ is a sentence of that language and that, in ϕ, the distinct odd-subscripted individual constants $\beta_1, \beta_2, \ldots, \beta_n$ (and no other odd-subscripted individual constants) occur. Then we can give the traditional explanation of satisfaction: ϕ is *satisfied* by a θ-$[\beta_1/б_1, \beta_2/б_2, \ldots, \beta_n/б_n]$-sequence iff ϕ would have been true had $\beta_1, \beta_2, \ldots, \beta_n$ named $б_1, б_2, \ldots, б_n$, respectively.[9]

For the case in which no odd-subscripted individual constants occur in ϕ, ϕ is satisfied by every θ-sequence iff ϕ is true; ϕ is satisfied by no θ-sequence iff ϕ is false.

[8] See Mates 1972: ch. 2, sect. 4.

[9] Here, I give what is more or less the standard explication given by Quine in many of his logical works, modified to take account of the fact that I will be frequently making modal assertions of satisfaction. See, for example, Quine 1970: 36–8, or his even earlier Quine 1959: 90.

Note

(1) If a θ-$[\beta_1/\text{б}_1, \beta_2/\text{б}_2, \ldots, \beta_n/\text{б}_n]$-sequence satisfies a sentence ϕ, then every θ-$[\beta_1/\text{б}_1, \beta_2/\text{б}_2, \ldots, \beta_n/\text{б}_n]$-sequence that exists satisfies ϕ.

(2) It should be emphasized that, from the Actualist's perspective, no sequence $(\text{б}_1, \text{б}_2, \ldots, \text{б}_n, \ldots)$ actually exists if each of the elements $\text{б}_1, \text{б}_2, \ldots, \text{б}_n, \ldots$ does not actually exist—there are no merely possible objects to serve as the required elements of the sequence. Similarly, just because there could have been objects $\text{б}_1, \text{б}_2, \ldots, \text{б}_n$, which satisfied some condition ϕ, one cannot conclude, according to the Actualist's principles, that there exist any objects that satisfy ϕ. In this respect, the Actualist's logic is weaker than the Possibilist's. The Actualist's principles prevent one from making certain existential inferences that the Possibilist's position licenses. In my proof of the Fundamental Theorem, to follow, I shall restrict my reasoning so as to satisfy the above restraints. In this respect, my reasoning will be *consistent with Actualism*, although it will not presuppose the truth of Actualism. Thus, in the reasoning to follow, from the hypothesis that there could have been objects $\text{б}_1, \text{б}_2, \ldots, \text{б}_n$ which satisfied some condition ϕ, one will not be able to infer that there exist any objects which satisfy ϕ. Of course, in thus restricting the reasoning to Actualistically acceptable principles, one is not contravening Possibilistic principles of reasoning.

(3) Now, suppose that $\text{б}_1, \text{б}_2, \ldots, \text{б}_n, \ldots$ actually exist. Can we infer that the sequence $(\text{б}_1, \text{б}_2, \ldots, \text{б}_n, \ldots)$ actually exists? From the perspective of mathematical Platonism (or Set Theoretic Realism), we could. From my point of view, sequences are fictions. Strictly speaking, sequences do not exist. (Indeed, no mathematical objects exist.) However, as I indicated earlier, I make use of the fiction of mathematical objects in this section of the book for purposes of simplicity. (Later, I will explain the rationale for reasoning in this way.) And it is part of this fiction that when the objects $\text{б}_1, \text{б}_2, \ldots, \text{б}_n, \ldots$ actually exist, so also does the sequence $(\text{б}_1, \text{б}_2, \ldots, \text{б}_n, \ldots)$. Thus, in the reasoning to follow, we have the principle that the sequence $(\text{б}_1, \text{б}_2, \ldots, \text{б}_n, \ldots)$ actually exists iff the elements $\text{б}_1, \text{б}_2, \ldots, \text{б}_n, \ldots$ actually exist. Similarly, if the world could have been such that the objects $\text{б}_1, \text{б}_2, \ldots, \text{б}_n, \ldots$ existed, then we are allowed to conclude that the world could have been such that the sequence $(\text{б}_1, \text{б}_2, \ldots, \text{б}_n, \ldots)$ existed.

NL Proto-Interpretations

These are specified by interpreting the connectives and operators in the way required for NL proto-interpretations of **MSC**:

[i] The logical connectives are to have their standard truth functional sense;

and

[ii] the modal operators are to be understood according to the intuitive analysis given earlier by [T-d] and [T-b].

In addition, the meanings of the quantifiers, individual constants, and predicates are given in the way described above by:

(1) selecting an English predicate of degree one to delimit the domain of the quantifiers to just those things that satisfy the predicate;
(2) assigning to each even-subscripted individual constant the sense of a name or rigidly denoting definite description of an actual object that satisfies the English predicate that gives sense to the quantifiers;

and

(3) designating the meaning of each predicate θ to be that of an assigned English predicate of degree n (where n is the degree of θ).

Comment. NL proto-interpretations enable us to use **M*** to express thoughts directly. The Barwise–Etchemendy computer program "Tarski's World" provides us with a concrete example of one such "interpreted" formal language which can be used to express thoughts directly: beginning logic students learn very quickly to use the "interpreted" formal language of "Tarski's World" to describe three-dimensional geometric "worlds", and many students are able to achieve sufficient proficiency with the language to be able to describe "worlds" without having to go through a process of translating English sentences into the formal language. This language of "Tarski's World" can be regarded as a first-order language that has been given an NL proto-interpretation.

Some Notation

If $\mathcal{I}$ is an NL proto-interpretation, then I shall use the notation '$\mathcal{I}(\mathbf{d})$' to refer to the English predicate assigned by $\mathcal{I}$ to delimit the domain of the quantifiers. Another notational convention I use here—one which is an extension of a convention I made use of in the two previous chapters—is '$[\theta/\mathcal{I}]$': thus, if θ is a predicate of **M***, then '$[\theta/\mathcal{I}]$' refers to the predicate θ, understood to have the meaning given to it by $\mathcal{I}$. It follows that $[\theta/\mathcal{I}]$ is not the same predicate as the English predicate assigned by $\mathcal{I}$ to θ, but the two predicates will have the same meaning. I shall use 'ext$[\theta/\mathcal{I}]$' to refer to the extension of $[\theta/\mathcal{I}]$. Similarly, if α is an individual constant, then '$[\alpha/\mathcal{I}]$' refers to α, understood to have the meaning assigned to it by $\mathcal{I}$, and 'den$[\alpha/\mathcal{I}]$' will refer to the denotation of $[\alpha/\mathcal{I}]$. Also in line with the conventions of the preceding chapter, if θ is a sentence of **M***, then '$[\theta/\mathcal{I}]$' refers to θ, understood to have the meaning given to it by $\mathcal{I}$.

To illustrate these notational devices, suppose that we have an NL proto-interpretation, say $\mathcal{H}$, with the following assignments:

> **d**: ① is an actual person
> M^1: ① is male
> b: Bob Dole
> c: the 1996 Republican presidential nominee

Then $\mathcal{H}(\mathbf{d})$ is the predicate '① is an actual person'; $[M^1/\mathcal{H}]$ is the unary predicate 'M^1' interpreted to have the same meaning as '① is male'; ext$[M^1/\mathcal{H}]$ is the set of all males; $[b/\mathcal{H}]$ is the individual constant 'b', interpreted to have the sense of the name 'Bob Dole'; den$[b/\mathcal{H}]$ = den$[c/\mathcal{H}]$ = the 1996 Republican presidential nominee; $[M^1b/\mathcal{H}]$ is the atomic sentence 'M^1b', interpreted to have the meaning of 'Bob Dole is male'; and $[(\exists x)M^1x/\mathcal{H}]$ is true iff there is an actual person who is male.

Representations via Bijections

Suppose that $\mathcal{I}$ is an NL proto-interpretation and that $\mathbf{I} = \langle \mathbf{W}, \mathbf{D}, \mathbf{w^*}, \mathbf{d}, \mathbf{r}, \mathbf{e} \rangle$ is an S5 interpretation of $\mathbf{M^*}$. Suppose also that, for some $w \in \mathbf{W}$, the world had been such that there was a bijection f_w, from $\mathbf{d}(w)$ to what was the extension of $\mathcal{I}(\mathbf{d})$, such that:

(1) For every predicate θ of any degree n, and for every n-tuple of members of $\mathbf{d}(w)$, $\langle d_1, d_2, \ldots, d_n \rangle$:

$$\langle d_1, d_2, \ldots, d_n \rangle \in \mathbf{e}(\theta, w)$$
$$\text{iff } \langle \mathrm{f}_w(d_1), \mathrm{f}_w(d_2), \ldots, \mathrm{f}_w(d_n) \rangle \in \text{ext}[\theta/\mathcal{I}];$$

and

(2) for every even-subscripted individual constant α such that $\mathbf{r}(\alpha) \in \mathbf{d}(w)$:

$$\mathrm{f}_w(\mathbf{r}(\alpha)) = \text{den}[\alpha/\mathcal{I}].$$

In that case, it would be natural to regard $\langle \mathbf{d}(w), \mathbf{r}(\alpha), \mathbf{e}(\phi, w), \mathcal{I} \rangle$ as representing, via f_w, how the world was, *in so far as what was of concern were the domains of the quantifiers, the denotations of the individual constants, and the extensions of the predicates*, when these are interpreted as specified by $\mathcal{I}$. Thus, I shall write:

$$\langle \mathbf{d}(w), \mathbf{r}(\alpha), \mathbf{e}(\phi, w) \rangle_{\mathcal{I}} \text{ represents the world via } \mathrm{f}_w$$

as short for:

f_w is a bijection, from $\mathbf{d}(w)$ to the extension of $\mathcal{I}(\mathbf{d})$, such that

(1) for every predicate θ of any degree n, and for every n-tuple of members of $\mathbf{d}(w)$, $\langle d_1, d_2, \ldots, d_n \rangle$:

$<d_1, d_2, \ldots, d_n> \in \mathbf{e}(\theta, w)$
iff $<\mathrm{f}_w(d_1), \mathrm{f}_w(d_2), \ldots, \mathrm{f}_w(d_n)> \in \mathrm{ext}[\theta/\mathcal{I}]$;

and

(2) for every even-subscripted individual constant α such that $\mathbf{r}(\alpha) \in \mathbf{d}(w)$:

$\mathrm{f}_w(\mathbf{r}(\alpha)) = \mathrm{den}[\alpha/\mathcal{I}]$.[10]

Hence, the expression 'the world could have been such that there was a function f_w via which $<\mathbf{d}(w), \mathbf{r}(\alpha), \mathbf{e}(\xi, w)>_{\mathcal{I}}$ represented the world' is to mean:

the world could have been such that there was a bijection f_w, from $\mathbf{d}(w)$ to the extension of $\mathcal{I}(\mathbf{d})$, such that

(1) for every predicate θ of any degree n, and for every n-tuple of members of $\mathbf{d}(w)$, $<d_1, d_2, \ldots, d_n>$:

$<d_1, d_2, \ldots, d_n> \in \mathbf{e}(\theta, w)$
iff $<\mathrm{f}_w(d_1), \mathrm{f}_w(d_2), \ldots, \mathrm{f}_w(d_n)> \in \mathrm{ext}[\theta/\mathcal{I}]$;

and

(2) for every even-subscripted individual constant α such that $\mathbf{r}(\alpha) \in \mathbf{d}(w)$:

$\mathrm{f}_w(\mathbf{r}(\alpha)) = \mathrm{den}[\alpha/\mathcal{I}]$.

With these definitions and conventions, I can now explain what it is for an NL proto-interpretation to conform to an S5 interpretation.

NL Proto-Interpretations that Conform to S5 Interpretations

$\mathcal{I}$ will be an NL proto-interpretation of **M*** conforming to S5 interpretation **I = <W, D, w*, d, r, e>** iff $\mathcal{I}$ is an NL proto-interpretation of **M*** which satisfies four conditions to be specified below. To prepare the reader for the complicated statements of conditions to follow, a few introductory explanatory comments will be given.

In order that an NL proto-interpretation conform to an S5 interpretation, the S5 interpretation must appropriately represent the logically relevant extensional elements that can be abstracted from the NL proto-interpretation. Recall the two conditions given to ensure conformity in the case of **MSC** of the previous chapter. Since the structures of **M*** are significantly more complicated than those of **MSC**, I needed more than two conditions to ensure conformity for the case at hand.

[10] When I say that $<\mathbf{d}(w), \mathbf{r}(\alpha), \mathbf{e}(\phi, w)>_{\mathcal{I}}$ represents the world, I mean to be saying, to be more accurate, that $<\mathbf{d}(w), \mathbf{r}(\alpha), \mathbf{e}(\phi, w)>_{\mathcal{I}}$ represents how the world is.

The following should give you some rough idea of what these conditions require: the first condition requires the NL proto-interpretation, $\mathscr{I}$, to be such that every possibility that the S5 interpretation represents is indeed a possibility. The second condition requires $\mathscr{I}$ to be such that there be no possibility that the S5 interpretation fails to represent. The other conditions are complicated and will be explained later. Here, then, are the four conditions:

[C-1] For every $w \in \mathbf{W}$, the world could have been such that there was a function via which $<\mathbf{d}(w), \mathbf{r}(\alpha), \mathbf{e}(\xi, w)>_{\mathscr{I}}$ represented the world.

[C-2] The world could not have been such that, for no $w \in \mathbf{W}$, was there a function via which $<\mathbf{d}(w), \mathbf{r}(\alpha), \mathbf{e}(\xi, w)>_{\mathscr{I}}$ represented the world.

[C-3] For all $\mathscr{o}_1, \mathscr{o}_2, \ldots, \mathscr{o}_n \in \mathbf{D}$ and for every $w, w_1, w_2, \ldots, w_n \in \mathbf{W}$ such that $\mathscr{o}_1 \in \mathbf{d}(w_1), \mathscr{o}_2 \in \mathbf{d}(w_2), \ldots, \mathscr{o}_n \in \mathbf{d}(w_n)$,

if every $\mathscr{o}_1, \mathscr{o}_2, \ldots, \mathscr{o}_n \in \mathbf{d}(w)$, then it is not the case that the world could have been such that there was a function f_1 via which $<\mathbf{d}(w_1), \mathbf{r}(\alpha), \mathbf{e}(\xi, w_1)>_{\mathscr{I}}$ represented the world; the world could have been such that there was a function f_2 via which $<\mathbf{d}(w_2), \mathbf{r}(\alpha), \mathbf{e}(\xi, w_2)>_{\mathscr{I}}$ represented the world; . . .; and the world could have been such that there was a function f_n via which $<\mathbf{d}(w_n), \mathbf{r}(\alpha), \mathbf{e}(\xi, w_n)>_{\mathscr{I}}$ represented the world, such that, had the world been such that there was a function via which $<\mathbf{d}(w), \mathbf{r}(\alpha), \mathbf{e}(\xi, w)>_{\mathscr{I}}$ represented the world, it is not the case that there would have also been a function g via which $<\mathbf{d}(w), \mathbf{r}(\alpha), \mathbf{e}(\xi, w)>_{\mathscr{I}}$ represented the world such that $g(\mathscr{o}_1) = f_1(\mathscr{o}_1)$, $g(\mathscr{o}_2) = f_2(\mathscr{o}_2), \ldots, g(\mathscr{o}_n) = f_n(\mathscr{o}_n)$;

and if, for some k, $\mathscr{o}_k \notin \mathbf{d}(w)$, then it is not the case that the world could have been such that there was a function f_k via which $<\mathbf{d}(w_k), \mathbf{r}(\alpha), \mathbf{e}(\xi, w_k)>_{\mathscr{I}}$ represented the world such that had the world been such that there was a function via which $<\mathbf{d}(\mathrm{w}), \mathbf{r}(\alpha), \mathbf{e}(\xi, w)>_{\mathscr{I}}$ represented the world, there would have been a function g via which $<\mathbf{d}(w), \mathbf{r}(\alpha), \mathbf{e}(\xi, w)>_{\mathscr{I}}$ represented the world such that, for some $x \in \mathbf{d}(w)$, $g(x) = f_k(\mathscr{o}_k)$.

[C-4] For every $w, w_1, w_2, \ldots, w_n \in \mathbf{W}$, for any atomic sentence θ consisting of a predicate of degree m immediately followed by m occurrences of individual constants, among which are occurrences of the odd-subscripted individual constants $\beta_1, \beta_2, \ldots, \beta_n$, and for all $\mathscr{o}_1, \mathscr{o}_2, \ldots, \mathscr{o}_n \in \mathbf{D}$, if there is a $u \in \mathbf{W}$ such that, had the world been such that there was a function via which $<\mathbf{d}(u), \mathbf{r}(\alpha), \mathbf{e}(\xi, u)>_{\mathscr{I}}$ represented the world, there would have been a function g via which $<\mathbf{d}(u), \mathbf{r}(\alpha), \mathbf{e}(\xi, u)>_{\mathscr{I}}$ represented the world such that some $\mathscr{I}(\mathbf{d})$-$[\beta_1/g(\mathscr{o}_1), \beta_2/g(\mathscr{o}_2), \ldots, \beta_n/g(\mathscr{o}_n)]$-sequence satisfied the condition:

> [$\theta/\mathcal{J}$] truly describes what would have been the case, had the world had been w,

> then had the world been such that there was a function h via which $<\mathbf{d}(w), \mathbf{r}(\alpha), \mathbf{e}(\xi, w)>_{\mathcal{J}}$ represented the world, for each $j \in \{1, 2, \ldots, n\}$, there exists an $x \in \mathbf{d}(w)$ such that $h(x) = f_j(\mathscr{b}_j)$.

As before, an NL proto-interpretation that conforms to an S5 interpretation will be said to be an *NL interpretation*. In other words, the NL interpretations are just the NL proto-interpretations that conform to some S5 interpretation.

Some Explanations

Consider the phrase:

> the world could have been such that there was a function f via which $<\mathbf{d}(w), \mathbf{r}(\alpha), \mathbf{e}(\xi, w)>_{\mathcal{J}}$ represented the world.

Such phrases occur many times in the above statements of the C-conditions. The displayed phrase functions like a quantifier. Like a quantifier, it has a term, 'f', occurring in it that operates like a bound variable, and it has something like a scope. To make explicit the quantifier-like nature of such phrases, one could always append to them the phrase 'such that' to signal the fact that what follows is logically linked to the quantifier-like phrase. Frequently, I omit the 'such that' primarily for the sake of brevity.

The following comments should aid the reader to appreciate the significance of the above conditions. [C-1] and [C-2] are clearly analogues of conditions [i] and [ii] of the definition of 'NL proto-interpretation that conforms to an S5 interpretation of **MSC**'. [C-2] could also be expressed: *no matter how the world might have been, for some* $w \in \mathbf{W}$, *there would have been a function* f_w *via which* $<\mathbf{d}(w), \mathbf{r}(\alpha), \mathbf{e}(\xi, w)>_{\mathcal{J}}$ *represented the world.*

Conditions [C-3] and [C-4] are somewhat difficult to grasp and remember on first reading. It is useful to have an abbreviated version to facilitate understanding and retention. A more perspicuous statement of this condition can be given by making use of the following definitions and abbreviations.

[Definition]

> The expression '*representing function*' will be used to refer to any bijection f which is such that, for some $w \in \mathbf{W}$, $<\mathbf{d}(w), \mathbf{r}(\alpha), \mathbf{e}(\xi, w)>_{\mathcal{J}}$ represents the world via f. f will be called a '*representing function for w*'.

[Abbreviation 1]

> the world could have been such that there was a function f via which $<\mathbf{d}(w), \mathbf{r}(\alpha), \mathbf{e}(\xi, w)>_{\mathcal{J}}$ represented the world

is to be abbreviated:

there could have been a representing function f for w.

[Abbreviation 2]

had the world been such that there was a function via which $<\mathbf{d}(w), \mathbf{r}(\alpha), \mathbf{e}(\xi, w)>_{\mathscr{I}}$ represented the world

is to be abbreviated:

had the world been w.

[Abbreviation 3]

there would have been a function g via which $<\mathbf{d}(w), \mathbf{r}(\alpha), \mathbf{e}(\xi, w)>_{\mathscr{I}}$ represented the world

is to be abbreviated:

some g would have been a representing function for w.

[Abbreviation 4]

it is not the case that there would have been a function g via which $<\mathbf{d}(w), \mathbf{r}(\mathrm{a}), \mathbf{e}(\mathrm{x}, w)>_{\mathscr{I}}$ represented the world

is to be abbreviated:

no g would have been a representing function for w.

Then condition [C-3] can be written:

[C-3] (The Consistency Condition): For all $б_1, б_2, \ldots, б_n \in \mathbf{D}$ and for every $w, w_1, w_2, \ldots, w_n \in \mathbf{W}$ such that $б_1 \in \mathbf{d}(w_1), б_2 \in \mathbf{d}(w_2), \ldots, б_n \in \mathbf{d}(w_n)$,

if $\{б_1, б_2, \ldots, б_n\} \subseteq \mathbf{d}(w)$, then it is not the case that there could have been a representing function f_1 for w_1; . . .; there could have been a representing function f_n for w_n; such that, had the world been w, no g would have been a representing function for w such that $\mathrm{g}(б_1) = \mathrm{f}_1(б_1)$, $\mathrm{g}(б_2) = \mathrm{f}_2(б_2), \ldots, \mathrm{g}(б_n) = \mathrm{f}_n(б_n)$;

and if, for some k, $б_k \notin \mathbf{d}(w)$, then it is not the case that there could have been a representing function f_k for w_k such that, had the world been w, there would have been a representing function g for w such that, for some $x \in \mathbf{d}(w)$, $\mathrm{g}(x) = \mathrm{f}_k(б_k)$.

[C-4] (The Serious Actualism Condition): For every $w, w_1, w_2, \ldots, w_n \in \mathbf{W}$, for any atomic sentence θ consisting of a predicate of degree m immediately followed by m occurrences of individual constants, among which are occurrences of the odd-subscripted individual constants $\beta_1, \beta_2, \ldots, \beta_n$, and for all $б_1, б_2, \ldots, б_n \in \mathbf{D}$, if there is a $u \in \mathbf{W}$ such that, had the world been u, there would have been a representing function g for u such that some $\mathcal{I}(\mathbf{d})$-[$\beta_1/g(\mathit{б}_1)$, $\beta_2/g(\mathit{б}_2)$, . . ., $\beta_n/g(\mathit{б}_n)$]-sequence satisfied the condition:

> [$\theta/\mathcal{I}$] truly describes what would have been the case, had the world had been w,

then for any representing function h for w, and for each $j \in \{1, 2, \ldots, n\}$, there exists an $x \in \mathbf{d}(w)$ such that $h(x) = f_j(\mathit{б}_j)$.

These abbreviated versions of the above conditions are somewhat misleading, since they may make it appear that worlds are being presupposed, but I am confident that they will make it easier to grasp and retain the basic ideas.

As was indicated above, [C-3] is a consistency condition of sorts. This condition can be more easily grasped by way of contrast with another consistency condition that I hoped to use as one of the conditions that NL proto-interpretations had to satisfy to be an NL interpretation. This condition was:

> [C-3*] For all $\mathit{б}_1, \mathit{б}_2, \ldots, \mathit{б}_n \in \mathbf{D}$ and for any $w_1, w_2, \ldots, w_n \in \mathbf{W}$ such that $\mathit{б}_1 \in \mathbf{d}(w_1)$, $\mathit{б}_2 \in \mathbf{d}(w_2)$, . . ., $\mathit{б}_n \in \mathbf{d}(w_n)$, there could have been a representing function f_1 for w_1; . . .; there could have been a representing function f_n for w_n; such that, had the world been w, some g would have been a representing function for w such that, if every $\mathit{б}_1, \mathit{б}_2, \ldots, \mathit{б}_n \in \mathbf{d}(w)$, $g(\mathit{б}_1) = f_1(\mathit{б}_1)$, $g(\mathit{б}_2) = f_2(\mathit{б}_2)$, . . ., $g(\mathit{б}_n) = f_n(\mathit{б}_n)$; and if, for some k, $\mathit{б}_k \notin \mathbf{d}(w)$, then for no $x \in \mathbf{d}(w)$ is $g(x) = f_k(\mathit{б}_k)$.

To see how [C-3*] would function, let us consider the situation from the Possibilist perspective. Thus, I wanted to require that, in order for an NL proto-interpretation to conform to **I**, there would have to be a bijection from **D** to the set of all possible objects that satisfy the English predicate specifying the domain. Such a requirement, of course, would make no sense from the point of view of Actualism. So to obtain a kind of Actualistic substitute for such a bijection, I wanted to require, in effect, that, for each $w \in \mathbf{W}$, there would have been a representing function that mapped $\mathbf{d}(w)$ one–one with the set of those things that would have satisfied the predicate specifying the domain, had the world been such that $\langle \mathbf{d}(w), \mathbf{r}(\alpha), \mathbf{e}(\xi, w)\rangle_{\mathcal{I}}$ represented the world via this representing function. But this substitute would not be adequate, since these different representing functions are not required to be *consistent with each other*: One representing function might map an element of **D** to one thing, and another might map this same element to a different thing. Condition [C-3*] addresses this possibility. Thus, given representing functions f_k for w_k, $k = 1, 2, \ldots, n$, [C-3*] requires that, had the world been w, there would also have been a representing function g that is "consistent with f_k" in the sense that, if all the $\mathscr{b}_k \in \mathbf{d}(w)$, then g maps each $\mathscr{b}_k$ to the same thing that f_k would have mapped $\mathscr{b}_k$ to (had the world been such that f_k represented w_k). On the other hand, if some $\mathscr{b}_k \notin \mathbf{d}(w)$, [C-3*] requires that, for that case, nothing in $\mathbf{d}(w)$ be mapped by g to what f_k would have mapped $\mathscr{b}_k$ to.

For an illustration of how condition [C-3*] works, consider an S5 interpretation **I**, only part of which is given by Table 2.

Table 2

	w_1	w_2	w_3	w_4	w_5
d:	{1, 2, 3}	{1, 2, 3}	{1}	{2}	∅
F^2:	{<1, 2>}	{<2, 1>}	∅	∅	∅
			d: 3		

Let $\mathscr{I}$ be an NL proto-interpretation, the essentials of which are given below:

> **d**: ① is one of the triple, Bill Clinton, Hillary Rodham Clinton, or Bob Dole
> M^2: ① is Bill Clinton and ② is Hillary Rodham Clinton
> *d:* Bob Dole

The question to be answered is whether $\mathscr{I}$ is an NL proto-interpretation conforming to **I**. Clearly **I** can be filled out so as to have Condition [C-2] satisfied. [C-1] is also clearly satisfied. But how about [C-3*]? Notice that there could have been a representing function for w_1, namely, the function that takes 1 to Bill Clinton, 2 to Hillary Clinton, and 3 to Bob Dole; and there could have been a representing function for w_2, namely, the function that takes 1 to Hillary Clinton, 2 to Bill Clinton, and 3 to Bob Dole. However, had the world been w_2, there would not have been any representing function for w_2 that was "consistent" with any representing function for w_1, that is, no representing function g for w_2 would have been such that g(1) = Bill Clinton and g(2) = Hillary Clinton. Hence, [C-3*] is not satisfied, and $\mathscr{I}$ is not an NL proto-interpretation conforming to **I**.

Unfortunately, $\mathscr{I}$ was unable to find a reasonably straightforward proof of the sort of connecting theorem I wanted when I used [C-3*]. So I replaced [C-3*] with [C-3]. [C-3] is a much more stringent condition than is [C-3*]. It requires that for any representing function f_1 for w_1, . . ., any representing function f_n for w_n, there is a representing function g for w such that, if every $\mathscr{b}_1, \mathscr{b}_2, \ldots, \mathscr{b}_n \in \mathbf{d}(w)$, $g(\mathscr{b}_1) = f_1(\mathscr{b}_1)$, $g(\mathscr{b}_2) = f_2(\mathscr{b}_2)$, . . ., $g(\mathscr{b}_n) = f_n(\mathscr{b}_n)$. Since this stronger condition makes it much more difficult to construct NL interpretations, using [C-3*] instead of [C-3] would be preferable. However, weakening the condition in this way would undoubtedly require a much more complicated proof than what is to be given below. My use of condition [C-3] instead of [C-3*] will be discussed in more detail in the next chapter.

I turn now to the role that condition [C-4] is playing. Since the semantics of **M*** has been specified to conform to Serious Actualism, we need to require that any NL proto-interpretation conforming to **I** reflect this feature of the formal semantics. Condition [C-4] expresses such a requirement. Thus, any NL proto-interpretation conforming to an S5 interpretation **I** must reflect the Serious Actualistic perspective that the semantics of **M*** requires of **I**. More specifically, [C-4] requires, in effect, that, had the world been such that some member of the set $\{f_1(\mathscr{b}_1), f_2(\mathscr{b}_2), \ldots, f_n(\mathscr{b}_n)\}$ did not exist, then $[\theta/\mathscr{I}]$ would not have been satisfied by any $\mathscr{I}(\mathbf{d})$-$[\beta_1/f_1(\mathscr{b}_1), \beta_2/f_2(\mathscr{b}_2), \ldots, \beta_n/f_n(\mathscr{b}_n)]$-sequence. Since θ is an atomic sentence, [C-4] requires that, in the envisaged situation, the interpreted version of the predicate **Φ** be satisfied by only *n*-tuples of objects that would have existed.

Serious Actualism motivated the feature of the semantics of **M*** that distinguishes it from **M**. The reasoning to follow is consistent with Serious Actualism in so far as it does not conflict with Serious Actualism. But the proofs themselves do not presuppose Serious Actualism; the Possibilist should find the reasoning acceptable. Condition [C-4] is a requirement for NL proto-interpretations to *conform to* S5 interpretations of **M***—a Serious Actualistic system.

An Example of an NL Interpretation

Consider the S5 interpretation **J** given by Table 3.

Table 3

	w_1	w_2	w_3	w_4	w_5
d:	{1, 2}	{1, 2}	{1}	{2}	∅
M^2:	{<1, 2>, <2, 1>}	∅	∅	∅	∅
		b: 1			
		h: 2			

It can be seen that the sentence '$(\exists x)M^2b_3x$' is satisfied at w_1 under **J** by some **D**-$[b_3/1]$-sequence and also by some **D**-$[b_3/2]$-sequence. However, that sentence is not satisfied at w_2 under **J** by any **D**-$[b_3/1]$-sequence.

What follows gives (in outline form) an NL proto-interpretation $\mathscr{I}$:

d: ① is one of the pair, Bill Clinton or Hillary Rodham Clinton
M^2: ① is married to ②
b: Bill Clinton
h: Hillary Rodham Clinton.

To determine that all four conditions are satisfied by this proto-interpretation (and hence that the interpretation is an NL interpretation) may prove to be a helpful exercise. The reader can also verify that some $\mathcal{J}(\mathbf{d})$-[c_3/Hillary Rodham Clinton, d_3/Bill Clinton]-sequence satisfies $[M^2c_3d_3/\mathcal{J}]$.

5. The Fundamental Theorem

The Completeness and Soundness Theorems of first-order logic link two fundamental notions: model theoretic consequence and derivability. The fundamental theorem of the integral calculus links the indefinite integral with the definite integral. In this chapter, two quite distinct kinds of interpretations have been analysed. On the one hand, there are the S5 interpretations, which are purely mathematical in nature. They are mathematical models—objects of pure set theory—which can be regarded as representing how the world could have been in so far as the domains of the quantifiers, extensions of the predicates, and the referents of the individual constants are of concern. Natural Language interpretations, on the other hand, are much closer to what we ordinarily have in mind when we speak of "interpreting the symbols of a language": they provide the symbols of our logical language with meanings. However, as will be shown in this section, these two kinds of interpretations are related in an important and logically useful way. The crucial link between these two kinds of interpretations will be given by a theorem, which does for modal quantificational logic what Theorem [2] did for modal sentential logic. The theorem to follow is the fundamental theorem of the semantical system developed in this work.

The proof to follow will be given in detail. Why, it may be asked, do I give so much detail, when it is quite common for mathematical logicians to provide only sketches or bare outlines of their proofs? Here it needs to be mentioned that my proof will not be a proof in some generally accepted field of mathematics, such as set theory or algebra, in which the standards of proof are both widely known and accepted. I certainly cannot assume that most readers could reconstruct my reasoning on the basis of a few hints (e.g. "by mathematical induction"), especially given the easily overlooked subtleties involving subjunctive counterfactuals that this reasoning requires. Furthermore, given the aims of this work, there is a special reason for laying out the details of this proof. This proof shows the reader what modal principles I use to support the Fundamental Theorem—something that the reader needs to know in order to judge whether I, in some way, assume or presuppose the existence of possible worlds, in developing my anti-realistic view of modal logic. However, as was suggested in the previous chapter, some readers who find lengthy logical reasoning tedious and boring, will undoubtedly wish to skip the proof to follow and proceed directly to Section 5, where the consequences of this theorem are taken up.

The proof of the Fundamental Theorem proceeds by way of five lemmas. In what follows, every lemma and theorem should be understood to have the phrase 'For every S5 interpretation **I** = <**W**, **D**, **w***, **d**, **r**, **e**> of **M***, and for every NL proto-interpretation $\mathscr{I}$ conforming to **I**', prefixed.

Before beginning the proofs, it is useful to note the following "mini-lemma":

> For every $w \in \mathbf{W}$, and for every n-ary predicate ϕ, and for all $\delta_1, \delta_2, \ldots, \delta_n \in \mathbf{d}(w)$, had the world been w, then it would have been the case that, for any representing function f of w,
>
> either $<\mathrm{f}(\delta_1), \mathrm{f}(\delta_2), \ldots, \mathrm{f}(\delta_n)> \in \mathrm{ext}[\phi/\mathscr{I}]$
> or $<\mathrm{f}(\delta_1), \mathrm{f}(\delta_2), \ldots, \mathrm{f}(\delta_n)> \notin \mathrm{ext}[\phi/\mathscr{I}]$,
>
> and also
>
> either $<\mathrm{f}(\delta_1), \mathrm{f}(\delta_2), \ldots, \mathrm{f}(\delta_n)> \in \mathrm{ext}[\phi/\mathscr{I}]$
> or $<\mathrm{f}(\delta_1), \mathrm{f}(\delta_2), \ldots, \mathrm{f}(\delta_n)> \in \mathrm{ext}[-\phi/\mathscr{I}]$.

(I leave it as a simple exercise for the reader to verify that this is indeed the case.)

> **Lemma 1**. For every $w \in \mathbf{W}$, for every n-ary predicate ϕ and for all $\delta_1, \delta_2, \ldots, \delta_n \in \mathbf{d}(w)$, had the world been w, then it would have been the case that, for any f and g which were representing functions of w,
>
> $<\mathrm{f}(\delta_1), \mathrm{f}(\delta_2), \ldots, \mathrm{f}(\delta_n)> \in \mathrm{ext}[\phi/\mathscr{I}]$ iff
> $<\mathrm{g}(\delta_1), \mathrm{g}(\delta_2), \ldots, \mathrm{g}(\delta_n)> \in \mathrm{ext}[\phi/\mathscr{I}]$.

Lemma 1 tells us that, in the above described situation in which the world is w, in so far as one is only concerned about the extensions of predicates, it would not make any difference what function was the representing function of w. Roughly, in that situation, if f mapped an n-tuple of elements of $\mathbf{d}(w)$ to an n-tuple of objects that satisfied $[\phi/\mathscr{I}]$, then any representing function of w would do the same thing.

Proof. Had the world been w, then since $<\mathbf{d}(w), \mathbf{r}(\alpha), \mathbf{e}(\phi, w)>_{\mathscr{I}}$ represented the world via both f and g, it would have been the case that:

$<\delta_1, \delta_2, \ldots, \delta_n> \in \mathbf{e}(\phi, w)$ iff
$<\mathrm{f}(\delta_1), \mathrm{f}(\delta_2), \ldots, \mathrm{f}(\delta_n)> \in \mathrm{ext}[\phi/\mathscr{I}]$;

and:

$<\delta_1, \delta_2, \ldots, \delta_n> \in \mathbf{e}(\phi, w)$ iff
$<\mathrm{g}(\delta_1), \mathrm{g}(\delta_2), \ldots, \mathrm{g}(\delta_n)> \in \mathrm{ext}[\phi/\mathscr{I}]$.

Hence, had the world been w, it would have been the case that:

$$<f(\delta_1), f(\delta_2), \ldots, f(\delta_n)> \in ext[\phi/\mathscr{I}] \text{ iff}$$
$$<g(\delta_1), g(\delta_2), \ldots, g(\delta_n)> \in ext[\phi/\mathscr{I}].$$

Comment. In the following discussion, the reader will come across such statements as:

[A] some $\mathscr{I}(\mathbf{d})$-$[\beta_1/g(\delta_1), \beta_2/g(\delta_2), \ldots, \beta_n/g(\delta_n)]$-sequence satisfied the condition: $[\theta/\mathscr{I}]$ truly describes what would have been the case, had the world been w;

and:

[B] some $\mathscr{I}(\mathbf{d})$-$[\beta_1/g(\delta_1), \beta_2/g(\delta_2), \ldots, \beta_n/g(\delta_n)]$-sequence would have satisfied $[\theta/\mathscr{I}]$, had the world been w.

Are these just two different ways of saying the same thing? If not, what is the difference? Statement [B] implies that, had the world been w, there would have existed a $\mathscr{I}(\mathbf{d})$-$[\beta_1/g(\delta_1), \beta_2/g(\delta_2), \ldots, \beta_n/g(\delta_n)]$-sequence—after all, how could a $\mathscr{I}(\mathbf{d})$-$[\beta_1/g(\delta_1), \beta_2/g(\delta_2), \ldots, \beta_n/g(\delta_n)]$-sequence satisfy $[\theta/\mathscr{I}]$ if it didn't exist? But [A] implies no such thing. Thus, consider the following statement:

> I could have had a son who would have satisfied 'x does not exist', had the world been such that there was no oxygen.

That may seem to be a paradoxical thing to say, since it strongly suggests that there could be someone who satisfied the condition: 'x does not exist.' On the other hand, the statement

> I could have had a son who was such that, taking the variable 'x' (in the quoted open-sentence to follow) to stand for this son, 'x does not exist' truly describes what would have been the case, had the world been such that there was no oxygen at all

does not have the paradoxical implications of the earlier one. Similarly, [A] does not imply that, had the world been w, there would have been a $\mathscr{I}(\mathbf{d})$-$[\beta_1/g(\delta_1), \beta_2/g(\delta_2), \ldots, \beta_n/g(\delta_n)]$-sequence that satisfied some condition. What [A] comes to is simply:

> $[\theta/\mathscr{I}]$ is such that, taking β_1 to denote $g(\delta_1)$, β_2 to denote $g(\delta_2)$, ..., β_n to denote $g(\delta_n)$, results in a sentence that truly describes what would have been the case, had the world been w.

And there is nothing paradoxical in that.[11]

[11] In Chapter 8, Section 3, I explain the distinction between [A] and [B] in terms of possible worlds semantics.

Notice that, in those cases in which, had the world been w, there would have existed $f(d_1)$, $f(d_2)$, . . ., $f(d_n)$—as is the case in Lemmas 2 and 4—we need not distinguish forms [A] and [B]: either form will do. Why then do I use form [A] for these two lemmas? The Fundamental Theorem is proved by induction on the indexes of the formulas of the formal language, and Lemma 4 gives us the index 0 case. Of course, for the higher index cases, I have to use form [A], so I use the form [A] version for Lemma 4 to make the formulation consistent with the other cases.

> **Lemma 2**. For all $w \in \mathbf{W}$, for all $d_1, d_2, \ldots, d_n \in \mathbf{d}(w)$, and for every *atomic* sentence θ, if $\beta_1, \beta_2, \ldots, \beta_n$ are the n odd-subscripted individual constants that occur in θ, then the following holds:
>
> *If* θ is satisfied at w under $\mathbf{I}$ by some $\mathbf{D}$-$[\beta_1/d_1, \beta_2/d_2, \ldots, \beta_n/d_n]$-sequence, *then*, for some $z \in \mathbf{W}$, had the world been z, there would have been a representing function g for z such that some $\mathcal{I}(\mathbf{d})$-$[\beta_1/g(d_1), \beta_2/g(d_2), \ldots, \beta_n/g(d_n)]$-sequence satisfied the condition: $[\theta/\mathcal{I}]$ truly describes what would have been the case, had the world been w.
>
> *If* θ is not satisfied at w under $\mathbf{I}$ by some $\mathbf{D}$-$[\beta_1/d_1, \beta_2/d_2, \ldots, \beta_n/d_n]$-sequence, *then*, for some $z \in \mathbf{W}$, had the world been z, there would have been a representing function g for z such that some $\mathcal{I}(\mathbf{d})$-$[\beta_1/g(d_1), \beta_2/g(d_2), \ldots, \beta_n/g(d_n)]$-sequence satisfied the condition: $[-\theta/\mathcal{I}]$ truly describes what would have been the case, had the world been w.

Proof: By hypothesis, θ is atomic, say $\Phi\alpha_1\alpha_2 \ldots \alpha_m$, where each odd-subscripted individual constant β_k must occur among the $\alpha_1, \alpha_2, \ldots, \alpha_m$. Now, had the world been w, then, for every m-tuple of objects from $\mathbf{d}(w)$, say $\langle d_1, d_2, \ldots, d_m\rangle$, and any f that was a representing function for w, it would have been the case that:

$$\langle d_1, d_2, \ldots, d_m\rangle \in \mathbf{e}(\Phi, w) \text{ iff}$$
$$\langle f(d_1), f(d_2), \ldots, f(d_m)\rangle \in \text{ext}[\Phi/\mathcal{I}].$$

Also, from condition [C-1], we know that, had the world been w, there would have been a representing function f for w. So assume that

[A] θ is satisfied at w under $\mathbf{I}$ by some $\mathbf{D}$-$[\beta_1/d_1, \beta_2/d_2, \ldots, \beta_n/d_n]$-sequence.

We can infer that:

[B] Had the world been w, there would have been a representing function g for w such that $[\theta/\mathcal{I}]$ was satisfied by some $\mathcal{I}(\mathbf{d})$-$[\beta_1/g(d_1), \beta_2/g(d_2), \ldots, \beta_n/g(d_n)]$-sequence.

(Notice here the role of the requirement that every representing function map every even-subscripted individual constant α, which is such that $\mathbf{r}(\alpha) \in \mathbf{d}(w)$, to den[$\alpha/\mathcal{I}$]—without this requirement, [A] could not be inferred.)

It follows (by existential generalization) that

> [C] For some $z \in \mathbf{W}$, had the world been z, there would have been a representing function g for z such that some $\mathcal{I}(\mathbf{d})$-[β_1/g(δ_1), β_2/g(δ_2), . . ., β_n/g(δ_n)]-sequence satisfied the condition: [$\theta/\mathcal{I}$] truly describes what would have been the case, had the world been w.

Suppose this time that

> [D] θ is not satisfied at w under **I** by some **D**-[β_1/δ_1, β_2/δ_2, . . ., β_n/δ_n]-sequence.

Then we can infer (using the mini-lemma) that

> [E] Had the world been w, there would have been a representing function g for w such that [$-\theta/\mathcal{I}$] was satisfied by some $\mathcal{I}(\mathbf{d})$-[β_1/g(δ_1), β_2/g(δ_2), . . ., β_n/g(δ_n)]-sequence.

Thus,

> [F] For some $z \in \mathbf{W}$, had the world been z, there would have been a representing function g for z such that some $\mathcal{I}(\mathbf{d})$-[β_1/g(δ_1), β_2/g(δ_2), . . ., β_n/g(δ_n)]-sequence satisfied the condition: [$-\theta/\mathcal{I}$] truly describes what would have been the case, had the world been w.

> **Lemma 3**. For every $w_1, w_2 \in \mathbf{W}$, if there could be representing functions f_1 and f_2 of w_1 and w_2 respectively, then if $\delta_1 \in \mathbf{d}(w_1) \cap \mathbf{d}(w_2)$, $f_1(\delta_1) = f_2(\delta_1)$; and if $\delta_1, \delta_2 \in \mathbf{d}(w_1)$, where $\delta_1 \neq \delta_2$, then $f_1(\delta_1) \neq f_2(\delta_2)$.

Proof: Suppose that the antecedent is the case and that $\delta_1 \in \mathbf{d}(w_1) \cap \mathbf{d}(w_2)$. Then, had the world been w_2, there would have been a representing function g for w_2 such that $g(\delta_1) = f_1(\delta_1)$ and $g(\delta_1) = f_2(\delta_1)$. It follows that $f_1(\delta_1) = f_2(\delta_1)$.

Suppose this time that the antecedent is the case and that $\delta_1, \delta_2 \in \mathbf{d}(w_1)$, where $\delta_1 \neq \delta_2$. Then, had the world been w_1, there would have been a representing function g for w_1 such that $g(\delta_1) = f_1(\delta_1)$ and $g(\delta_2) = f_2(\delta_2)$. Since $\delta_1 \neq \delta_2$, it follows that $g(\delta_1) \neq g(\delta_2)$ and hence that $f_1(\delta_1) \neq f_2(\delta_2)$.

Comment. Lemma 3 makes it clear that, for any $w \in \mathbf{W}$, there could not have been more than one representing function for w. It also tells us that the representing functions of w_1 and w_2 respectively must agree on what they assign to any element the two domains have in common. From this, the first two corollaries of the lemma can easily be established:

Corollary 1. For all $w \in \mathbf{W}$, for all $\mathscr{b}_1, \mathscr{b}_2, \ldots, \mathscr{b}_n \in \mathbf{D}$, and for every sentence θ, if $\beta_1, \beta_2, \ldots, \beta_n$ are the n odd-subscripted individual constants that occur in θ, then the following holds:

If, for some $z \in \mathbf{W}$, had the world been z, there would have been a representing function g for z such that some $\mathscr{I}(\mathbf{d})$-$[\beta_1/g(\mathscr{b}_1), \beta_2/g(\mathscr{b}_2), \ldots, \beta_n/g(\mathscr{b}_n)]$-sequence satisfied the condition: $[\theta/\mathscr{I}]$ truly describes what would have been the case, had the world been w;

then it is not the case that, for some $z \in \mathbf{W}$, had the world been z, there would have been a representing function g for z such that some $\mathscr{I}(\mathbf{d})$-$[\beta_1/g(\mathscr{b}_1), \beta_2/g(\mathscr{b}_2), \ldots, \beta_n/g(\mathscr{b}_n)]$-sequence did not satisfy the condition: $[\theta/\mathscr{I}]$ truly describes what would have been the case, had the world been w.

Comment 1. The above corollary does not require θ to be atomic, nor does it require $\mathscr{b}_1, \mathscr{b}_2, \ldots, \mathscr{b}_n$ to be elements of $\mathbf{d}(w)$. For such cases, it is important to distinguish claims like:

some $\mathscr{I}(\mathbf{d})$-$[\beta_1/g(\mathscr{b}_1), \beta_2/g(\mathscr{b}_2), \ldots, \beta_n/g(\mathscr{b}_n)]$-sequence satisfied the condition: $[\theta/\mathscr{I}]$ truly describes what would have been the case, had the world been w

from claims like:

some $\mathscr{I}(\mathbf{d})$-$[\beta_1/g(\mathscr{b}_1), \beta_2/g(\mathscr{b}_2), \ldots, \beta_n/g(\mathscr{b}_n)]$-sequence would have satisfied $[\theta/\mathscr{I}]$, had the world been w.

As was pointed out earlier, the latter implies that the elements of the $\mathscr{I}(\mathbf{d})$-$[\beta_1/g(\mathscr{b}_1), \beta_2/g(\mathscr{b}_2), \ldots, \beta_n/g(\mathscr{b}_n)]$-sequence would have existed, had the world been w; not so, the former.

Corollary 2. For all $w \in \mathbf{W}$, for all $\mathscr{b}_1, \mathscr{b}_2, \ldots, \mathscr{b}_n \in \mathbf{D}$, and for every sentence θ, if $\beta_1, \beta_2, \ldots, \beta_n$ are the n odd-subscripted individual constants that occur in θ, then the following holds:

If, for some $z \in \mathbf{W}$, had the world been z, there would have been a representing function g of z such that some $\mathscr{I}(\mathbf{d})$-$[\beta_1/g(\mathscr{b}_1), \beta_2/g(\mathscr{b}_2), \ldots, \beta_n/g(\mathscr{b}_n)]$-sequence satisfied the condition: $[\theta/\mathscr{I}]$ truly describes what would have been the case, had the world been w;

then it is not the case that, for some $z \in \mathbf{W}$, had the world been z, there would have been a representing function g for z such that some $\mathscr{I}(\mathbf{d})$-$[\beta_1/g(\mathscr{b}_1), \beta_2/g(\mathscr{b}_2), \ldots, \beta_n/g(\mathscr{b}_n)]$-sequence satisfied the condition: $[-\theta/\mathscr{I}]$ truly describes what would have been the case, had the world been w.

Comment 2. A word should be said in explanation of expressions of the form 'had the world been w, then ϕ'. To state that *the world had been* w is to give a certain amount of information about how the world had been—much about how the world had been is thus fixed—but not everything. As I use the expression 'had the world been w, then ϕ', it is true iff, had the world been w, then no matter how else the world might have been, so long as it was w, ϕ would have been the case.

Lemma 4. For all $w \in \mathbf{W}$, for all $\delta_1, \delta_2, \ldots, \delta_n \in \mathbf{d}(w)$, and for every *atomic* sentence θ, if $\beta_1, \beta_2, \ldots, \beta_n$ are the n odd-subscripted individual constants that occur in θ, then the following holds:

θ is satisfied at w under $\mathbf{I}$ by some $\mathbf{D}$-$[\beta_1/\delta_1, \beta_2/\delta_2, \ldots, \beta_n/\delta_n]$-sequence iff, for some $z \in \mathbf{W}$, had the world been z, there would have been a representing function g for z such that some $\mathscr{I}(\mathbf{d})$-$[\beta_1/\mathrm{g}(\delta_1), \beta_2/\mathrm{g}(\delta_2), \ldots, \beta_n/\mathrm{g}(\delta_n)]$-sequence satisfied the condition: $[\theta/\mathscr{I}]$ truly describes what would have been the case, had the world been w.

Proof: This follows directly from Lemma 2 and Corollary 2 to Lemma 3.

Lemma 5. For every $w \in \mathbf{W}$, for every $\delta_1, \delta_2, \ldots, \delta_n \in \mathbf{D}$, and for every sentence θ, if $\beta_1, \beta_2, \ldots, \beta_n$ are the n odd-subscripted individual constants that occur in θ, the following holds:

Either, for some $z \in \mathbf{W}$, had the world been z, there would have been a representing function g for z such that some $\mathscr{I}(\mathbf{d})$-$[\beta_1/\mathrm{g}(\delta_1), \beta_2/\mathrm{g}(\delta_2), \ldots, \beta_n/\mathrm{g}(\delta_n)]$-sequence satisfied the condition: $[\theta/\mathscr{I}]$ truly describes what would have been the case, had the world been w;

or, for some $z \in \mathbf{W}$, had the world been z, there would have been a representing function g for z such that some $\mathscr{I}(\mathbf{d})$-$[\beta_1/\mathrm{g}(\delta_1), \beta_2/\mathrm{g}(\delta_2), \ldots, \beta_n/\mathrm{g}(\delta_n)]$-sequence satisfied the condition: $[-\theta/\mathscr{I}]$ truly describes what would have been the case, had the world been w.

Comment. The above lemma is an implication of what is called 'conditional excluded middle', $(\phi \Rightarrow \chi) \vee (\phi \Rightarrow -\chi)$, the validity of which has been hotly disputed.[12] It needs to be emphasized that I am not presupposing that controversial law.

Proof (by induction on the index of θ). I start with the case in which θ is a sentence of index 0. Then θ is atomic, say $\boldsymbol{\Phi}\alpha_1\alpha_2 \ldots \alpha_m$, where each odd-subscripted individual constant β_k must occur among the $\alpha_1, \alpha_2, \ldots, \alpha_m$. Either $\{\delta_1, \delta_2, \ldots, \delta_n\} \subseteq \mathbf{d}(w)$ or $\{\delta_1, \delta_2, \ldots, \delta_n\}$ is not a subset of $\mathbf{d}(w)$.

I begin by supposing the former. Now excluded middle applies to θ in

[12] See Harper's Introduction to Harper, Stalnaker, and Pearce 1981: 4–11. Cf. Hajek and Ned 1994: 83–4.

so far as either θ is satisfied at w under **I** by some **D**-$[\beta_1/б_1, \beta_2/б_2, \ldots, \beta_n/б_n]$-sequence or θ is not satisfied at w under **I** by some **D**-$[\beta_1/б_1, \beta_2/б_2, \ldots, \beta_n/б_n]$-sequence. Suppose θ is satisfied at w under **I** by some **D**-$[\beta_1/б_1, \beta_2/б_2, \ldots, \beta_n/б_n]$-sequence. Then, by Lemma 2, we have:

> For some $z \in \mathbf{W}$, had the world been z, there would have been a representing function g for z such that some $\mathscr{I}(\mathbf{d})$-$[\beta_1/g(б_1), \beta_2/g(б_2), \ldots, \beta_n/g(б_n)]$-sequence satisfied the condition: $[\theta/\mathscr{I}]$ truly describes what would have been the case, had the world been w.

Suppose θ is not satisfied at w by such a sequence. Then, again by Lemma 2, we have:

> For some $z \in \mathbf{W}$, had the world been z, there would have been a representing function g for z such that some $\mathscr{I}(\mathbf{d})$-$[\beta_1/g(б_1), \beta_2/g(б_2), \ldots, \beta_n/g(б_n)]$-sequence satisfied the condition: $[-\theta/\mathscr{I}]$ truly describes what would have been the case, had the world been w.

Suppose, this time, that some member $б_k$ of the set $\{б_1, б_2, \ldots, б_n\}$ is not a member of $\mathbf{d}(w)$. I shall begin by assuming that:

[A] It is not the case that, for some $z \in \mathbf{W}$, had the world been z, there would have been a representing function g for z such that some $\mathscr{I}(\mathbf{d})$-$[\beta_1/g(б_1), \beta_2/g(б_2), \ldots, \beta_n/g(б_n)]$-sequence satisfied the condition: $[\theta/\mathscr{I}]$ truly describes what would have been the case, had the world been w.

Then,

[B] For every $z \in \mathbf{W}$, had the world been z, there would have been a representing function g for z such that there was no $\mathscr{I}(\mathbf{d})$-$[\beta_1/g(б_1), \beta_2/g(б_2), \ldots, \beta_n/g(б_n)]$-sequence that satisfied the condition: $[\theta/\mathscr{I}]$ truly describes what would have been the case, had the world been w.

But we also know that

[C] For some $z \in \mathbf{W}$, had the world been z, there would have been a representing function g for z such that some $\mathscr{I}(\mathbf{d})$-$[\beta_1/g(б_1), \beta_2/g(б_2), \ldots, \beta_n/g(б_n)]$-sequence existed.

From [B] and [C], we can infer

[D] For some $z \in \mathbf{W}$, had the world been z, there would have been a representing function g for z such that some $\mathscr{I}(\mathbf{d})$-$[\beta_1/g(б_1), \beta_2/g(б_2), \ldots, \beta_n/g(б_n)]$-sequence did not satisfy the condition: $[\theta/\mathscr{I}]$ truly describes what would have been the case, had the world been w.

Since θ is atomic, it is clear that such a $\mathcal{I}(\mathbf{d})$-[$\beta_1/g(\mathscr{b}_1)$, $\beta_2/g(\mathscr{b}_2)$, . . ., $\beta_n/g(\mathscr{b}_n)$]-sequence did *not* satisfy the condition

> [θ/$\mathcal{I}$] truly describes what would have been the case, had the world been *w*

only if it satisfied the condition

> [–θ/$\mathcal{I}$] truly describes what would have been the case, had the world been *w*.

Therefore, if [A] is the case, the lemma holds. On the other hand, if [A] is not the case, the lemma clearly holds. In either case, the lemma holds; so it holds for $k = 0$.

I now suppose that the lemma holds for all sentences of index less than *k*, where $k > 0$. Then, for any sentence θ of index *k*, θ must be a sentence of one of the following nine forms: –ψ, (φ & χ), (φ ∨ χ), (φ → χ), (φ ↔ χ), (∃α)φ, (α)φ, ◇φ, □φ. I shall consider each of these possibilities in the above order.

Case (1). θ = –ψ.

Then ψ is of index *k*-1. Hence, by the inductive hypothesis, we have:

[A] *Either*, for some $z \in \mathbf{W}$, had the world been *z*, there would have been a representing function g for *z* such that some $\mathcal{I}(\mathbf{d})$-[$\beta_1/g(\mathscr{b}_1)$, $\beta_2/g(\mathscr{b}_2)$, . . ., $\beta_n/g(\mathscr{b}_n)$]-sequence satisfied the condition: [ψ/$\mathcal{I}$] truly describes what would have been the case, had the world been *w*;

or, for some $z \in \mathbf{W}$, had the world been *z*, there would have been a representing function g for *z* such that some $\mathcal{I}(\mathbf{d})$-[$\beta_1/g(\mathscr{b}_1)$, $\beta_2/g(\mathscr{b}_2)$, . . ., $\beta_n/g(\mathscr{b}_n)$]-sequence satisfied the condition: [–ψ/$\mathcal{I}$] truly describes what would have been the case, had the world been *w*.

Suppose that the upper disjunct of [A] holds. Then:

[B] For some $z \in \mathbf{W}$, had the world been *z*, there would have been a representing function g for *z* such that some $\mathcal{I}(\mathbf{d})$-[$\beta_1/g(\mathscr{b}_1)$, $\beta_2/g(\mathscr{b}_2)$, . . ., $\beta_n/g(\mathscr{b}_n)$]-sequence satisfied the condition: [– –ψ/$\mathcal{I}$] truly describes what would have been the case, had the world been *w*.

On the other hand, if the lower disjunct of [A] holds, we have what is wanted. Thus, in either case, we have what is wanted.

Case (2). θ = (φ & χ).

Both φ and χ have indexes less than *k*. Hence, we can infer:

[A] *Either*, for some $z \in \mathbf{W}$, had the world been z, there would have been a representing function g for z such that some $\mathcal{I}(\mathbf{d})$-$[\beta_1/g(\mathit{6}_1), \beta_2/g(\mathit{6}_2), \ldots, \beta_n/g(\mathit{6}_n)]$-sequence satisfied the condition: $[\phi/\mathcal{I}]$ truly describes what would have been the case, had the world been w;

or, for some $u \in \mathbf{W}$, had the world been u, there would have been a representing function h for u such that some $\mathcal{I}(\mathbf{d})$-$[\beta_1/h(\mathit{6}_1), \beta_2/h(\mathit{6}_2), \ldots, \beta_n/h(\mathit{6}_n)]$-sequence satisfied the condition: $[-\phi/\mathcal{I}]$ truly describes what would have been the case, had the world been w.

and

[B] *Either*, for some $z \in \mathbf{W}$, had the world been z, there would have been a representing function g for z such that some $\mathcal{I}(\mathbf{d})$-$[\beta_1/g(\mathit{6}_1), \beta_2/g(\mathit{6}_2), \ldots, \beta_n/g(\mathit{6}_n)]$-sequence satisfied the condition: $[\chi/\mathcal{I}]$ truly describes what would have been the case, had the world been w;

or, for some $u \in \mathbf{W}$, had the world been u, there would have been a representing function h for u such that some $\mathcal{I}(\mathbf{d})$-$[\beta_1/h(\mathit{6}_1), \beta_2/h(\mathit{6}_2), \ldots, \beta_n/h(\mathit{6}_n)]$-sequence satisfied the condition: $[-\chi/\mathcal{I}]$ truly describes what would have been the case, had the world been w.

I suppose, first, that both upper disjuncts of [A] and [B] hold. Then,

[C] For some $z \in \mathbf{W}$, had the world been z, there would have been a representing function g for z such that some $\mathcal{I}(\mathbf{d})$-$[\beta_1/g(\mathit{6}_1), \beta_2/g(\mathit{6}_2), \ldots, \beta_n/g(\mathit{6}_n)]$-sequence satisfied the condition: $[\phi/\mathcal{I}]$ truly describes what would have been the case, had the world been w;

and also:

for some $u \in \mathbf{W}$, had the world been u, there would have been a representing function h for u such that some $\mathcal{I}(\mathbf{d})$-$[\beta_1/h(\mathit{6}_1), \beta_2/h(\mathit{6}_2), \ldots, \beta_n/h(\mathit{6}_n)]$-sequence satisfied the condition: $[\chi/\mathcal{I}]$ truly describes what would have been the case, had the world been w.

In view of Lemma 3, it can be seen that

[D] For some $z \in \mathbf{W}$, had the world been z, there would have been a representing function g for z such that some $\mathcal{I}(\mathbf{d})$-$[\beta_1/g(\mathit{6}_1), \beta_2/g(\mathit{6}_2), \ldots, \beta_n/g(\mathit{6}_n)]$-sequence satisfied the condition:

$[\phi/\mathcal{I}]$ truly describes what would have been the case, had the world been w;

and also:

$[\chi/\mathcal{I}]$ truly describes what would have been the case, had the world been w;

and hence the condition:

[ϕ & $\chi/\mathcal{I}$] truly describes what would have been the case, had the world been w.

I suppose, second, that the upper disjunct of [A] and the lower disjunct of [B] hold. Then, by essentially the above reasoning, we can infer:

[E] For some $z \in \mathbf{W}$, had the world been z, there would have been a representing function g for z such that some $\mathcal{I}(\mathbf{d})$-[$\beta_1/g(\mathscr{b}_1)$, $\beta_2/g(\mathscr{b}_2)$, . . ., $\beta_n/g(\mathscr{b}_n)$]-sequence satisfied the condition:

[$\phi/\mathcal{I}$] truly describes what would have been the case, had the world been w

and also the condition:

[$-\chi/\mathcal{I}$] truly describes what would have been the case, had the world been w

and hence the condition:

[$-(\phi$ & $\chi)/\mathcal{I}$] truly describes what would have been the case, had the world been w.

Similarly, in the remaining two cases, we can repeat, basically, the above reasoning to infer that the right (or lower) disjunct of the lemma holds. So we obtain the desired result for this case.

Cases (3)–(5) can be proved in essentially the above way *mutatis mutandis*, so I shall proceed to:

Case (6). $\theta = (\exists\alpha)\phi$.

Let β be the first odd-subscripted individual constant not occurring in ϕ. Then,

For some $z \in \mathbf{W}$, had the world been z, there would have been a representing function g for z such that some $\mathcal{I}(\mathbf{d})$-[$\beta_1/g(\mathscr{b}_1)$, $\beta_2/g(\mathscr{b}_2)$, . . ., $\beta_n/g(\mathscr{b}_n)$]-sequence satisfied the condition: [$\theta/\mathcal{I}$] truly describes what would have been the case, had the world been w,

iff

for some $\mathscr{b} \in \mathbf{d}(w)$, for some $z \in \mathbf{W}$, had the world been z, there would have been a representing function g for z such that some $\mathcal{I}(\mathbf{d})$-[$\beta_1/g(\mathscr{b}_1)$, $\beta_2/g(\mathscr{b}_2)$, . . ., $\beta_n/g(\mathscr{b}_n)$, $\beta/g(\mathscr{b})$]-sequence satisfied the condition: [$\phi\alpha/\beta/\mathcal{I}$] truly describes what would have been the case, had the world been w.

But $\phi\alpha/\beta$ is a sentence of index k-1. Hence, for every $\mathscr{b} \in \mathbf{d}(w)$, we have:

[A] *Either*, for some $z \in \mathbf{W}$, had the world been z, there would have been a representing function g for z such that some $\mathscr{I}(\mathbf{d})$-$[\beta_1/g(\mathscr{b}_1), \beta_2/g(\mathscr{b}_2), \ldots, \beta_n/g(\mathscr{b}_n), \beta/g(\mathscr{b})]$-sequence satisfied the condition: $[\phi\alpha/\beta/\mathscr{I}]$ truly describes what would have been the case, had the world been w;

or, for some $z \in \mathbf{W}$, had the world been z, there would have been a representing function g for z such that some $\mathscr{I}(\mathbf{d})$-$[\beta_1/g(\mathscr{b}_1), \beta_2/g(\mathscr{b}_2), \ldots, \beta_n/g(\mathscr{b}_n), \beta/g(\mathscr{b})]$-sequence satisfied the condition: $[-\phi\alpha/\beta/\mathscr{I}]$ truly describes what would have been the case, had the world been w.

Suppose that, for some $\mathscr{b} \in \mathbf{d}(w)$, the upper disjunct holds. That is,

[B] For some $z \in \mathbf{W}$, had the world been z, there would have been a representing function g for z such that some $\mathscr{I}(\mathbf{d})$-$[\beta_1/g(\mathscr{b}_1), \beta_2/g(\mathscr{b}_2), \ldots, \beta_n/g(\mathscr{b}_n), \beta/g(\mathscr{b})]$-sequence satisfied the condition: $[\phi\alpha/\beta/\mathscr{I}]$ truly describes what would have been the case, had the world been w.

Then,

[C] For some $z \in \mathbf{W}$, had the world been z, there would have been a representing function g for z such that some $\mathscr{I}(\mathbf{d})$-$[\beta_1/g(\mathscr{b}_1), \beta_2/g(\mathscr{b}_2), \ldots, \beta_n/g(\mathscr{b}_n)]$-sequence satisfied the condition: $[\theta/\mathscr{I}]$ truly describes what would have been the case, had the world been w.

Suppose this time that, for no $\mathscr{b} \in \mathbf{d}(w)$, is it the case that [B]. Then, for every $\mathscr{b} \in \mathbf{d}(w)$, the lower disjunct of [A] must hold. We can infer that:

[D] For every $\mathscr{b} \in \mathbf{d}(w)$, for some $z \in \mathbf{W}$, had the world been z, there would have been a representing function g for z such that some $\mathscr{I}(\mathbf{d})$-$[\beta_1/g(\mathscr{b}_1), \beta_2/g(\mathscr{b}_2), \ldots, \beta_n/g(\mathscr{b}_n), \beta/g(\mathscr{b})]$-sequence satisfied the condition: $[-\phi\alpha/\beta/\mathscr{I}/\mathscr{I}]$ truly describes what would have been the case, had the world been w.

Hence,

[E] For some $z \in \mathbf{W}$, had the world been z, there would have been a representing function g for z such that some $\mathscr{I}(\mathbf{d})$-$[\beta_1/g(\mathscr{b}_1), \beta_2/g(\mathscr{b}_2), \ldots, \beta_n/g(\mathscr{b}_n)]$-sequence satisfied the condition: $[(\alpha)-\phi/\mathscr{I}]$ truly describes what would have been the case, had the world been w.

It follows that:

[F] For some $z \in \mathbf{W}$, had the world been z, there would have been a representing function g for z such that some $\mathscr{I}(\mathbf{d})$-$[\beta_1/g(\mathscr{b}_1), \beta_2/g(\mathscr{b}_2), \ldots, \beta_n/g(\mathscr{b}_n)]$-sequence satisfied the condition: $[-\theta/\mathscr{I}]$ truly describes what would have been the case, had the world been w.

Since Case (7) can be proved in essentially the way (6) was, I turn now to:

Case (8). $\theta = \Diamond\phi$.

Since ϕ is of index k-1, it follows by the inductive hypothesis that, for every $v \in \mathbf{W}$:

[A] *Either*, for some $z \in \mathbf{W}$, had the world been z, there would have been a representing function g for z such that some $\mathcal{S}(\mathbf{d})$-$[\beta_1/g(b_1), \beta_2/g(b_2), \ldots, \beta_n/g(b_n)]$-sequence satisfied the condition: $[\phi/\mathcal{S}]$ truly describes what would have been the case, had the world been v;

or, for some $z \in \mathbf{W}$, had the world been z, there would have been a representing function g for z such that some $\mathcal{S}(\mathbf{d})$-$[\beta_1/g(b_1), \beta_2/g(b_2), \ldots, \beta_n/g(b_n)]$-sequence satisfied the condition: $[-\phi/\mathcal{S}]$ truly describes what would have been the case, had the world been v.

I shall begin by assuming that, for some $v \in \mathbf{W}$,

[B] |For some $z \in \mathbf{W}$, had the world been z, there would have been a representing function g for z such that some $\mathcal{S}(\mathbf{d})$-$[\beta_1/g(b_1), \beta_2/g(b_2), \ldots, \beta_n/g(b_n)]$-sequence satisfied the condition: $[\phi/\mathcal{S}]$ truly describes what would have been the case, had the world been v.

Then, we can infer

[C] For some $z \in \mathbf{W}$, had the world been z, there would have been a representing function g for z such that some $\mathcal{S}(\mathbf{d})$-$[\beta_1/g(b_1), \beta_2/g(b_2), \ldots, \beta_n/g(b_n)]$-sequence satisfied the condition: $[\Diamond\phi/\mathcal{S}]$ truly describes what would have been the case, had the world been v.

But

$[\Diamond\phi/\mathcal{S}]$ truly describes what would have been the case, had the world been v,

only if

$[\Diamond\phi/\mathcal{S}]$ truly describes what would have been the case, no matter how the world may have been, and so, in particular, had the world been w.

So we can infer what is wanted.

Let us assume this time that, for no $v \in \mathbf{W}$, do we have

[D] |For some $z \in \mathbf{W}$, had the world been z, there would have been a representing function g for z such that some $\mathcal{S}(\mathbf{d})$-$[\beta_1/g(b_1), \beta_2/g(b_2), \ldots, \beta_n/g(b_n)]$-sequence satisfied the condition: $[\phi/\mathcal{S}]$ truly describes what would have been the case, had the world been v.

In that case, we can infer using the inductive hypothesis that, for all $v \in \mathbf{W}$:

[E] For some $z \in \mathbf{W}$, had the world been z, there would have been a representing function g for z such that some $\mathcal{S}(\mathbf{d})$-$[\beta_1/g(b_1), \beta_2/g(b_2), \ldots,$

> $\beta_n/g(\mathscr{d}_n)$]-sequence satisfied the condition: [$-\phi/\mathcal{I}$] truly describes what would have been the case, had the world been v.

We can conclude, in view of condition [C-2], that:

> [F] For some $z \in \mathbf{W}$, had the world been z, there would have been a representing function g for z such that some $\mathcal{I}(\mathbf{d})$-[$\beta_1/g(\mathscr{d}_1)$, $\beta_2/g(\mathscr{d}_2)$, . . ., $\beta_n/g(\mathscr{d}_n)$]-sequence satisfied the condition: [$-\Diamond\phi/\mathcal{I}$] truly describes what would have been the case, no matter how the world might have been, and hence had the world been w.

This completes Case (8). Case (9) can be proved in essentially the same way *mutatis mutandis.* So the lemma follows by induction.

The Fundamental Theorem

> **Theorem [3]**. For every $w \in \mathbf{W}$, for every $\mathscr{d}_1, \mathscr{d}_2, \ldots, \mathscr{d}_n \in \mathbf{D}$, and for every sentence θ, if $\beta_1, \beta_2, \ldots, \beta_n$ are the n odd-subscripted individual constants that occur in θ, then the following holds:
>
> θ is satisfied at w under **I** by some **D**-[$\beta_1/\mathscr{d}_1$, $\beta_2/\mathscr{d}_2$, . . ., $\beta_n/\mathscr{d}_n$]-sequence iff for some $z \in \mathbf{W}$, had the world been z, there would have been a representing function g for z such that some $\mathcal{I}(\mathbf{d})$-[$\beta_1/g(\mathscr{d}_1)$, $\beta_2/g(\mathscr{d}_2)$, . . ., $\beta_n/g(\mathscr{d}_n)$]-sequence satisfied the condition: [$\theta/\mathcal{I}$] truly describes what would have been the case, had the world been w.

Comment. Notice that n gives the number of odd-subscripted individual constants occurring in θ. When $n = 0$, θ has no occurrences of odd-subscripted individual constants at all, and hence is a closed sentence. In that case, θ is satisfied at a world (under an interpretation) by a **D**-sequence iff it is true at that world (under the interpretation), and since a closed condition containing such a sentence (with the meaning it has under $\mathcal{I}$) is true iff it is satisfied by any $\mathcal{I}(\mathbf{d})$-sequence, we could state the content of the indented part of the theorem as follows:

> For $w \in \mathbf{W}$, any closed-sentence θ is true at w under **I** iff [$\theta/\mathcal{I}$] truly describes what would have been the case, had the world been w.[13]

Proof: The proof of the theorem is by induction on the *index k* of the sentences of **M***. I begin with the case in which $k = 0$.

θ is atomic, say $\Phi\alpha_1\alpha_2 \ldots \alpha_m$, where each odd-subscripted individual constant β_k must occur among the $\alpha_1, \alpha_2, \ldots, \alpha_m$. Now, either all the mem-

[13] This is what I shall give as the principal corollary of the Fundamental Theorem.

bers of the set $\{\delta_1, \delta_2, \ldots, \delta_n\}$ are members of $\mathbf{d}(w)$ or at least one member of the set $\{\delta_1, \delta_2, \ldots, \delta_n\}$ is not a member of $\mathbf{d}(w)$.

Suppose the former. In that case, the theorem holds by Lemma 4.

Suppose the latter. That is, suppose that some member δ_k of the set $\{\delta_1, \delta_2, \ldots, \delta_n\}$ is not a member of $\mathbf{d}(w)$. Then since θ is atomic, by the Serious Actualism feature of **M***, θ cannot be satisfied at w under **I** by any **D**-$[\beta_1/\delta_1, \beta_2/\delta_2, \ldots, \beta_n/\delta_n]$-sequence. So suppose, for reductio, that:

> For some $z \in \mathbf{W}$, had the world been z, there would have been a representing function g for z such that some $\mathcal{J}(\mathbf{d})$-$[\beta_1/\mathrm{g}(\delta_1), \beta_2/\mathrm{g}(\delta_2), \ldots, \beta_n/\mathrm{g}(\delta_n)]$-sequence satisfied the condition: $[\theta/\mathcal{J}]$ truly describes what would have been the case, had the world been w.

In that case, by [C-4], had the world been w, any representing function h of w would be such that, for some $x \in \mathbf{d}(w)$, $\mathrm{h}(x) = \mathrm{g}(\delta_k)$. By [C-3], noting that $\delta_k \in \mathbf{d}(z)$, this is impossible.

I have shown that the theorem holds for the case in which $k = 0$. So I now suppose that the theorem holds for all sentences of index less than k, where $k > 0$. Then, for any sentence θ of index k, θ must be a sentence of one of the nine forms given in the proof of Lemma 5. I shall consider each of these nine possibilities in the above order.

Case (1). $\theta = -\psi$.

Then ψ is of index k-1. Hence, by the inductive hypothesis, we have:

> [A] ψ is satisfied at w under **I** by some **D**-$[\beta_1/\delta_1, \beta_2/\delta_2, \ldots, \beta_n/\delta_n]$-sequence iff for some $z \in \mathbf{W}$, had the world been z, there would have been a representing function g for z such that some $\mathcal{J}(\mathbf{d})$-$[\beta_1/\mathrm{g}(\delta_1), \beta_2/\mathrm{g}(\delta_2), \ldots, \beta_n/\mathrm{g}(\delta_n)]$-sequence satisfied the condition: $[\psi/\mathcal{J}]$ truly describes what would have been the case, had the world been w.

Hence,

> [B] ψ is not satisfied at w under **I** by some **D**-$[\beta_1/\delta_1, \beta_2/\delta_2, \ldots, \beta_n/\delta_n]$-sequence iff it is not the case that for some $z \in \mathbf{W}$, had the world been z, there would have been a representing function g for z such that some $\mathcal{J}(\mathbf{d})$-$[\beta_1/\mathrm{g}(\delta_1), \beta_2/\mathrm{g}(\delta_2), \ldots, \beta_n/\mathrm{g}(\delta_n)]$-sequence satisfied the condition: $[\psi/\mathcal{J}]$ truly describes what would have been the case, had the world been w.

From [B] and Lemma 5, we can infer:

> [C] If $-\psi$ is satisfied at w under **I** by some **D**-$[\beta_1/\delta_1, \beta_2/\delta_2, \ldots, \beta_n/\delta_n]$-sequence, then for some $z \in \mathbf{W}$, had the world been z, there would have been a representing function g for z such that some $\mathcal{J}(\mathbf{d})$-

$[\beta_1/g(\mathcal{b}_1), \beta_2/g(\mathcal{b}_2), \ldots, \beta_n/g(\mathcal{b}_n)]$-sequence satisfied the condition: $[-\psi/\mathcal{J}]$ truly describes what would have been the case, had the world been w.

To prove the other direction, suppose that:

[D] For some $z \in \mathbf{W}$, had the world been z, there would have been a representing function g for z such that some $\mathcal{J}(\mathbf{d})$-$[\beta_1/g(\mathcal{b}_1), \beta_2/g(\mathcal{b}_2), \ldots, \beta_n/g(\mathcal{b}_n)]$-sequence satisfied the condition: $[-\psi/\mathcal{J}]$ truly describes what would have been the case, had the world been w.

Then, in view of Lemma 3, we can infer:

[E] It is not the case that for some $z \in \mathbf{W}$, had the world been z, there would have been a representing function g for z such that some $\mathcal{J}(\mathbf{d})$-$[\beta_1/g(\mathcal{b}_1), \beta_2/g(\mathcal{b}_2), \ldots, \beta_n/g(\mathcal{b}_n)]$-sequence satisfied the condition: $[\psi/\mathcal{J}]$ truly describes what would have been the case, had the world been w.

From [B] and [E], we can infer:

[F] ψ is not satisfied at w under $\mathbf{I}$ by some $\mathbf{D}$-$[\beta_1/\mathcal{b}_1, \beta_2/\mathcal{b}_2, \ldots, \beta_n/\mathcal{b}_n]$-sequence,

and hence:

[G] $-\psi$ is satisfied at w under $\mathbf{I}$ by some $\mathbf{D}$-$[\beta_1/\mathcal{b}_1, \beta_2/\mathcal{b}_2, \ldots, \beta_n/\mathcal{b}_n]$-sequence.

So the theorem holds for Case (1).

Case (2). $\theta = (\phi \,\&\, \chi)$.

Both ϕ and χ have indexes less than k. Hence, we can infer:

[A] ϕ is satisfied at w under $\mathbf{I}$ by some $\mathbf{D}$-$[\beta_1/\mathcal{b}_1, \beta_2/\mathcal{b}_2, \ldots, \beta_n/\mathcal{b}_n]$-sequence iff, for some $z \in \mathbf{W}$, had the world been z, there would have been a representing function g for z such that some $\mathcal{J}(\mathbf{d})$-$[\beta_1/g(\mathcal{b}_1), \beta_2/g(\mathcal{b}_2), \ldots, \beta_n/g(\mathcal{b}_n)]$-sequence satisfied the condition: $[\phi/\mathcal{J}]$ truly describes what would have been the case, had the world been w;

and

[B] χ is satisfied at w under $\mathbf{I}$ by some $\mathbf{D}$-$[\beta_1/\mathcal{b}_1, \beta_2/\mathcal{b}_2, \ldots, \beta_n/\mathcal{b}_n]$-sequence iff, for some $z \in \mathbf{W}$, had the world been z, there would have been a representing function g for z such that some $\mathcal{J}(\mathbf{d})$-$[\beta_1/g(\mathcal{b}_1), \beta_2/g(\mathcal{b}_2), \ldots, \beta_n/g(\mathcal{b}_n)]$-sequence satisfied the condition: $[\chi/\mathcal{J}]$ truly describes what would have been the case, had the world been w.

Now

[C] $(\phi \,\&\, \chi)$ is satisfied at w under **I** by some **D**-$[\beta_1/\text{б}_1, \beta_2/\text{б}_2, \ldots, \beta_n/\text{б}_n]$-sequence iff ϕ is satisfied at w under **I** by some **D**-$[\beta_1/\text{б}_1, \beta_2/\text{б}_2, \ldots, \beta_n/\text{б}_n]$-sequence and χ is satisfied at w under **I** by some **D**-$[\beta_1/\text{б}_1, \beta_2/\text{б}_2, \ldots, \beta_n/\text{б}_n]$-sequence.

So we can infer that:

[D] $(\phi \,\&\, \chi)$ is satisfied at w under **I** by some **D**-$[\beta_1/\text{б}_1, \beta_2/\text{б}_2, \ldots, \beta_n/\text{б}_n]$-sequence iff

for some $z \in \mathbf{W}$, had the world been z, there would have been a representing function g for z such that some $\mathscr{I}(\mathbf{d})$-$[\beta_1/g(\text{б}_1), \beta_2/g(\text{б}_2), \ldots, \beta_n/g(\text{б}_n)]$-sequence satisfied the condition: $[\phi/\mathscr{I}]$ truly describes what would have been the case, had the world been w,

and

for some $z \in \mathbf{W}$, had the world been z, there would have been a representing function g for z such that some $\mathscr{I}(\mathbf{d})$-$[\beta_1/g(\text{б}_1), \beta_2/g(\text{б}_2), \ldots, \beta_n/g(\text{б}_n)]$-sequence satisfied the condition: $[\chi/\mathscr{I}]$ truly describes what would have been the case, had the world been w.

Because of Lemma 3, it can be seen that the right side of the biconditional [D] holds iff:

[E] For some $z \in \mathbf{W}$, had the world been z, there would have been a representing function g for z such that some $\mathscr{I}(\mathbf{d})$-$[\beta_1/g(\text{б}_1), \beta_2/g(\text{б}_2), \ldots, \beta_n/g(\text{б}_n)]$-sequence satisfied the condition: $[(\phi \,\&\, \chi)/\mathscr{I}]$ truly describes what would have been the case, had the world been w.

So, we obtain the desired result.

Cases (3)–(5) can be proved in ways quite similar to the above, so I shall proceed to:

Case (6). $\theta = (\exists\alpha)\phi$.

From the definition of 'true at w under **I**', and letting β be the first odd-subscripted individual constant not occurring in ϕ, we can infer that:

[A] $(\exists\alpha)\phi$ is satisfied at w under **I** by some **D**-$[\beta_1/\text{б}_1, \beta_2/\text{б}_2, \ldots, \beta_n/\text{б}_n]$-sequence iff, for some element б of $\mathbf{d}(w)$, $\phi\alpha/\beta$ is satisfied at w under **I** by some **D**-$[\beta_1/\text{б}_1, \beta_2/\text{б}_2, \ldots, \beta_n/\text{б}_n, \beta/\text{б}]$-sequence.

Since $\phi\alpha/\beta$ is a sentence of index $< k$, we know, by the inductive hypothesis, that:

[B] For every element б of $\mathbf{d}(w)$, $\phi\alpha/\beta$ is satisfied at w under **I** by some **D**-$[\beta_1/\text{б}_1, \beta_2/\text{б}_2, \ldots, \beta_n/\text{б}_n, \beta/\text{б}]$-sequence iff, for some $z \in \mathbf{W}$, had the world been z, there would have been a representing function g for z

such that some $\mathcal{I}(\mathbf{d})$-$[\beta_1/g(\sigma_1), \beta_2/g(\sigma_2), \ldots, \beta_n/g(\sigma_n), \beta/g(\sigma)]$-sequence satisfied the condition: $[\phi\alpha/\beta/\mathcal{I}]$ truly describes what would have been the case, had the world been w.

By a simple logical inference of first-order logic, we can obtain from [B]:

[C] For some element σ of $\mathbf{d}(w)$, $\phi\alpha/\beta$ is satisfied at w under **I** by some **D**-$[\beta_1/\sigma_1, \beta_2/\sigma_2, \ldots, \beta_n/\sigma_n, \beta/\sigma]$-sequence iff, for some element σ of $\mathbf{d}(w)$, and for some $z \in \mathbf{W}$, had the world been z, there would have been a representing function g for z such that some $\mathcal{I}(\mathbf{d})$-$[\beta_1/g(\sigma_1), \beta_2/g(\sigma_2), \ldots, \beta_n/g(\sigma_n), \beta/g(\sigma)]$-sequence satisfied the condition: $[\phi\alpha/\beta/\mathcal{I}]$ truly describes what would have been the case, had the world been w.

And it can also be seen that:

[D] For some element σ of $\mathbf{d}(w)$, and for some $z \in \mathbf{W}$, had the world been z, there would have been a representing function g for z such that some $\mathcal{I}(\mathbf{d})$-$[\beta_1/g(\sigma_1), \beta_2/g(\sigma_2), \ldots, \beta_n/g(\sigma_n), \beta/g(\sigma)]$-sequence satisfied the condition: $[\phi\alpha/\beta/\mathcal{I}]$ truly describes what would have been the case, had the world been w

iff

for some $z \in \mathbf{W}$, had the world been z, there would have been a representing function g for z such that some $\mathcal{I}(\mathbf{d})$-$[\beta_1/g(\sigma_1), \beta_2/g(\sigma_2), \ldots, \beta_n/g(\sigma_n)]$-sequence satisfied the condition: $[(\exists\alpha)\phi/\mathcal{I}]$ truly describes what would have been the case, had the world been w.

So from [A], [C], and [D], we get what is wanted.

Since Case (7) can be proved in essentially the way (6) was proved, I turn now to:

Case (8). $\theta = \Diamond\phi$.

From the definition of 'true at a world under an interpretation', we can conclude:

[A] $\Diamond\phi$ is satisfied at w under **I** by some **D**-$[\beta_1/\sigma_1, \beta_2/\sigma_2, \ldots, \beta_n/\sigma_n]$-sequence iff, for some $u \in \mathbf{W}$, ϕ is satisfied at u under **I** by some **D**-$[\beta_1/\sigma_1, \beta_2/\sigma_2, \ldots, \beta_n/\sigma_n]$-sequence.

Since ϕ is a sentence of index $< k$, it follows from the inductive hypothesis that:

[B] For every $u \in \mathbf{W}$, ϕ is satisfied at u under **I** by some **D**-$[\beta_1/\sigma_1, \beta_2/\sigma_2, \ldots, \beta_n/\sigma_n]$-sequence iff, for some $z \in \mathbf{W}$, had the world been z, there would have been a representing function g for z such that some $\mathcal{I}(\mathbf{d})$-$[\beta_1/g(\sigma_1), \beta_2/g(\sigma_2), \ldots, \beta_n/g(\sigma_n)]$-sequence satisfied the condition: $[\phi/\mathcal{I}]$ truly describes what would have been the case, had the world been u.

By a simple logical inference of first-order logic, we can obtain from [B]:

[C] For some $u \in \mathbf{W}$, ϕ is satisfied at u under **I** by some $\mathbf{D}$-$[\beta_1/\delta_1, \beta_2/\delta_2, \ldots, \beta_n/\delta_n]$-sequence iff, for some $u \in \mathbf{W}$, and for some $z \in \mathbf{W}$, had the world been z, there would have been a representing function g for z such that some $\mathcal{I}(\mathbf{d})$-$[\beta_1/g(\delta_1), \beta_2/g(\delta_2), \ldots, \beta_n/g(\delta_n)]$-sequence satisfied the condition: $[\phi/\mathcal{I}]$ truly describes what would have been the case, had the world been u.

We also have:

[D] For some $u \in \mathbf{W}$, for some $z \in \mathbf{W}$, had the world been z, there would have been a representing function g for z such that some $\mathcal{I}(\mathbf{d})$-$[\beta_1/g(\delta_1), \beta_2/g(\delta_2), \ldots, \beta_n/g(\delta_n)]$-sequence satisfied the condition: $[\phi/\mathcal{I}]$ truly describes what would have been the case, had the world been u

iff

for some $z \in \mathbf{W}$, had the world been z, there would have been a representing function g for z such that some $\mathcal{I}(\mathbf{d})$-$[\beta_1/g(\delta_1), \beta_2/g(\delta_2), \ldots, \beta_n/g(\delta_n)]$-sequence satisfied the condition: $[\Diamond\phi/\mathcal{I}]$ truly describes what would have been the case, no matter how the world might have been.

Since some $\mathcal{I}(\mathbf{d})$-$[\beta_1/g(\delta_1), \beta_2/g(\delta_2), \ldots, \beta_n/g(\delta_n)]$-sequence would satisfy the condition:

$[\Diamond\phi/\mathcal{I}]$ truly describes what would have been the case, no matter how the world might have been

iff that sequence would satisfy the condition:

$[\Diamond\phi/\mathcal{I}]$ truly describes what would have been the case, had the world been w,

we can infer from [D]:

[E] For some $u \in \mathbf{W}$, and for some $z \in \mathbf{W}$, had the world been z, there would have been a representing function g for z such that some $\mathcal{I}(\mathbf{d})$-$[\beta_1/g(\delta_1), \beta_2/g(\delta_2), \ldots, \beta_n/g(\delta_n)]$-sequence satisfied the condition: $[\phi/\mathcal{I}]$ truly describes what would have been the case, had the world been u

iff

for some $z \in \mathbf{W}$, had the world been z, there would have been a representing function g for z such that some $\mathcal{J}$(**d**)-[β_1/g(δ_1), β_2/g(δ_2), . . ., β_n/g(δ_n)]-sequence satisfied the condition: [$\Diamond\phi/\mathcal{J}$] truly describes what would have been the case, had the world been w.

Thus, from [A], [C], and [E], we get what is wanted.

This completes Case (8). Case (9) can be proved in essentially the same way *mutatis mutandis*. So the theorem follows by induction.

6. Some Consequences

I shall now give, without proof, some easy consequences of the Fundamental Theorem.

Principal Corollary. For $w \in$ **W**, any closed-sentence θ is true at w under **I** iff [$\theta/\mathcal{J}$] truly describes what would have been the case, had the world been w.

Comment 1. Suppose that $\mathcal{J}$ is an NL proto-interpretation and that **I** and **J** are S5 structures to which $\mathcal{J}$ conforms. It can be seen from the Principal Corollary that any sentence true at w under one of these structures will be true at w under the other one. It follows that all the S5 structures to which $\mathcal{J}$ conforms will satisfy exactly the same sentences. In short, there will be no significant differences between the S5 structures to which a proto-interpretation conforms.

Comment 2. From the Principal Corollary, we can obtain the analogues of the corollaries of the previous chapter. In the following, I shall state the corollaries in abbreviated form, dropping the universal quantifiers. For every S5 interpretation **I** = <**W**, **D**, **w***, **d**, **r**, **e**> of **M***, for every NL proto-interpretation $\mathcal{J}$ conforming to **I**:

Corollary [a]: If the world *is* such that there is a function via which <**d**(**w***), **r**(α), **e**(ϕ, **w***)>$_{\mathcal{J}}$ represents the world, then any closed-sentence ϕ is true under **I** iff [$\phi/\mathcal{J}$] is true.

Corollary [b]: For any closed-sentence ϕ, there is a $w \in$ **W** such that ϕ is true at w under **I** iff there is a $u \in$ **W** such that, had the world been such that there was a function via which <**d**(u), **r**(α), **e**(ϕ, u)>$_{\mathcal{J}}$ represented the world, then [$\phi/\mathcal{J}$] would have been true.

From the conditions that must be satisfied by an NL proto-interpretation conforming to **I**, it follows that the right side of the above biconditional holds iff the world could have been such that [$\phi/\mathcal{J}$] was true. And the latter is the case iff it is possible that [$\phi/\mathcal{J}$] is true. The left side of the biconditional of Corollary [b] holds iff $\Diamond\phi$ is true under **I**. Hence, we obtain:

Corollary [c]: For any closed-sentence ϕ, $\Diamond\phi$ is true under **I** iff it is possible that $[\phi/\mathcal{I}]$ is true.

It then follows that:

Corollary [d]: For any closed-sentence ϕ, $\Box\phi$ is true under **I** iff it is necessary that $[\phi/\mathcal{I}]$ is true.

Let us interrupt the flow of corollaries for a definition. I define an *adequate NL interpretation* to be an NL interpretation that conforms to some S5 structure. Then, we can state:

Corollary [e]: For every adequate NL interpretation $\mathcal{I}$ and any closed-sentence ϕ, if ϕ is valid, then it is necessary that $[\phi/\mathcal{I}]$ is true.

Corollary [f]: For every adequate NL interpretation $\mathcal{I}$ and every set of closed-sentences Γ, and every closed-sentence ϕ, if $\Gamma \models \phi$, then if each member θ of Γ is such that $[\theta/\mathcal{I}]$ is true, then $[\phi/\mathcal{I}]$ is true.

Corollary [g]: For every adequate NL interpretation $\mathcal{I}$ and every set of closed-sentences Γ, and every closed-sentence ϕ, if $\Gamma \vdash \phi$, then if each member θ of Γ is such that $[\theta/\mathcal{I}]$ is true, then $[\phi/\mathcal{I}]$ is true.

In other words, we can use our modal logic to conclude that, if all the members of Γ (interpreted according to $\mathcal{I}$) are true, then ϕ (so interpreted) must also be true.[14]

In the following chapter, I shall explore the philosophical implications of these results.

[14] The results of this chapter have certain points in common with the work of Menzel (Menzel 1990)—especially so, in the light of Ray 1996. One reason I have not taken the Menzel–Ray route is because its central claim (the link between truth and truth in a model) is not justified. Neither Menzel nor Ray produces a proof of something like my Fundamental Theorem. Another feature of their method of eliminating possible worlds from the semantics of modal logic that I find unacceptable is their presupposition that there is a set of all things and that it makes sense to quantify over everything that there is (recall the discussion in Section 4 of Chapter 7).

8

Modality without Worlds III

Explorations, Developments, and Defences

In this chapter, I explore, develop, and defend some of the major features of the anti-Realist position presented in earlier chapters.

1. The Anti-Realist Position

One feature of Forbes's anti-Realistic analysis of modal logic that aroused my suspicions from my very first reading of the account was the absence of any substantial discussion of the role of possible worlds *models* or *structures*. Forbes wrote as if the only philosophically significant semantical connection between the formal language of box and diamond and the language of possible worlds was to be given by his reverse translation doctrine. It seemed to me then, as it does now, that any plausible anti-realistic account of contemporary modal logic must assign an important role to the possible worlds structures themselves. The history of mathematics and logic is filled with great advances achieved as a result of work done with such structures; and it is hard to believe that these models of modal logic are not somehow intimately connected with the semantical features of the formal sentences that are "treated" by the logic. (In the discussion to follow, the important role that these S5 structures can play in modal reasoning will be emphasized.)

On the other hand, we have seen that the possible worlds structures cannot be understood in the full-blown literal fashion advocated by the Modal Realist without incurring a huge ontological burden. So it has been my aim to steer a course between these two unacceptable positions. The S5 models, the structures themselves, play a key role in my analysis of the semantics of the system. On the other hand, the structures themselves are only mathematical objects that represent—the use of these models does not require belief in possible worlds.

The semantical system developed here requires two sorts of "interpretations" of the logical language: on the one hand, we have S5 structures—the purely mathematical entities—which, as was made evident in the previous chapters, provide us with what can be regarded as mathematical representations of how the world could have been in so far as the domains of the quantifiers, the referents of even-subscripted individual constants, and the extensions of the predicates are of concern; and on the other hand, there are the NL interpretations, which give meaning to the symbols of the logical language so as to yield sentences that directly express statements. Thus, when the language **M*** is given an NL interpretation, it can be used to formulate modal theories and modal beliefs. The Fundamental Theorem of the previous chapter shows how these two kinds of "interpretations" are so logically related that (1) the derivational system specified in that chapter can be legitimately employed to draw conclusions and make inferences from such theories and beliefs; and (2) the model theory of S5 structures can be used to draw a variety of conclusions about the logical relationships that obtain among NL interpreted sentences of **M***, as well as of their truth or falsity. And we can do so without assuming that there is some "intended interpretation" or "canonical model" that requires the existence of the set of all possible worlds or that presupposes a set of all possible objects. *The applications of the modal system developed in the previous chapters do not require the S5 structures to be anything more than purely mathematical objects.* In particular, S5 modal quantificational logic can be applied to conclude, from the truth of the members of a set of sentences of **M***, that some NL interpreted sentence must be true—without supposing that the S5 models are anything more than pure sets.

I indicated above how our system of modal quantification logic can be used to assess and validate reasoning carried out in **M***, suitably interpreted. What about modal reasoning carried out in a natural language? How can this system play a role in the assessment and validation of such reasoning? But, here, the logical situation is a familiar one. How do we apply the logical theory of first-order logic to reasoning carried out in a natural language? We first *translate* the relevant sentences expressed in the natural language into the first-order formal language, suitably interpreted, and then apply the logical theory to the translated sentences. In short, the assessment is carried out by means of translations. Of course, translations can be more or less adequate—more or less logically accurate. Since we have no set of effectively calculable rules for translating a natural language into the formal language, the logical assessments we carry out will be, of necessity, relative—relative to the adequacy of the translation. But this is not to say that one cannot accurately evaluate translations that are made. One can have a great deal of confidence that some translation is, for logical purposes, more than adequate. Thus, there is no question that first-order logic has proved to be valuable in the evaluation of logical reasoning carried out in a natural language.

According to Plantinga and Lewis, it is only in the arena of applied semantics that a commitment to possible worlds is incurred. So it is appropriate, at this point, to investigate what ontological commitments are incurred when the system of possible worlds semantics developed in this work is applied to draw lessons and conclusions about conceptual possibility.

2. Analysing the Forbes and Plantinga Examples

The Forbes Example

Let us begin with a re-examination of the sort of example which Forbes thought raised problems for the anti-Realist. The inference schema

$$\frac{\Diamond P \quad \Diamond Q}{\Diamond(P \,\&\, Q)}$$

can be easily seen to be invalid when the modal sentences are translated into sentences that involve quantification over worlds. The problem for the anti-Realist, as Forbes sees it, is to explain why the explanation in terms of quantification over worlds seems natural and right, even though there are no such things as worlds.

Given the anti-Realist account developed in the previous chapters, it is easy to give the sort of explanation desired. The worlds sentences in question,

$$\frac{(\exists w)P(w) \quad (\exists w)Q(w)}{(\exists w)(P(w) \,\&\, Q(w))}$$

which were thought to "translate", in some sense, the modal sentences in the above schema, are obtained almost directly from the model theory of **MSC**. Thus, we can analyse the left premiss of the modal schema into:

> There is a $w \in \mathbf{W}$ such that, had the world been as it is represented as being by $\mathbf{v}(\sigma, w)$, then P would have been true.

And this sentence can be abbreviated: '$(\exists w)(\mathbf{v}(P, w) = 0)$'. Doing the same for the other premiss and the conclusion of the modal schema, we obtain:

$$\frac{(\exists w)(\mathbf{v}(P, w) = 0) \quad (\exists w)(\mathbf{v}(Q, w) = 0)}{(\exists w)(\mathbf{v}(P, w) = 0 \,\&\, \mathbf{v}(Q, w) = 0)}$$

Now simply shorten the notation of the above, writing 'P(w)' for '$\mathbf{v}(P, w) = 0$', and we obtain Forbes's schema

$$\frac{(\exists w)P(w)\ (\exists w)Q(w)}{(\exists w)(P(w)\ \&\ Q(w))}$$

So we can explain why the "translation" seems natural and intuitively correct and also why it is reasonable to use our acquaintance with the properties of quantification to conclude that the schema is invalid. However, this time there is little temptation to suppose that the quantification in the above schema is literally quantification over possible worlds. It is clear that the quantifiers in question can be taken to be ranging over some set **W** of an interpretation **I** = <**W**, **w***, **v**>, and that **W** can be regarded as simply an index set—whose elements are pure sets. Thus, in quantifying over **W**, we do not commit ourselves to possible worlds.

To say that the above inference schema is invalid is, of course, not to say that no "instance" of the schema will be a valid argument. Let $\mathcal{I}$ be the following (partial) NL proto-interpretation of **MSC**:

> *P*: There are seven consecutive sevens in the decimal expansion of π.
> *Q*: There are eight consecutive eights in the decimal expansion of π.

It would be perfectly reasonable to infer the truth of $[\Diamond(P\ \&\ Q)/\mathcal{I}]$ from the truth of the two premisses:

$$[\Diamond P/\mathcal{I}] \text{ and } [\Diamond Q/\mathcal{I}]$$

We can make such an inference because of the logical features of the sentences involved that are not captured by the semantical system of **MSC** (a system which focuses on the logical constants of the modal language).

More generally, suppose that it is shown, using model theory, that ϕ is not a model-theoretic consequence of Γ. Can we conclude that the truth of $[\phi/\mathcal{I}]$ cannot be legitimately inferred from the truth of the set of $[\theta/\mathcal{I}]$ such that $\theta \in \Gamma$? As the above example illustrates, we cannot. What is clear in the imagined case is that, since we can conclude from the soundness of our rules of derivation that θ cannot be proven from Γ by our rules of modal quantificational logic, the truth of $[\phi/\mathcal{I}]$ can be inferred from the truth of the set of $[\theta/\mathcal{I}]$ such that $\theta \in \Gamma$, only by appeal to logical features of the interpreted sentences that are not captured by the semantics of **MSC** (and hence an appeal would have to be made to logical features of parts of the sentences other than those of the logical constants of **MSC**). So it is clear that such model-theoretic results as the Soundness Theorem would be highly relevant to our assessment of the validity of modal reasoning, even if we are not able to draw the conclusion, from what has been developed in this work, that the truth of some sentence cannot be inferred from certain premisses.

Plantinga's Examples

Consider now the two ways of understanding the sentence

> Whatever is seen to be sitting is necessarily sitting

which, you will recall, played a role in Plantinga's defence of Modal Realism. There is, on the one hand, the *de dicto* reading:

> In every possible world, whatever is seen to be sitting is sitting.

And, on the other hand, we have the *de re* reading:

> Whatever is seen (in the actual world) to be sitting is sitting in every possible world.

Plantinga doubted, you will recall, that one could grasp the distinction between the *de dicto* and the *de re* readings if one adopted the view that all our talk of possible worlds is dispensable.

What we have, in this example, are some modal sentences of a natural language. How can the anti-Realist account of this chapter be used to analyse such sentences? The answer is to be found in replies Quine has made to an analogous worry about his criterion of ontological commitment. This system of modal logic can be used in assessing a modal sentence or a modal argument expressed in a natural language in so far as the relevant sentences can be translated into, or "approximated by", sentences of the formal language.[1]

To express the two key sentences of Plantinga's example in a formal modal language of the sort I have been discussing in this chapter, let us expand the vocabulary of **M*** to include an operator $\mathcal{S}$ to express the sense of the English phrase 'it is seen that' (I shall leave the technical developments to the reader's imagination). Let us interpret the monadic predicate 'S^1' to be:

> ① is sitting.

Then, in this NL interpreted expanded language, we can express the *de dicto* reading with the sentence:

$$\Box(x)(\mathcal{S}S^1x \rightarrow S^1x).$$

The *de re* reading can be written:

$$(x)(\mathcal{S}S^1x \rightarrow \Box S^1x).$$

Armed with the account of the previous chapter, it is easily seen that the

[1] Cf. Quine 1960: 242–3.

anti-Realist will be able to discern a big difference between the *de dicto* and the *de re* readings. One can use the semantical theory of **M*** to show that the logical implications of the two sentences are considerably different and easily distinguished. For example, the latter, but not the former, implies

$$E^1a \rightarrow (\mathcal{S}S^1a \rightarrow \Box S^1a)$$

It can be seen that the fundamental distinction between these two readings can be captured without having to quantify over, or refer to, possible worlds. Indeed, the *de dicto* reading can be given in English as:

> No matter how the world might have been, it still would have been the case that whatever would have been seen to be sitting would have been sitting.

And the *de re* reading can be given as:

> The world is such that whatever is seen to be sitting would have been sitting, no matter how the world might have been.

I see no reason at all to think that these English sentences can only be understood by making reference to possible worlds.

As for the other example of Plantinga's we considered earlier, 'Johan is essentially a person, but contingently a philosopher', the reader will recall that he analysed this sentence by referring to possible worlds. I, on the other hand, can analyse this sentence using **M***, when it is given an NL proto-interpretation in which the following assignments are made:

H: ① is a person;
P: ① is a philosopher;
j: Johan.

Then the sentence:

$$(Hj \,\&\, Pj) \,\&\, (\Box(E^1j \rightarrow Hj) \,\&\, -\Box(E^1j \rightarrow Pj)$$

gives a translation of the target English sentence which is decent enough for our purposes. We can then disambiguate the modal features of the sentence in question as follows:

> Johan is both a person and a philosopher, and no matter how the world might have been, so long as it had been such that Johan existed, Johan would have been a person; but it is not the case that, no matter how the world might have been, so long as it had been such that Johan existed, Johan would have been a philosopher.

Thus, the anti-Realist can grasp and explicate the relevant modal features of the target sentence without making reference to possible worlds.

3. The Plantinga–Konyndyk Objection

Let us reconsider the Actualistic objection to such modal languages as **M** (considered at the end of Chapter 1). Recall that it was objected that such languages as **M** cannot express the truths

> There could have been an object different from every actual object

and

> There could have been more objects than there are

without presupposing the existence of non-existent objects or ascribing properties to objects at worlds in which they do not exist. Does this objection apply to the kind of system developed in the previous chapter?

For the sake of simplicity, I shall concentrate on the first of the above two modal sentences. Furthermore, I shall consider the objection in terms of the sort of expanded language discussed in Chapter 1, in which an actuality operator has been added to the vocabulary. Thus, the sentence in question can be expressed in the expanded language by:

$$\Diamond(\exists x)\text{-}\mathcal{A}E^1 x$$

The question is: must such a modal language presuppose the existence of non-existent objects? Not at all. It is easy to see that this language would differ from **M*** only slightly and that a Fundamental Theorem could be easily proven for this language, using only reasoning that is compatible with Actualism. In other words, an Actualistic quantificational modal language could be constructed along these lines: it could express the target sentence in question without presupposing non-existent objects or "ascribing properties to objects at worlds in which they do not exist"; and the language would not require the existence of possible worlds.

4. The Problem of Transworld Identity Revisited

It is time to reconsider *the problem of transworld identity* discussed in Chapter 2, where it was suggested that the fact that Modal Realism allows such nice responses to the problem may be regarded as a strong argument for adopting some version of Realism. I now wish to argue that we do not have to take on the onerous ontological burdens of the Modal Realist in order to avoid the problem. From my point of view, there is no flock of possible worlds to give rise to the sort of considerations that generate the problem. Since possible worlds are not being presupposed at all, there is no problem of coming up with a qualitatively distinguishing essential property of Socrates, by means of which we can determine whether some person in another possible world is or is not Socrates. We can thus agree with Plantinga that "There is no such thing as "looking into" another possible world to see what is going on there. There is no such thing as inspecting the inhabitants of another possible world with a view to deciding which, if any, is Socrates" (Plantinga 1974: 96). With the conception of possible worlds models developed in this work, the problem of coming up with a *criterion* of transworld identity simply does not arise; and we can avoid the frustrating task of attempting to develop a "method of individuation" that can be applied across possible worlds.

The position being described here is compatible with the position on identity sketched in Chapter 2, Section 3, according to which the concept of *same organism* or of *same person* may be adequate for all practical purposes, given the general framework of regularities and natural laws within which the concept actually functions, but it is simply not sharp enough to be definitely applicable or definitely not applicable for a whole range of cases when this general framework is removed, as in effect it is when we are asked, by those in the grip of the problem of transworld identity, to try to apply the concept in fantastic situations.

It might be thought, however, that a problem arises for my account that is analogous to the problem of transworld identity. For example, condition [C-3], in Section 4 of Chapter 7, states (among other things) that:

> if $\{\sigma_1, \sigma_2, \ldots, \sigma_n\} \subseteq \mathbf{d}(w)$, then it is not the case that there could have been a representing function f_1 for w_1; . . . ; there could have been a representing function f_n for w_n; such that, had the world been w, no g would have been a representing function for w such that $g(\sigma_1) = f_1(\sigma_1)$, $g(\sigma_2) = f_2(\sigma_2), \ldots, g(\sigma_n) = f_n(\sigma_n)$.

One might wonder if the identity statements above presuppose some kind of transworld identity.

Let us examine the Consistency Condition with this question in mind. For our purposes, it would be enough to consider the case in which $n = 1$, so that we can restrict our consideration (dropping the subscript '1') to:

> it is not the case that there could have been a representing function f for w such that, had the world been u, no g would have been a representing function for u such that $g(\sigma) = f(\sigma)$.

Let us focus on what follows the phrase 'it is not the case that', since that is the part that would presuppose transworld identity, if anything would.

> there could have been a representing function f for w such that, had the world been u, no g would have been a representing function for u such that $g(\sigma) = f(\sigma)$.

I see no reason to suppose that this statement presupposes transworld identity. The underlying logical form of this part of [C-3] can be seen to be little different from that of the sentence:

> There could have been a woman who was able to high-jump six feet and who was such that, had she been given a thorough training in high-jumping techniques from the time she was ten years old, there would have been no other woman in the world who could have high-jumped as high as she could.

Here, there is no temptation to suppose that some sort of "criterion of transworld identity" is required to understand such a statement.

5. Problems, Worries, and Uses

Worries About Structures

We saw above that, for the case in which **M*** has been given an NL proto-interpretation $\mathcal{I}$ conforming to some S5 interpretation **I**, where Γ is a set of closed-sentences of **M*** and ϕ is a closed-sentence of **M***, if a derivation tells us that $\Gamma \models \phi$, we can conclude that if each member θ of Γ is such that $[\theta/\mathcal{I}]$ is true, then $[\phi/\mathcal{I}]$ must also be true. However, in order to apply the system in the way described above, we would need to have an NL proto-interpretation that conforms to some S5 interpretation. Thus, given some NL proto-interpretation, how can we determine whether there is some S5 interpretation to which the proto-interpretation conforms? Indeed, given some S5 interpretation **I**, and some NL proto-interpretation $\mathcal{G}$, might we not have difficulty determining whether or not $\mathcal{G}$ conforms to **I**? More worrisome yet, in view of the stringency of condition [C-3], might we not have great difficulties coming up with appropriate NL interpretations with respect to which we can apply the Fundamental Theorem?

In response to these difficulties, I should like to suggest that the situation is not as problematic as it may appear. Suppose that we wish to use modal logic to reason about some living human beings; to be specific, let us imagine that we are concerned with all the people *who actually live in the United States and who have one and only one Social Security number*. This allows us to construct an English predicate that gives us $\mathcal{I}$(**d**) of an NL proto-interpretation $\mathcal{I}$ (just prefix '① is a person' to the italicized portion of the previous sentence and make the appropriate changes in grammar). Let us imagine that we have assigned, in the appropriate way, referents for the even-subscripted individual constants and also English predicates for the predicates of **M***, thus completing the specification of $\mathcal{I}$. The problem is to get an appropriate S5 interpretation to which $\mathcal{I}$ conforms.

Let us start by specifying that the second coordinate **D** of any S5 interpretation we shall consider is to be the set of natural numbers x such that x is the Social Security number of a person who satisfies $\mathcal{J}(\mathbf{d})$. Suppose that $\mathbf{J} = \langle \mathbf{W}, \mathbf{D}, \mathbf{w}^*, \mathbf{d}, \mathbf{r}, \mathbf{e} \rangle$ is some S5 interpretation which is such that, with respect to the pair $\langle \mathcal{J}, \mathbf{J} \rangle$, conditions [C-1], [C-2], and [C-4] are satisfied. It follows that, for every w in **W**, had the world been w, there would have been at least one representing function for w. Now, in typical cases of this sort, among the representing functions for w, there will always be one that can be singled out as special, namely that function that maps each member n of $\mathbf{d}(w)$ to that person who has Social Security number n. Let us call this function 'the preferred representing function'. Then, in typical cases of this sort, if we counted only preferred representing functions as representing functions, the consistency condition [C-3] would be satisfied. Thus, if we replace each occurrence of the term 'representing function', in the four conditions that must be satisfied in order for an NL proto-interpretation to conform to an S5 structure, with occurrences of the term 'preferred representing function', the resulting set of conditions can easily be satisfied in cases of this sort, so that an appropriate S5 interpretation to which $\mathcal{J}$ conforms* can be found (or at least reasonably postulated to exist)—where 'conforms*' has this new sense of conform. Then, we can apply a version of the Fundamental Theorem (call it 'the Fundamental* Theorem'), obtained by replacing all occurrences of the terms 'representing function' and 'conforms' (as well as the different grammatical forms of these two terms) in the statement and proof of the original theorem with occurrences of the terms 'preferred representing function' and 'conforms*' (and the corresponding grammatical forms of these two terms) respectively. It will now be plausible to maintain that there is an S5 structure to which $\mathcal{J}$ conforms*, so that the Fundamental* Theorem is applicable. Since, for purposes of assessing the validity of modal arguments and explaining modal reasoning, the Fundamental* Theorem has all the useful implications of the Fundamental Theorem, we can freely use, in assessing modal arguments and modal principles expressed by means of $\mathcal{J}$, the sort of analyses described at the end of Chapter 7.

One may not always be able to specify the S5 structure to which $\mathcal{J}$ conforms*, but, in most cases in which the system is being applied, it is not necessary that one specify any specific S5 interpretation to which the relevant NL proto-interpretation is to conform (or conform*). In assessing some inference or chain of inferences, it may be necessary that *there be* some S5 interpretation to which one's NL proto-interpretation conforms (or conforms*), but generally, one need not know which one.

It should also be noted that, in general, this [C-3] problem should not arise for the case in which we are dealing with interpretations of the modal sentential calculus **MSC**. Nor should there be great difficulties, in general, with NL proto-interpretations of **M*** that restrict the domains of the quantifiers to countable (finite or countably infinite) totalities, as will become apparent from the following considerations.

Let us suppose that we wish to reason (modally) about some countable set S of (actually existing) individual things, each of which in theory could be denoted by some English definite description. And assume that the English predicates one wishes to use can be so chosen that the resulting NL proto-interpretation would not only satisfy the Serious Actualism and the compossibility conditions, but it would be such that, for appropriate S5 interpretations, conditions [C-1], [C-2], and [C-4] would be satisfied. Thus, we would have an NL proto-interpretation, say $\mathcal{J}$, such that each element of S is denoted by some even-subscripted individual constant. Corresponding to $\mathcal{J}$, there will be an S5 interpretation $\mathbf{I} = \langle \mathbf{W}, \mathbf{D}, \mathbf{w}^*, \mathbf{d}, \mathbf{r}, \mathbf{e} \rangle$ of the sort sketched above, which is such that **D** is the set of even natural numbers. Then, in theory, there should be an NL proto-interpretation, say $\mathcal{J}$, such that each element of S is denoted by some even-subscripted individual constant. Corresponding to $\mathcal{J}$, there will be an S5 interpretation $\mathbf{I} = \langle \mathbf{W}, \mathbf{D}, \mathbf{w}^*, \mathbf{d}, \mathbf{r}, \mathbf{e} \rangle$ such that **D** is the set of even natural numbers. Then, for every $w \in \mathbf{W}$, had the world been w, there would have been only one representing function of w, since any representing function f of w would have to be such that $\mathrm{f}(\mathbf{r}(\alpha_n)) = \mathrm{den}[\alpha_n/\mathcal{J}]$. Thus, condition [C-3] will be satisfied, and $\mathcal{J}$ will be an NL interpretation conforming to **I**. So for this case, there will be no barrier to applying the Fundamental Theorem.

Still, it must be admitted that such applications presuppose domains of relatively small cardinality. What is to be done when one is dealing with huge infinite domains? Consider, to take an extreme example, Descartes's claim in the *Meditations* that the soul is essentially indivisible. Suppose that we wish to analyse this claim and that we are not concerned with analysing the modal term 'indivisible'. We can then use **M*** to analyse what Descartes was claiming by giving the language the following (partial) NL proto-interpretation:

S^1: ① is a soul;
I^1: ① is indivisible.

We should investigate the two claims:

$$\text{(a) } \Box(x)(S^1x \rightarrow I^1x);$$

and

$$\text{(b) } \Box(x)(S^1x \rightarrow \Box(E^1x \rightarrow I^1x)).$$

But how can we be confident that there is any S5 interpretation of **M*** to which any NL proto-interpretation with such predicates will conform* (assuming that we could come up with some way of selecting preferred representing functions)? In particular, think of how many souls there might have been? Is there any cardinal limit to what the totality of souls (in Descartes's sense of that term) might have been? At this point, one might begin to wonder whether any set theoretical structure could capture what we are trying to express.

Troublesome as this problem is, it should be noted that similar problems arise for the Modal Realist. We saw earlier, in investigating Plantinga's views, that the Realist's totality of possible worlds and her domain of possible objects (or essences) would have to be enormous—so much so, in fact, that some philosophers would maintain that they could not be sets at all. Yet, many Modal Realists make use of the results of model theory in their applications of modal logic, without worrying about the fact that the structures dealt with by the mathematical modal logician are taken to be sets.[2] In those cases, the modal logician may be simply assuming that the set theoretical structures of her semantical system are sufficiently close to the "reality" she intends to be reasoning about to allow such applications. Strictly speaking, however, that assumption is far from trivial. It is an assumption that is similar (in one respect) to the sort of assumption that may be needed to make certain applications of my anti-Realist system, e.g. that there is an S5 interpretation of **M*** to which some NL proto-interpretation conforms*.

One way of dealing with the problem for this case may be to analyse Descartes's claim in terms of a reduced domain. If we suppose, with Descartes, that there actually are such things as souls, then we can specify the domain of our NL proto-interpretation by means of some such English predicate as:

> ① is one of the triple, Bill Clinton's soul, Hillary Rodham Clinton's soul, or Bob Dole's soul.

It can be plausibly argued that the NL proto-interpretation one gets when one fills out the assignment is an NL interpretation and that there will be no problem with coming up with a set theoretical S5 structure to which it conforms. We can then use the above sentences, (a) and (b), in our analysis, with the conviction that, even though the NL proto-interpretation we have given does not yield everything that Descartes had in mind, the reduction in the domain will produce an analysis that is sufficient for our purposes.

It can also be seen that, for many philosophical purposes, there is no necessity to apply the Fundamental Theorem. In the case of Descartes's

[2] Plantinga, for example, does not concern himself in Plantinga 1974 with the question of whether the totality of essences is or is not a proper class. There are philosophers who have concerned themselves with this question. See Lear 1977. Vann McGee discusses some aspects of this problem in his McGee 1992. Cf. my comments on McGee's views in this connection in Chihara 1998.

claim, what one wants is clarification and disambiguation. One wants to know: did Descartes mean this or that? For that sort of purpose, one does not need an NL proto-interpretation of **M***that conforms to some S5 interpretation. Suppose that one believed that the notion of soul is just too vague or unclear to yield any sort of set theoretical model. Even so, one can use the two sentences of **M*** as guides for the construction of English sentences that express the two alternatives in question. Thus, we can take alternative (a) to say:

> No matter how the world might have been, every soul would have been indivisible,

and take alternative (b) to say:

> No matter how the world might have been, every soul would have been essentially indivisible.

The reason is this. One knows how a sentence of an NL interpreted language **M*** "works" when there is an S5 interpretation of **M*** to which the NL proto-interpretation conforms. In that case, we have the set theoretical machinery to model its logical features; we have a syntactic system of rules of proof with which to derive its consequences; and we can study the modal system by placing it among a whole family of related logical systems. The fact that we can model the semantics of the language for such clear cut cases of NL interpretations, gives us grounds for having confidence in our grasp of the basic logical features of the language, so that when we use NL proto-interpreted sentences in situations where we are unable to apply the set theoretical machinery, we are still reasonably confident of our grasp of the basic logical features of these sentences. Thus, we can be confident that we have distinguished above two quite distinct but acceptable English readings of the Cartesian sentence, based upon our grasp of the two sentences of **M*** given earlier. In summary, even in cases where there is so much vagueness, unclarity, excessive size, or complexity in an NL proto-interpretation of **M*** that there seems to be no set theoretical structure to which the interpretation can conform (as undoubtedly is the case in the Cartesian example), one can still use the NL proto-interpretated language to make distinctions, bring out ambiguities, and offer explanations—which, after all, is one of the principal uses of modal logic in philosophical investigations. (Recall the discussion of Plantinga's examples earlier in this chapter.)

Heuristic and Didactic Uses of S5 Structures

Closely related to the above uses of modal logic in philosophy is the use of S5 structures as heuristic devices for investigating systems of modal reasoning or, alternatively, as didactic devices for clarifying the logic of some conceptual modal notions. It was just such uses of S5 structures in Chihara 1990 that gave rise to the objection, by reviewers, that I thereby committed myself to the existence of possible worlds.

Suppose that **I** = <**W**, **D**, **w***, **d**, **r**, **e**> is an S5 interpretation of **M*** and that $\mathcal{I}$ is an NL proto-interpretation conforming to **I**. Let w be some specific member of **W**. Then, the world could have been such that there was a bijection f_w from $\mathbf{d}(w)$ to what was the extension of $\mathcal{I}(\mathbf{d})$ such that

(1) for every predicate θ of any degree n, and for every n-tuple of members of $\mathbf{d}(w)$, $<d_1, d_2, \ldots, d_n>$:

$$<d_1, d_2, \ldots, d_n> \in \mathbf{e}(\theta, w) \text{ iff } <f_w(d_1), f_w(d_2), \ldots, f_w(d_n)> \in \text{ext}[\theta/\mathcal{I}];$$

and

(2) for every even-subscripted individual constant α such that $\mathbf{r}(\alpha) \in \mathbf{d}(w)$,

$$f_w(\mathbf{r}(\alpha)) = \text{den}[\alpha/\mathcal{I}].$$

In this way, $\mathbf{d}(w)$ tells us what the extension of $\mathcal{I}(\mathbf{d})$ would have been, and $\mathbf{e}(\theta, w)$ tells us what ext$[\theta/\mathcal{I}]$ would have been. Thus, the ordered triple $<\mathbf{d}(w), \mathbf{r}(\alpha), \mathbf{e}(\xi, w)>$ functions as a mathematical representation of how the world could have been, in so far as the domain $\mathcal{I}(\mathbf{d})$ of the quantifiers, the referents of the even-subscripted individual constants, den$[\alpha/\mathcal{I}]$, and the extensions of the predicates, ext$[\theta/\mathcal{I}]$, are of concern.

The S5 structure, **I** itself, gives us a set of such mathematical representations—it gives us a kind of total picture—the totality of representations of how the world could have been, in so far as . . . Thus, once we realize that S5 structures can provide us with such sets of mathematical representations, it is easy to see why it can be useful and enlightening to reason in terms of these "possible worlds" structures. For one thing, most philosophers are much more adept at reasoning about such mathematical entities than about modal concepts. We have a precise and well developed theory of such things; and we have received much more training in dealing with sets and set theoretical relationships than we have in dealing with modal statements and entailments. Furthermore, when we place modal reasoning within the framework of set theory—more specifically, model theory—we can then comfortably operate in a familiar way in a relatively familiar field. Thus, we can see why such structures can be useful in explaining the logic of the constructibility quantifiers I set forth in Chihara 1990 and why I would find such S5 structures of heuristic value in devising proofs involving these constructibility quantifiers. In view of Chapters 6 and 7 above, it can now be seen that *such didactic and heuristic uses of S5 structures do not presuppose, entail, or require the existence of possible worlds, as was suggested by the reviewers* (this will be argued in detail in the following chapter).

What needs to be emphasized, in any case, is that heuristic and didactic uses of possible worlds semantics do not, in general, presuppose that all S5 structures have NL proto-interpretations that conform to them. Consider, in this connection, how our system of possible worlds structures can aid in the explanation of some refined modal distinction. Take the distinction, made in the previous chapter, between the statement:

> [A] Some $\mathcal{I}(\mathbf{d})$-[$\beta_1/g(\sigma_1)$, $\beta_2/g(\sigma_2)$, . . ., $\beta_n/g(\sigma_n)$]-sequence *satisfies* the condition: [$\theta/\mathcal{I}$] truly describes what would have been the case, had the world been *w*,

and

> [B] Some $\mathcal{I}(\mathbf{d})$-[$\beta_1/g(\sigma_1)$, $\beta_2/g(\sigma_2)$, . . ., $\beta_n/g(\sigma_n)$]-sequence would have satisfied [$\theta/\mathcal{I}$], had the world been *w*.

This distinction can be clarified, in terms of possible worlds structures, by noting simply that, in the case of statement [A], one takes the elements of the sequence from the actual world **w*** but then determines what is true of these elements at world *w*, whereas, in the case of statement [B], one both takes the elements of the sequence from world *w* and also determines what is true of these elements at that world. Notice that, in order to understand this explanation, the hearer need not know anything specific about the particular S5 interpretation **I** in question. All that is required is general knowledge of how the ordered triples $<\mathbf{d}(w), \mathbf{r}(\alpha), \mathbf{e}(\xi, w)>$ function as mathematical representations of how the world could have been and how truth at a world under an S5 interpretation is determined. It should be clear that this use of the system of possible worlds structures does not presuppose that all the S5 structures have NL proto-interpretations that conform to them.

These last two sections indicate just some of the options, techniques, and strategies available to the philosopher who wishes to avoid the commitment to possible worlds while making use of modal logic. Creative spirits can no doubt devise many others.

*Sentences that do not Translate into M**

Recall that Forbes ran into great difficulty finding convincing reverse translations in his formal language of some modal sentences. He felt compelled, by some examples advanced by Hazen, to expand the vocabulary of his modal language to include a number of novel modal operators. How, it might be asked, do I respond to such examples? Might I not be forced to expand my language **M*** to include new modal operators? Related to that problem, there are many things we say in ordinary everyday situations that seem to involve quantification over possibilities or possible situations. For example, we sometimes say such things as 'There is a possibility that I won't be able to make your dinner party'. How do I analyse such things?

I shall take up the latter problem first. For many typical cases like that of the example, such *existential* talk is simply not to be taken literally. For all practical purposes, one could just as well say 'It is possible that I won't be able to make your dinner party'. In other words, although we do say such things, they are not things an anti-Realist need assert when they are taken to be (literally) existential assertions. After all, we say all sorts of things, in everyday life, that ought not to be taken literally. 'I'll be damned', 'We'll slaughter them', and 'You'll eat those words' are just a few examples.

No doubt, however, there are sentences that one can construct, which really do express the existence of possibilities and which cannot be expressed in some ontologically innocent way. What is my response to such assertions? Well, why should we suppose that such assertions are true? It seems to me that the burden of justification rests upon the person making the assertion. If the claim is that there are ordinary everyday assertions that are both ontologically committed to the existence of possible worlds and also clearly true and justified, then I would like to see just what the assertions are and how they are justified. I can only say here that I know of no such assertion and that I doubt very much that any really convincing examples of such assertions can be given

Turning now to the case of modal sentences of natural languages that do not translate into the vocabulary of my formal modal language, I should first emphasize that Forbes found such examples troubling because of his acceptance of the reverse translation thesis. Recall that, because of his adoption of that thesis, Forbes felt it necessary to defend the view that "every meaningful possible worlds sentence is a rendering of some modal sentence" (Forbes 1985: 81). But I have rejected the reverse translation thesis. And there is nothing in my position that requires me to defend such an extreme view of worlds sentences. I see no reason at all to deny that the Modal Realist's sentence 'There is a possible world in which donkeys talk' is a meaningful sentence, the truth of which requires the existence of possible worlds. Of course, I would deny the truth of such a worlds sentence, but that is very different from claiming that the sentence makes no reference to worlds or is meaningless.

I should also emphasize that I have not claimed (nor do I see the need to claim) that every meaningful modal sentence any philosopher may think up can be expressed in the language **M***. Take the sentence

My car **a** could have been the same colour as your car **b** is actually

which I discussed in Chapter 4. To deal with this sentence, you will recall, Forbes had to expand the vocabulary of this modal language to include complex modal operators that were both subscripted and superscripted.

But I see no reason to revise my language **M*** to express this sentence. The above sentence already makes no reference to worlds. So I see no need to give a world-free analysis of the sentence. Furthermore, as of now, the sentence plays no important role in any important scientific or philosophical theory, which cries out for formalization in a formal modal language. Nor have I come across any philosophical arguments for the conclusion that possible worlds exist, the analysis of which requires a formalization of that sentence. So in the present situation, I see no pressing need to develop a formal language to express such sentences in an expanded version of **M***.

Circularity in my Use of Subjunctives?

In my own account of the semantics of S5 modal logic, I have made heavy use of (subjunctive) counterfactual conditionals. Now there are philosophers (Lewis and Stalnaker being outstanding examples) who believe that such conditionals need to be analysed as statements about possible worlds. Such a position could be thought to raise problems for me, since it can be argued that I cannot give my anti-Realist account of possible worlds semantics, without requiring, at the same time, the acceptance of the Modal Realist's possible worlds analyses of these conditionals. Such an argument would open my account to the charge of being viciously circular. So what am I to say in response to these advocates of a possible worlds analysis of counterfactual conditionals?

I cannot hope to tackle this enormously complex and deep topic in a relatively short space; it would take at least another book to treat this question adequately. So what I shall say here should be taken to be only suggestions of how a response could be constructed.

First, it is not obvious to me that a possible worlds analysis is the only acceptable way or even the most promising way of analysing counterfactual conditionals. There are promising competing analyses in the field, which do not presuppose possible worlds at all, but which nevertheless deserve serious consideration. In this regard, I would mention the metalinguistic approach of Nelson Goodman and Roderick Chisholm, the suppositional account of J. L. Mackie, and the inferential view of Igal Kvart.[3] Second, I see no problem at all in a philosopher giving explanations, clarifications, and analyses of counterfactual conditionals, in which talk about possible worlds is used as a heuristic or didactic device. Kvart, for example, finds it useful for certain purposes to make reference to (explicitly defined) "possible-world-descriptions" in some of his analyses:

> We have seen that the possible-world-description model was quite helpful for heuristic and motivational purposes . . . Here we have used not possible worlds, but

[3] See Goodman 1973, Chisholm 1946, Mackie 1973, and Kvart 1986.

possible-world *descriptions*, and so we are not committed to possible worlds in any way. (Kvart 1986: 55)

Finally, I do not find any of the possible worlds analyses of counterfactual conditionals I have seen to be so convincing that I am tempted to abandon my ontological scruples and accept belief in possible worlds in order to accept such an analysis.

6. My Modal Analysis does not Eliminate Modality

Since my analysis of S5 possible worlds semantics explicitly makes use of modal notions, it might be thought that I do not provide a "reduction" of modal notions to non-modal ones in the way that Lewis's analysis does. Admittedly, I do not show how to eliminate modal notions from one's "total theory". How destructive of my position is this feature? So far as I can see, no one has constructed such a "reduction". In what follows, I shall argue, in particular, that Lewis has not managed to effect a genuine reduction of modal principles and concepts to purely non-modal ones.

In order to explain my reasons for questioning the widely held view that Lewis has effected a genuine reduction of the modal to the non-modal, I shall examine, in more detail than I have provided thus far, the main features of Lewis's *Counterpart theory* which was set forth in Lewis 1979*a*. In what follows, I shall take it to be a deductive theory formalized in the first-order language **FL** described in Chapter 5. The vocabulary of *Pure Counterpart theory* will then be a set whose members are the following non-logical constants (I omit the superscripts):

$$W \; A \; I \; C$$

The first two displayed above are unary predicates; the latter two are binary predicates. Under the intended interpretation, they are to be understood as follows:

W: ① is a world
A: ① is actual
I: ① is in ②
C: ① is a counterpart of ②.

To be a sentence of pure Counterpart theory, a sentence of **FL** may contain occurrences of nonlogical constants only if the constants are members of the vocabulary. It will be convenient to expand the vocabulary of Counterpart theory in order to express other ideas, resulting in an *applied Counterpart theory*. The set of assertions of pure or applied Counterpart theory will consist of those sentences of the theory that are consequences of the following eight axioms:

A1: $(x)(y)(Ixy \rightarrow Wy)$.
A2: $(x)(y)(z)(Ixy \,\&\, Ixz) \rightarrow y = z)$.
A3: $(x)(y)(Cxy \rightarrow (\exists z)Ixz)$.
A4: $(x)(y)(Cxy \rightarrow (\exists z)Iyz)$.
A5: $(x)(y)(z)(Ixy \,\&\, Izy \,\&\, Cxz) \rightarrow x = z)$.
A6: $(x)(y)(Ixy \rightarrow Cxx)$.
A7: $(\exists x)(Wx \,\&\, (y)(Iyx \leftrightarrow Ay))$.
A8: $(\exists x)Ax$.

These axioms can be understood to be saying:

A1: It is only a world that a thing can be in.
A2: A thing can be in only one world.
A3: Everything that is a counterpart of something is in a world.
A4: Everything that has a counterpart is in a world.
A5: Of all the things that exist in the same world as something, the only one that is a counterpart of that thing is the thing itself.
A6: Everything in a world is a counterpart of itself.
A7: There is a world which has in it all and only actual things.
A8: Something is actual.

Now how are we supposed to get modal logic in this system? Lewis gives us a translation scheme that allows us to regard sentences of quantificational modal logic as translations of sentences of **FL**. Sentences of modal logic are not sentences of Counterpart theory: they are sentences of another language that are short-hand or abbreviated versions of sentences of **FL**. (This may bring to mind Russell's no-class theory, according to which sentences of set theory are merely abbreviations of sentences of Russell's theory of propositional functions.)[4] One advantage of developing modal logic in this way is that we do not need a special kind of logic: we simply use standard first-order logic to carry out our deductions. Here, then, is the effective set of rules of translation that he gives.

Lewis's Translation Scheme

Suppose that ϕ is a formula of first-order logic. And suppose that n individual symbols occur free in ϕ. (Remember: any occurrence of an individual constant in ϕ is a free occurrence.) And suppose that $\alpha_1, \alpha_2, \ldots, \alpha_n$ are n individual symbols, not necessarily distinct. Then $\phi\alpha_1 \ldots \alpha_n$ is the formula that results from replacing all free occurrences of the first individual symbol occurring in ϕ with α_1, replacing all free occurrences of the second individual symbol occurring in ϕ with $\alpha_2, \ldots$, replacing all free occurrences of the nth individual symbol occurring in ϕ with α_n.

[4] For a detailed discussion of Russell's no-class theory, see Chihara 1973, ch. 1, sect. 4.

1. The translation of ϕ is $\phi^{@}$ (ϕ holds in the actual world)
   i.e. $(\beta)((\alpha)(I\alpha\beta \leftrightarrow A\alpha) \,\&\, [\phi]^{\beta})$

where '$[\phi]^{\beta}$' is to be understood according to the following recursive definition:

2. $[\phi]^{\beta} = \phi$, if ϕ is atomic.
3. $[-\phi]^{\beta} = -[\phi]^{\beta}$
4. $[\phi \,\&\, \chi]^{\beta} = [\phi]^{\beta} \,\&\, [\chi]^{\beta}$
5. $[\phi \vee \chi]^{\beta} = [\phi]^{\beta} \vee [\chi]^{\beta}$
6. $[\phi \rightarrow \chi]^{\beta} = [\phi]^{\beta} \rightarrow [\chi]^{\beta}$
7. $[\phi \leftrightarrow \chi]^{\beta} = [\phi]^{\beta} \leftrightarrow [\chi]^{\beta}$
8. $[(\alpha)\phi]^{\beta} = (\alpha)(I\alpha\beta \rightarrow [\phi]^{\beta})$
9. $[(\exists\alpha)\phi]^{\beta} = (\exists\alpha)(I\alpha\beta \,\&\, [\phi]^{\beta})$

[In the following, $\alpha_1 \ldots \alpha_n$ are distinct individual symbols, and the variables $\gamma_1, \ldots, \gamma_n$ are the first n variables that do not occur within the scope of the quantifiers.]

10. $[\Box\phi\alpha_1 \ldots \alpha_n]^{\beta} = (\delta)(\gamma_1) \ldots (\gamma_n)((W\delta \,\&\, I\gamma_1\delta \,\&\, C\gamma_1\alpha_1 \,\&\, \ldots \,\&\, I\gamma_n\delta \,\&\, C\gamma_n\alpha_n) \rightarrow [\phi\gamma_1 \ldots \gamma_n]^{\delta})$
11. $[\Diamond\phi\alpha_1 \ldots \alpha_n]^{\beta} = (\exists\delta)(\exists\gamma_1) \ldots (\exists\gamma_n)((W\delta \,\&\, I\gamma_1\delta \,\&\, C\gamma_1\alpha_1 \,\&\, \ldots \,\&\, I\gamma_n\delta \,\&\, C\gamma_n\alpha_n) \,\&\, [\phi\gamma_1 \ldots \gamma_n]^{\delta})$

For heuristic purposes, it is useful to treat expressions of the form '$[—]^{\beta}$' as giving us conditions under which — is true at world β. Thus, clause 11 can be regarded as telling us that $\Diamond Rab$ is true at world β iff there is a world γ in which some counterpart x of a exists and also some counterpart y of b exists such that x is in the relation R to y.

Notice that we don't say that x is in the relation R to y at γ. According to the Lewis theory, properties and relations do not have extensions *at worlds* (as they do according to the standard "canonical" interpretation of **M**); they just have extensions. Recall that the sixth component of an S5 interpretation assigns an extension to a predicate at each world, the idea being that an individual, say Bill Clinton, might satisfy a predicate at one world, and not in another. This cannot happen according to Counterpart theory: Bill Clinton can exist in only one world, and he either satisfies the predicate or he doesn't.

Modal Principles

An examination of the modal principles that come out valid under this translation scheme will indicate the kind of modal logic that will result. It is easy to show that the following principle is valid:

$$\Box\phi \rightarrow \phi.$$

(This may surprise some readers, since the sentence '$\Box Fa \rightarrow Fa$' was found to be an invalid sentence of the counterpart modal semantics given in Chapter 2. The industrious reader may wish to investigate the differences in the two approaches that yield this striking difference in validity.) Furthermore, for any sentence ϕ containing no free occurrences of individual symbols,

$$\Diamond\phi \rightarrow \Box\Diamond\phi$$

is also valid. Indeed, it is not difficult to show that if one restricts the sentences of this short-hand theory to just sentences containing no free occurrences of individual symbols, one would get an S5 modal logic.

Consider, however, the following instance of the Brouwer axiom:

$$Fa \rightarrow \Box\Diamond Fa$$

is not inferable, in Counterpart theory, from the axioms. But, although the above is not valid, it would be valid if the counterpart relation were symmetric. Since the Brouwer axiom is a theorem of standard versions of S5 modal logic, it follows that Lewis's theory is not, strictly speaking, an S5 system.[5]

Has Lewis Given a Non-Modal Explanation of Modality?

Now initially, it may seem that Lewis has given a completely non-modal analysis of modality. After all, the modal terms appear as defined terms in Lewis's system, explicated in such a way that they do not themselves presuppose or involve modal notions. Thus, Lewis could plausibly claim that there are no modal terms among the primitives of his theory. Based upon an examination of the vocabulary of Counterpart theory, one could easily come to the conclusion that no modal concepts are used to define any of the non-logical constants of the theory. Upon reflection, however, suspicions may arise, especially concerning the terms 'actual' and 'world'. One might suppose that 'actual' is really a modal term since it seems to pick out one out of a multitude of possible worlds. In response, Lewis could argue that his indexical analysis of actuality provides an adequate resolution of this worry.

How about 'world'? Here, Lewis's mereological definition of 'world' is crucial. A world, according to Lewis's view, is a mereological sum of all the things that are spatiotemporally related. None of the notions used in this explanation seems to be modal, so Lewis can argue that his notion of world is non-modal.

[5] Indeed, it does not yield S4 either, since the characteristic axiom of S4, namely $\Box\phi \rightarrow \Box\Box\phi$, is not valid, although it would be if the counterpart relation were required to be transitive. This and the previous point are made by Lewis in Lewis 1979*a*: 121–2.

As for the relation 'in', which is another primitive of Counterpart theory, it can be defined mereologically: x is in β iff x is a part of a mereological sum of all the things that are spatiotemporally related. Finally, we have the counterpart relation, and although this notion has not been given a clear definition, it is supposed to be some sort of similarity relation—which does not seem to be modal.

Shalkowski's Doubts

Despite the above considerations in favour of the thesis that Lewis has effected a reduction of modal notions to non-modal ones, Scott Shalkowski has argued (in Shalkowski 1994) that Lewis's analysis of modal notions presupposes modal principles. He maintains that the Modal Realist's ontology must satisfy the following two conditions: (1) the worlds must be sufficiently abundant so as to exhaust all the possibilities—they must not leave any gaps in the space of possibilities; and (2) the worlds must not be so abundant so as to include impossible worlds. He then argues: "If the modal realist's ontology meets these conditions, a reduction of modality in terms of possible worlds and their constituents is circular" (Shalkowski 1994: 80).

Why is this so? Of course, if the reducing theory literally asserts that *the worlds are all possible* and *that the worlds exhaust all the possibilities,* then the reducing theory surely does not reduce modality to the non-modal, since the reducing theory explicitly makes assertions containing modal terms. But the question that needs to be answered is this: *must the reducing theory make those assertions or at least presuppose or imply these assertions in order to meet these conditions*? To put it another way, could not the reducing theory meet the above conditions, even though none of the above assertions (nor any assertions that imply them) is included among its assertions? Might not the worlds satisfy the two conditions even though the theory makes no such modal assertion, as is claimed by Shalkowski?

Let us suppose that the theory makes no such modal assertion. Then, to that extent, whether the above two conditions are met will not be decided by the theory itself. Should we then conclude that the theory is inadequate? Shalkowski thought so, claiming that, if the theory does not assert or imply that these two conditions are satisfied, the reduction would be seen to be as "arbitrary a reduction as the reduction in terms of impossibilia or bottle caps" (Shalkowski 1994: 80). Shalkowski evidently thought that the reducer would be in the untenable position of having no idea of whether the things in the domain of the theory could do the job they are being asked to do.

To obtain some perspective on this conclusion of Shalkowski's, let us consider the following question: is it legitimate for the Modal Realist to ask if his postulated domain of worlds is adequate for the job of representing all the possibilities, and then determine that it is—without being involved in a circularity? Lewis's answer to this question is to be found in his discussion of a problem he poses for the ersatzist. He supposes that the ersatzist has specified his axioms *guided by his "pre-existing modal opinions"*. Then, he considers the question: even if the ersatzist ends with a correct definition of necessity and possibility, wouldn't the procedure be circular? No, not at all, is the reply Lewis gives.

> Circularity is a matter of what you analyse in terms of what; *it is not a matter of why you think your analysis is right*. It is not circularity if you build your analysis to give the answers it ought to, exercising your understanding of the *analysandum* as you go. You might as well say that an analysis of 'cousin' in terms of 'parent' was circular, because you wouldn't have given that analysis if you hadn't already a working command of the word 'cousin'! The object of analysis is to reduce our burden of primitive notions, and to make tacit understanding explicit—not to bootstrap ourselves into understanding what we didn't understand at all beforehand. (Lewis 1986: 154, italics mine)

Consider, in the light of the above quote, Lewis's world-analysis of the modal operators. It is clear that he makes use of a fund of modal opinions in devising his analysis. But this use would not undermine his claim to be giving a purely non-modal account of modality, if these modal opinions were merely used to confirm that his world analysis was right. Such a use would not be circular since he would only be exercising his understanding of the *analysandum* in order to test his analysis.

A Comparison to Logicism

To appreciate Lewis's position on this score, let us compare

> [1] Lewis's attempt to reduce the assertions and concepts of modal logic to the assertions and concepts of a theory of worlds

with

> [2] the Logicist's attempts to reduce the assertions and concepts of classical mathematics to the assertions and concepts of logic.

Certainly, Logicists began their project in possession of a fund of mathematical beliefs and mathematical concepts. Not even their most severe critics objected to the Logicist's analyses on that score.[6]

Poincaré understood the Logicist's programme in the following way: "Can mathematics be reduced to logic without having to appeal to principles peculiar to itself?" (Poincaré 1953: 143). Later, he put it this way:

[6] See, for example, Poincaré 1953: 143 ff.

Well, what I want to find out is, whether it is true that once the principles of logic are admitted we can, I will not say discover, but demonstrate all mathematical truths without making a fresh appeal to [mathematical] intuition.[7]

Russell and Whitehead are widely thought to have effected a reduction of classical mathematics to the logical theory of *Principia Mathematica*. The reduction supposedly was carried out in the following way. One has two theories, Classical Mathematics, henceforth 'CM', on the one hand, and *the logical theory of Principia Mathematica*, or 'PM' for short, on the other. The authors produce a translation scheme whereby sentences of CM get understood as abbreviations of sentences of PM. Then it is claimed that the assertions of CM are just abbreviations of theorems of PM. One can conclude, they thought, that mathematical theorems are just disguised theorems of logic.

Unfortunately, there was a huge flaw in this reduction. Russell and Whitehead couldn't prove all the theorems of CM in PM—at least with the logical laws that they were sure of. In desperation, they had to include, among their logical laws, the infamous Axiom of Reducibility, which they attempted to justify by saying that it was needed to prove certain (mathematical) theorems of CM that they could not prove without this axiom.[8] In other words, one could not justify this axiom by appealing only to *logical intuitions*. Our logical sense did not tell us that the Axiom of Reducibility was true—the plausibility of this axiom came from the fact that it had certain consequences, such as the least upper bound theorem, that are theorems of Classical Mathematics and that we wanted to assert. Thus, it was our *mathematical sense* that supported this conviction. In other words, the reduction did not work by giving the translation scheme and then relying just on logic to prove all the theorems of classical analysis. On the contrary, it was Classical Mathematics that was used to justify accepting as a law of logic a proposition that did not seem to be true even to Russell.

Needless to say, this way of reducing mathematics to logic was highly controversial. There was no objection to using preanalytic beliefs, about what are or should be the theorems of classical mathematics, *to test the claim that classical mathematics can be derived in PM*; but there was an objection to using our knowledge of classical mathematics to infer what should be a principle of logic. Using *mathematical beliefs to justify including the axiom of reducibility among the principles of logic* engendered a great deal of scepticism about the purported reduction of mathematics to logic.

To see how this example from the philosophy of mathematics is relevant

[7] Poincaré 1953: 149. For a discussion of Poincaré's objections to Logicism, see Chihara 1996.

[8] See Chihara 1973: ch. 1, sect. 9.

to the question of whether Lewis has reduced modality to the non-modal, let us judge whether the following analysis of modal operators effects an acceptable reduction of modality to the non-modal.

Thoughts in the Mind of God: An Analysis of Modality

According to this analysis, God has thoughts about other universes. Being God, His thoughts about other universes are complete in the following way: the universe is conceived in complete detail, down to the smallest of minutiae (no detail is left undetermined by His thoughts). Now these *complete thoughts of universes* will be said to have thoughts as parts. The idea is this: if the thought is of a universe in which Smith has eggs for breakfast on Thursday, 28 September 1995, then part of this complete thought is the thought that, in this universe, Smith has eggs for breakfast on Thursday, 28 September 1995. (I am not suggesting, here, that this notion of "part" is the part relation that is the subject of mereology.)

A rough idea of this analysis can now be gleaned from the following two formulas:

> $\Diamond\phi$ iff there is a *complete thought of a universe* C such that the thought that, in this universe, ϕ is a part of C;
>
> $\Box\phi$ iff every *complete thought of a universe* C is such that the thought that, in this universe, ϕ is a part of C.

What about *de re* sentences with nested operators? Take the sentence

$$\Diamond(\exists x)\Box(x \text{ exists} \rightarrow x \text{ is rational}).$$

This sentence is true iff there is a *complete thought* C_1 *of a universe* such that a part of C_1 is the thought that, in this universe, there is some individual x and this latter thought is such that, every *complete thought of a universe* C_2 which is of a universe in which x exists, has as a part the thought that, in this universe, x is rational.

Now *de re* sentences are supposed to bring on problems with transworld identity. Do we have an analogous problem here? If God has a *complete thought* T_1 of a universe in which an individual x has property F and He has a *complete thought* T_2 of a universe in which an individual y has property G, what makes T_1 and T_2 thoughts about the same individual (x and y)? Remember, *complete thoughts of a universe* are complete down to the smallest detail; and one detail of a complete thought will be whether the individual x that has property F in the universe in question is or is not the same individual as the individual y that has the property G in the universe thought of in complete thought T_2. Speaking very roughly here, one might say that the complete thought T_1 will contain the thought that x is the very same individual as the individual y thought of in T_2.

I will not continue to develop this analysis of modality. The main outlines of this analysis should have emerged in enough detail for my purposes. Consider now how the proponents of this analysis might respond to criticisms. The objection is raised: "This is an absurd analysis—how would you even know that you could have had eggs for breakfast this morning? You don't know what is in the mind of God." We can imagine the reply: "I do know what is in the mind of God. I have the preanalytic certitude that I could have had eggs for breakfast this morning, so I conclude (on the basis of my analysis) that there is a *complete thought of a universe* a part of which is the thought that, in this universe, I have eggs for breakfast." Here, the proponent of this analysis used his preanalytic beliefs about modality to guide him in deciding what thoughts he should attribute to the mind of God. This is not a case of using preanalytic beliefs as a test to see if the analysis works. The preanalytic beliefs are functioning to determine what the reducing theory should assert. And they form the ground for the claim that it is known that there is a complete thought of the appropriate sort.

A Re-examination of Lewis's Reduction of Modality to Worlds

Let us grant that what was claimed in the above quote from Lewis 1986 about analysing the meaning of 'cousin' in terms of parents is true enough. If I were attempting to give an analysis of the word 'cousin', I could use certain preanalytic truths of which I am certain, such as that Harry, Don, and Sally could be correctly said to be cousins of mine. Suppose now that I ask myself: "How are they related to me?" I know that Harry, Don, and Sally have a mother who is my mother's sister. By analysing other such examples, I can come to the conviction that a cousin is a child of a sibling of a parent. And there is no circularity in using my fund of preanalytic beliefs about cousins to test my analysis. But notice that, in this case, I establish on independent grounds that Harry has a mother, who this mother is, and that this mother is a sister of my mother. I do not use my beliefs about "cousinhood" to discover these things or to convince myself that I know such things.

This is where Lewis's procedure is very different. Lewis has no independent way of discovering what worlds there are, what the nature of these worlds are, or anything about these worlds (apart from "our world"). So how does Lewis arrive at his general principles of world existence?

The Principle of Plenitude

Consider Lewis's discussion (which occurs in Lewis 1986: section 1.8) of the principle that *absolutely every way a world could possibly be is a way that some world is*. Of course, stated in the above manner, this "principle of plenitude" is clearly modal in character and in need of analysis in a way that would eliminate the modal terms. After considering various possible analyses, Lewis concludes that none will do (Lewis 1986: 87). So he sets out to find some other way of expressing the idea that there are no gaps in possibility space. (Notice that Lewis is here, in effect, attempting to show that his worlds satisfy the sufficiency condition that Shalkowski discusses.) He then postulates a *principle of recombination*, inspired by Hume's denial of necessary connections between distinct existents, according to which *patching together parts of different worlds always yields another world* (Lewis 1986: 87). What this means is by no means clear on first reading. Lewis explains: "Roughly speaking, the principle is that anything can coexist with anything else, at least provided they occupy distinct spatiotemporal positions. Likewise, anything can fail to coexist with anything else" (Lewis 1986: 88).

This explanation of the principle is thoroughly modal in character; it enables us to see clearly enough what modal principle he is trying to capture. Can we not then translate this principle into Lewis's world language via the Counterpart translation scheme? Thus, could he not state that the counterpart of anything can coexist (or fail to coexist) with the counterpart of anything else? No. The problem is: what counts as a counterpart of an object *x* depends upon more than just whether the object is similar to *x*; it also depends upon their respective surroundings and their respective relations with other things. For this reason, Lewis resorts to stating the principle in terms of "duplicates": given various parts of different worlds, there is another world in which "duplicates" of these parts are to be found. Whether these duplicates are counterparts is another matter.

How can anyone know any such thing? In particular, how can anyone know that there is another world in which "duplicates" of my heart and lungs are to be found? It seems preposterous to suppose that anyone is in a position to know such a general fact about other worlds. The reason that Lewis thinks he can affirm his principle of recombination is because he believes this principle expresses the Humean modal principle he has in mind and he believes he can affirm the Humean principle. It should now be clear that modal beliefs are not only guiding the very formulation of this principle of recombination, but also modal beliefs form the very ground of Lewis's conviction that the principle of recombination is true.

We see that key assertions of Lewis's world theory are based on *modal intuitions*. Thus, when Lewis produces his analysis of modal notions, the *intuitive modal theory is never really dropped; it continues to function within the new framework*. Because of this, I do not believe that Lewis's "reduction of modality to worlds" is a genuine reduction in any serious sense.[9]

[9] The reader should study Bigelow and Pargetter 1987, especially sect. 4, for a quite different line of reasoning for questioning Lewis's reduction of modality to the non-modal.

7. Ontological Commitment to Worlds or Ways?

To those who have been influenced by Quine's views on ontology[10] it may seem that the analysis of the applications of modal logic given above presupposes the existence of strange entities called 'ways the world could have been'; this is because, in developing my analysis of possible worlds semantics in the previous chapters, I sometimes make use of language in a way that appears to make reference to such "ways". Since it has been argued by Lewis that such "ways" are, in fact, possible worlds[11] (the actual world supposedly being one of these possible worlds), there seem to be grounds for concluding that applications of modal logic, even according to my analysis, presuppose the existence of possible worlds.

Am I Committed to the Existence of Possible Worlds?

I shall argue that the apparent references to "ways" are not essential and that I could carry out the analysis I gave earlier without presupposing the existence of such strange entities. However, since one of the goals of this work is to explain how modal logic can be applied without presupposing the existence of *possible worlds*, I shall begin by arguing that the kind of "ways" the view apparently presupposes are not possible worlds.

It should be emphasized, first of all, that what are apparently being presupposed are not just "ways the world could have been", but rather "ways the world could have been, *in so far as the truth values of the sentential letters are of concern*" or "ways the world could have been, *in so far as the range of the quantifiers and the extensions of the predicates are of concern*". Such "ways" have much less structure than genuine possible worlds are supposed to have; these "ways" also have much less informational content than possible worlds are thought to possess. (This is fortunate, because otherwise one could not find any sort of adequate mathematical representation of such "ways".)

To grasp this last point, consider one of Lewis's objections to a rival view, according to which the "possible worlds" of modal logic are maximal consistent sets of sentences—a view that he calls "linguistic ersatzism".[12] Lewis notes that the language of the ersatzist will simply lack the linguistic resources to describe all the possibilities that there are:

> There is no way the theorist can specify his worldmaking language so that it will have words—names or predicates—for properties alien to his world. So any language he can specify will be inadequate to distinguish some of the possibilities there are.[13]

[10] See Chihara 1973: ch. III for a discussion of Quine's views on ontology.

[11] In Lewis 1979*b*: 182.

[12] See Lewis 1986: sect. 3.2.

[13] Lewis 1986: 139.

Lewis suggests, in this way, that in order to represent *every* possibility—even those that cannot even be expressed within the languages we have—we need genuine worlds possessing those features, from the most general to the minutest and most specific, that only a real world can supply. Clearly, the "ways" to which I apparently refer do not come close to matching real worlds for structure and informational content.

Take some "way" described by one of the **C** sentences. All that the **C** sentence tells us about the "way the world could have been" is what truth values some sentential letters would have had if the world had been that way. How could that "way" be a *possible world*, in either Lewis's or Plantinga's sense of that term? Recall that, according to Plantinga, any possible world contains not only the structure of set theory, property theory, and mathematics, but also all the structure of each and every possible world. How could a **C** sentence contain all that information? Clearly, it could not.

There is another way of attacking the above reasoning linking "ways that the world could have been" to possible worlds. Robert Stalnaker has argued that if "ways that the world could have been" are possible worlds, then the "way that the world is" should be the actual world. But the "way that the world is" is a property of the world, not the world itself.[14] Stalnaker's position is surely more plausible than Lewis's. In what ways could Socrates have been different? He could have lived a very different sort of life: he could have lived a solitary life upon a mountain top. He could also have been a carpenter instead of a philosopher. These are modal attributes Socrates possessed. Socrates possessed the property of possibly having been a carpenter. Socrates lacked the property of possibly having been a cabbage. Now it might be said that the actual world also has modal attributes. It could have been a world containing fewer stars than planets. The different ways in which the world could have been different than it is are modal attributes that it possesses; they are not possible worlds. Peter van Inwagen sums it up nicely, writing: "[S]urely the cosmos cannot itself be identical to any way the cosmos could have been: to say this would be like saying that Socrates is identical with the way Socrates is, which is bad grammar."[15] I thus conclude that my analysis does not presuppose the existence of possible worlds.

[14] Stalnaker 1984: 46.

[15] Van Inwagen 1980: 406. Lewis has responded to this quotation by claiming that "the choice whether to take a 'way' as a unit set or as its sole member seems to be of the utmost unimportance" (Lewis 1986: 87 n. 57). What lies behind this response is Lewis's explication of 'ways that the world could have been' as sets of possible worlds. A 'maximally specific way' then will be a unit set, so the difference between a maximally specific way that a world could have been and a world comes down to the difference between a unit set whose only member is a world and the world itself - a difference that Lewis regards as unimportant. But this re-

However, in so responding, I have not addressed another worry that may have struck the reader: it may appear that my analysis presupposes the existence of "ways that the world could have been", and even if these sorts of entities are not possible worlds, they still seem to be things that I should not at all be happy with. So to this worry I now turn.

Am I Committed to the Existence of Ways?

At this point, we need to examine just how these apparent references to "ways the world could have been" occur in my analyses. One such apparent reference occurs in sentences of the form:

> The world could have been such that, had the world in fact been that way, 'P' would have been true and 'Q' would have been false.

Such apparent references are not troubling since the phrase 'had the world in fact been that way' is not essential to the meaning of such sentences. I could write instead:

> It could have been the case that 'P' was true and 'Q' was false.

Or I could write:

> The world could have been such that 'P' was true and 'Q' was false.

Thus, the troubling phrases can be eliminated in an appropriate paraphrase.

Apart from such cases, an examination of my reasoning reveals that there are basically two other kinds of apparent references to "ways the world could have been". (1) I, sometimes, speak of things describing or representing ways that the world could have been. For example, in one place, I say of a mathematical function that it can be regarded as representing a way that the world could have been, in so far as the truth values of the sentential letters are of concern, and I also say that each **C**-sentence describes a way that the world could have been, in so far as the truth value of the sentential letters are of concern. (2) At one point, I suggest that we may think of the members of **W** as ways the world could have been, in so far as the truth values of the sentential letters are of concern.

Examining the first sort of apparent reference to "ways the world could have been", it can be seen that what are being presupposed are not "ways the world could have been" but rather "representations of ways the world could have been" or "descriptions of ways the world could have been". It may be objected, however, that a representation of a way the world could have been presupposes the existence of such a way. If there are no such

sponse is not an adequate defence of the ordinary language argument for the existence of possible worlds that van Inwagen was criticizing.

things as a "way the world could have been", how can something represent a way that the world could have been? How can something represent what does not exist? But surely there is no problem with that. A diagram can represent how a traffic accident could have taken place, even if no such accident ever took place. A diagram can represent a configuration of buildings none of which actually exists. So why cannot a **C** sentence describe a way that the world could have been, in so far as the truth values of the sentential letters are of concern, even if there are no such entities as "ways"?

As for the second sort of apparent reference to "ways the world could have been", it can be seen that I am only suggesting that one can think of the elements of **W** as ways the world could have been. To think of *A* as a *B* is not to imply that *B*s exist.

Might not some philosophers object to even *representations of ways of ways the world could have been*? After all, these representations are mathematical functions. Can it not be argued that I have presupposed the existence of mathematical entities? I shall take up this question in the following chapter.

9

Anti-Realism in Mathematics

As I noted in the Introduction, the initial impulse to explore the metaphysics of possible worlds semantics came to me from my work in the philosophy of mathematics. In this final chapter, I discuss the implications of the modal theory developed in this work for my view of mathematics. I shall begin with a discussion of Mathematical Realism, because my own position regarding the ontology of mathematics can only be appreciated by way of contrast with the metaphysical doctrines it rejects.

1. Mathematical Realism

Mathematical Realism (or Mathematical Platonism) is the view that mathematical objects exist and that mathematical terms, such as 'five', 'the null set', and 'π', refer to these mathematical objects. It should be noted, however, that Mathematical Realists do not always insist that all the standard mathematical terms refer. Some single out a special class of terms—say, the referring expressions of set theory—as the ones that genuinely refer. (The other mathematical objects are then taken to be "reducible" to the objects to which the special terms refer.)

Initially, the philosophical advantages of espousing Mathematical Realism appear to be quite impressive. First of all, "genuine mathematics", as it is actually practised by mathematicians and not some philosophical version of it, can be understood to be a straightforward theory about mathematical objects. For example, it can be plausibly maintained that set theory, as it is taught in mathematics departments throughout the world, is a theory about . . . SETS. Since the Mathematical Realist believes that mathematical objects truly exist, she can take the axioms of set theory to be just truths about sets. Secondly, the logical form of the statements of mathematics can be taken at face value, so that the semantics of the language used can be regarded as the standardly referential Tarskian one.

The most influential and well-known of the mathematical Platonists are Gödel, Quine, and, at least during one period, Putnam.[1] More recently, in

[1] I have discussed the views of these Mathematical Realists in Chihara 1973 and, more recently, in Chihara 1990.

a work that has received much attention in the literature (Maddy 1990), Penelope Maddy has developed a philosophically sophisticated version of Mathematical Realism by building on the writings of all three of the above mentioned philosophers.[2] She agrees with the traditional Mathematical Platonist that set theory is a theory about mathematical objects. She also agrees that the logical form of the sentences of set theory should be taken at face value and given the standardly referential Tarskian semantics. However, by making several daring moves, she attempts to repair what she takes to be serious defects of each of the earlier versions. Maddy calls her version of Realism "Set Theoretic Realism".

2. Maddy's Set Theoretic Realism

Maddy's Compromise

Maddy's quest for an adequate version of Mathematical Realism begins from the perspective of Quine's *epistemology naturalized.* This is the view that our best understanding of the world can be obtained from contemporary science and not from a priori canons of knowledge put forward by philosophers, so that epistemology is taken to be a part of science, namely a branch of psychology (Maddy 1990: 9). From this perspective, Maddy enthusiastically seconds the Quine–Putnam "indispensability argument" (discussed earlier in Chapter 3, Section 3) for the conclusion that mathematical objects—more specifically, sets—are presupposed, implied, or required by our best overall scientific theory;[3] a conclusion that gives rise to the dictum "mathematical entities are indispensable for our best theory of the world" (Maddy 1990: 59). If this argument is correct, it would seem that rationality would demand that we accept the existence of sets. For if

[2] I should also mention her reliance on George Pitcher's account of perception. Recently, Maddy has begun to have second thoughts about the positions taken in her book—see, for example, Maddy 1992 and Maddy 1995, where she raises a number of doubts about the indispensability argument. But compare Hellman 1995 for an interesting reply to some of Maddy's objections to indispensability arguments. In any case, in this chapter, I shall present the views she espouses in her book Maddy 1990, since this is the view that most sharply diverges from my own. Besides, even if she abandons the more extreme features of her set theoretical realism, assessing the view still has point: there are other philosophers who accept the view. See, for example, Gillies 1992. For the sake of simplicity of exposition, I shall write as if Maddy is the Set Theoretic Realist of her book, although (as I indicated above) she is clearly moving away from many of the positions she adopted there. I should mention that I find her more recent views much more in line with my own.

[3] See Section 7 below for more details on the indispensability argument. I have discussed Quine's indispensability argument in greater detail in Chihara 1990: ch. 1. Putnam's version of the indispensability argument is taken up in ch. 8, pp. 165 ff. Unlike Maddy (and Hartry Field), I take Putnam's version to be significantly different from Quine's.

our best theory of the world requires the existence of sets, and if it is rational to believe our best theory of the world, then it must be rational to believe in sets.

Although Maddy finds this argument compelling, she cannot accept an idea that Quine has linked to his indispensability argument, namely, the idea that *unapplied mathematics is without justification*. Maddy strenuously rejects Quine's claim that any mathematics that goes beyond what is applied in our best scientific theories is "on a par with uninterpreted theories"; and she argues that this Quinean view just does not do justice to mathematics "as it is practiced", pointing out that mathematicians do not await the verdict of the history of science to determine what is and what is not a mathematical truth (Maddy 1990: 31). "Rather," she claims, "mathematicians have a whole range of justificatory practices of their own, ranging from proofs and intuitive evidence, to plausibility arguments and defences in terms of consequences" (Maddy 1990: 31).

Maddy gives another reason for questioning the Quine–Putnam version of Mathematical Realism. As she understands their reasoning, the grounds they give for accepting any mathematical assertion as true are extremely theoretical, depending as they do upon what is required for our best scientific theories. Maddy believes that such a view of mathematical justification is questionable, writing: "But isn't it odd to think of '2 + 2 = 4' or 'the union of the set of even numbers with the set of odd numbers is the set of all numbers' as highly theoretical principles?" (Maddy 1990: 31). According to Maddy, then, the Quine–Putnam sort of justification does not provide us with an adequate account of actual mathematics, because the very general and highly theoretical kind of justification proposed by Quine and Putnam simply does not account for the obviousness of many truths of elementary mathematics

Maddy holds, on the other hand, that Gödel's views on the nature of mathematics fit our actual mathematical practices and our actual mathematical experiences much better than those of Quine and Putnam. Gödel held that mathematicians have something like a perception of the objects of set theory—something that he called 'mathematical intuition'—which allows us to "see" (in some metaphorical sense) certain things to be true of sets: certain axioms of set theory "force themselves upon us as being true".

However, not all axioms of set theory are of the above sort:

> Just as there are facts about physical objects that aren't perceivable, there are facts about mathematical objects that aren't intuitable. In both cases, our belief in such 'unobservable' facts is justified by their role in our theory, by their explanatory power, their predictive success, their fruitful interconnections with other well-confirmed theories, and so on. (Maddy 1990: 32)

So Maddy thinks that Gödel's epistemology is two-tiered. (a) Some of the basic axioms of set theory are just intuited to be true (these are the obvious or self-evident ones); whereas (b) others are justified by their consequences in the way that molecular theory was justified.

Although Maddy is sympathetic to much of Gödel's philosophy of mathematics, one part of the view she finds defective: she is not satisfied with Gödel's account of why we are justified in believing in the existence of mathematical objects. On this score, she finds the Quine–Putnam account more plausible. Thus, Maddy develops a version of Mathematical Realism that she characterizes as a compromise:

> From Quine/Putnam, this compromise takes the centrality of the indispensability arguments; from Gödel, it takes the recognition of purely mathematical forms of evidence and the responsibility for explaining them. Thus, it averts a major difficulty with Quine/Putnam—its unfaithfulness to mathematical practice—and a major difficulty with Gödelism—its lack of a straightforward argument for the truth of mathematics. (Maddy 1990: 35)

Maddy's Doctrine of the Perceivability of Sets

There are two major problems with traditional forms of Mathematical Realism. (1) It seems difficult to explain how we obtain knowledge of mathematical objects, since traditionally, they have been taken to be abstract objects that do not exist in space or time. How, it might be asked, have we humans arrived at our knowledge of the existence and properties of entities with which we are not in any sort of causal relationships? (2) It seems difficult to explain how we can even refer to these entities, since we seem to be so utterly isolated from them. How can I refer to objects that I cannot see, touch, smell or hear? And how can I tell if anyone else refers to the same thing that I do when I use such terms as 'seven' or 'null set'?

To meet these problems, Maddy advances the radical doctrine that we can and do causally interact with sets. We can see them, feel them, smell them, taste them, and even (it would seem) eat them. Maddy holds that a "mixed set", that is, a set whose transitive closure contains urelements (or non-sets), has a location in physical space (namely, the exact place in space at which the urelements are located) and can be perceived. The set of eggs in my refrigerator, for example, is to be found in exactly the place that the eggs are to be found. According to Maddy, the set occupies exactly the same space that the eggs do. By taking mixed sets to be locatable in physical space, Maddy believes that she will be able to give a reasonable explanation of how we can refer to sets and even of how we can gain knowledge of them and their properties. After all, with this doctrine, sets are held to be little different from physical objects.

To see why Maddy thinks it is reasonable to claim to perceive sets, we need to examine her views on perception. Relying on an analysis of George

Pitcher's, Maddy lays out her account of perception in the following way (in Maddy 1980: 171):

> A person, *P*, perceives an *x* (something that is an *x*) at location *L* if and only if
> (i) there is an *x* at *L*;
> (ii) *P* acquires perceptual beliefs about the *x*—in particular, the belief that there is an *x* at *L*;
> (iii) the *x* at *L* is involved in the generation of this perceptual belief state in an appropriate causal way . . .

Thus, in order that Steve perceive a tree before him, it is not enough that there be a tree before him or even that Steve have the perceptual belief that there is a tree before him: "[W]e must insist", she tells us, "that whatever makes the belief true—in this case, the tree in front of Steve—be responsible for his belief in some appropriate way" (Maddy 1990: 51). What is an "appropriate way"? She says:

> Appropriate participation is exemplified by the role of my hand in the generation of my perceptual belief that there is a hand before me [when I hold my hand in front of my eyes]. (Maddy 1990: 63)

In other words, the object perceived must be causally responsible for the perceptual belief, in the way my hand causes my perceptual belief that my hand is right in front of my eyes, when I see my hand right in front of my eyes.

With this view of perception, Maddy can argue that sets are perceivable. A person can open her refrigerator, and gain a perceptual belief about the set of eggs that is there, so that, if the set is responsible for her perceptual belief in the appropriate way, she will have perceived the set.

What about the properties that sets have? What about the relations into which sets enter? How do we gain knowledge of such things? Well, how do we know that the top of my desk is rectangular in shape? We can just see that it has such a feature. And how do we know that the desk is to the right of my computer (from the perspective of my couch)? Again, we can just see that it is. Now, Maddy suggests that we obtain our knowledge of the properties and relations of sets in just that way, that is, by perception. Thus, how do we know that, in situation *T*, there are *two* pieces of fruit on my desk? Maddy suggests that we obtain this knowledge by seeing the set John and seeing that John has the number property of *having two members*. She asserts: "Just as the perception of physical objects includes the perception of their properties, so the perception of sets does the same" (Maddy 1990: 88). In other words, the very act of perceiving John includes perceiving the cardinality property. Indeed, she seems to believe that the number property perceived is one and the same thing as the number itself, writing "these number properties are the numbers". And she disputes the arguments that were given by Benacerraf and Frege for denying that numbers are properties or concepts (Maddy 1990: 89–94).

Numbers, then, are held to be properties. Since number properties are perceivable and locatable in space, and since number properties are numbers, numbers are perceivable and locatable in space. For this reason, Maddy holds that the number two is located exactly where John is located. She tells us that "ten is located where the set of my fingers is located, in motion over the keys of my word processor" (Maddy 1990: 87).

Let us now turn to the question of how this view of set perception provides Maddy with the theoretical material to explain the "obviousness" of the basic axioms of set theory. Briefly and roughly, here is how her explanation runs. She extrapolates from the doctrines of the neurophysiologist Donald Hebb according to which there are, in the visual cortex, certain configurations of cells that he calls "cell-assemblies"—cells that are active during visual perception. In the case of the perception of triangles, Hebb holds that these cell-assemblies function as triangle detectors, which give rise automatically to various general beliefs about the figures that stimulate them, such as the belief that the figure being perceived is three-sided. Three-sidedness is, one might say, "built into" the concept of triangle. Hebb also maintains that there are cell-assemblies that function as physical object detectors and that these cell-assemblies (which involve higher-order cell-assemblies) give rise automatically to such beliefs as that the object being perceived is space-occupying and perception-independent (i.e. they exist independently of being perceived). Maddy takes the next step of postulating such cell-assemblies for the case of set perception, writing:

> What goes for physical objects should also go for sets: the development of higher-order cell-assemblies responsive to particular sets gives rise to an even higher-order assembly corresponding to the general concept of set. The structure of this general set assembly is then responsible for various intuitive beliefs about sets, for example . . . that they have various subsets, that they can be combined and so on. And these intuitions underlie the most basic axioms of our scientific theory of sets. (Maddy 1990: 70)

In Chihara 1982, I raised a number of objections to an article (Maddy 1980) in which Maddy presented and defended her doctrine that one can literally see sets. Her book (Maddy 1990) contains (implicit) responses to some of these objections, so it is appropriate to give some consideration to these responses.

One objection I raised in my article is quoted in full in Maddy's book. Here's the quotation:

> Imagine that I am sitting at my desk. Its surface has been cleared of everything except an apple. Now according to Maddy, we can literally perceive on the desk, in addition to the apple, the set of apples on my desk (which happens to be a unit set). It is claimed that this set has a location in space, the exact spot where the apple is.

Supposedly, it also came into existence at a particular time (when the apple did), and will go out of existence at a particular time (when the apple does). Obviously, Maddy thinks this set can be moved about in space. Now if we can perceive this set with the sense of sight, what does it look like? Evidently, it looks exactly like the apple itself. After all, I don't see anything at that exact region in space that looks different from the apple. One wonders how this object is to be distinguished (perceptually) from the apple, since it has exactly the same shape and color. Perhaps it feels different. Let's touch it. But I can't feel anything there other than the apple. Evidently, this strange entity feels no different from the apple. How about its smell or taste? Again, it would seem that the set must be identical in smell and taste to the apple. So it looks, feels, smells, and tastes exactly like the apple. . . yet it is a distinct entity. One would think that an entity with these properties would be of interest to the physicist.[4]

Maddy has responded to this passage by revising her position so that, now, *the apple is taken to be identical to its singleton* (Maddy, 1990: 153). More generally, she holds that every physical individual is identical with its singleton. Notice that she does not claim that every object is the same thing as its singleton. She is not committed to maintaining that, for example, a set x is identical to $\{x\}$. However, she does believe that any physical object is identical to its singleton. Quine had already advocated the identification of an individual with its singleton for the purpose of simplifying the formalism of his set theory, so this revision does not seem very radical to Maddy. She notes that such a revision would require only a minor modification of the axioms of Zermelo-Fraenkel set theory (Maddy, 1990: 153 n. 4).

Another objection I raised in Chihara 1982 concerned pure sets. Here, I must backtrack a bit to explain this objection. In an even earlier paper, I had raised a number of objections to attempts by mathematical Realists to justify their belief in the existence of sets. In particular, I questioned Gödel's claim that the epistemological position we are in *vis-à-vis* the existence of sets, is no different from ("an exact replica of") the epistemological position we are in *vis-à-vis* the existence of physical objects. Here is what I wrote:

We now have a highly developed and complex system of beliefs about the world, encompassing not only mathematics and physics, but also theories about ourselves and other people, and about the experiences we have and how we communicate with one another. With this rich background of beliefs, we are also pictured as having certain "mathematical experiences" that we share with others and communicate to one another; and the Gödelian justification is supposed to show us the rationality of extending our beliefs so as to explain these experiences by postulating

[4] Chihara 1982: 223–4. There are many other objections to Maddy's paper to be found in Chihara 1990: ch. 10. See especially pp. 210–14 for a detailed criticism of Maddy's claim that prehistoric people perceived and believed in sets. Unfortunately, I was not in a position to respond to the replies in Maddy 1990 when I wrote that chapter, since Maddy's book had not yet appeared.

the existence of mathematical objects and some sort of interaction between us and these objects. (Chihara 1982: 218)

But I questioned the idea that we were ever in a comparable position with respect to physical objects, writing:

What theories, what systems of beliefs, formed the foundations for the inference to physical objects? When were we in such a position?

Maddy responded to this question by admitting that "we were never in such a position with respect to physical objects". But she went on to deny that we were ever in such a position with respect to sets, claiming that people have perceived sets throughout history, going so far as to claim that even prehistoric people perceived and believed in sets.[5] This provided her with a basis for arguing that there is no need for a *theoretical justification* for our belief in sets, any more than there is a need for a theoretical justification for our belief in physical objects. Thus, Maddy claimed that my objection involves the error of substituting theoretical justification for perceptual justification (Maddy 1980: 191).

Recall, however, that she had conceded that *pure sets* do not have a location in physical space and cannot be perceived. I replied in Chihara 1982 that, even if it were granted that *mixed sets* are perceivable, Gödel (as she understood him) would still be faced with the problem of giving a theoretical justification for his belief in *pure sets*.

Now her book contains an implicit reply to this objection. Here's what she says:

In fact, the pure sets aren't really needed. The set theoretic realist . . . can take the subject matter of set theoretic science to be the radically impure hierarchy generated from the set of physical individuals by the usual power set operation, except that the empty set is omitted at each stage. On this picture, each set, no matter how exalted in rank, is located where the physical stuff in its transitive closure is located. The theory of this structure differs only trivially from that of the usual hierarchy with individuals. It can serve all the same purposes. (Maddy 1990: 156–7)

She concludes with the words: "I've suggested two minor alterations in the set theoretic realist's ontology: the identification of physical objects with their singletons and the elimination of pure sets" (Maddy 1990: 157).

3. A Gothic Tale

Although there is much more to Maddy's Set Theoretic Realism than I have indicated above, I trust that enough of her views have been sketched to provide you with an idea of the sort of ontological doctrine I oppose. In

[5] Maddy 1980: 191.

response to various objections that have been levelled at Mathematical Realism, Maddy has fashioned a theory that is downright Gothic.

Do Number Terms Refer?

An estimation of how far she has moved from what I first characterized as Mathematical Realism can be obtained from a consideration of the following thesis: having argued that numbers are located in space, she goes on to proclaim that if 'numbers' are supposed to be the referents of number words, then "there are no numbers" (Maddy 1990: 99). Such words as 'one', 'two', and 'the first odd prime number' may have the superficial grammar of referring expressions, she says, but we should resist the temptation to specify referents for them. On the one hand, then, the number ten is supposed to be located exactly where my fingers are. On the other hand, the word 'ten' is not supposed to refer to anything. What then is being claimed when she says that ten is located at exactly the place my fingers are? Is she not skirting paradox?

Do Sets Have Mass?

Another paradoxical feature of her views can be brought out by the question: are sets supposed to have mass, weight, kinetic energy, and momentum? More generally, are they supposed to have the common physical properties that ordinary physical objects possess? Maddy does not say; but consider the alternatives. If the answer is 'no', then she will be faced with the mystery of explaining how an object lacking such physical properties can generate perceptual beliefs in the appropriate *causal way* that her theory of perception requires. If the answer is 'yes', then consider a tightly sealed glass container in which there are two apples (and, according to the typical scientist, no other macroscopic object that is detectable). It would seem that, according to Maddy's theory, there would have to be infinitely many distinct macroscopic objects in that box (all the sets generated from the pair of apples), each with a significant amount of mass. With all that mass in the container, how could it be moved about so easily? She might reply that, in such situations, in making our scientific calculations, we should simply ignore the mass and other physical properties of the sets that are in the container. For some reason, she might say, the mass of a set does not have a physical effect. Again, we seem to have paradoxical entities.

Maddy's Hebbian Explanation of the Obviousness of the Elementary Laws of Mathematics

Consider now Maddy's attempt to give an "account of the obviousness of elementary mathematics". She had purported to "explain what intuition is and how it works" (Maddy 1990: 35). But consider some of the strange consequences of her Hebbian explanation. Practically everyone can see triangles and physical objects, so there may be some plausibility to Hebb's hypothesis that there are the triangle-detecting and physical-object-detecting cell-assemblies in the cortex. On the other hand, practically no one claims to see sets. If we humans have set detectors, as Maddy supposes, why is it that so few people are able to perceive (in Pitcher's "intentional" sense) sets? It might be argued that it takes special training or instruction to develop set detectors; but if so, then what happens to her account of the obviousness of mathematics? If only people who receive this special training develop these set detectors, then only these few people should have the intuitions that are, according to Maddy, built into these detectors. So she will not have explained what she needs to explain, namely the obviousness *to the general population* of the elementary and basic principles of mathematics.

Related to the above is another feature of Maddy's account that I find hard to accept. According to her Hebbian account of perception, those who can perceive triangles have the triangle detectors, and those who can perceive physical objects have the physical object detectors; and having the detectors is supposed to account for our intuitive beliefs about the obvious properties of triangles and physical objects—those being the beliefs that are "built into" the appropriate detectors. Now, according to Maddy, sets are space occupying, perception independent things. If, as Maddy believes, we have cell-assemblies that function as set detectors, then one would think that one of the principles of set theory would be that sets are space-occupying, perception-independent things. Since no widely accepted set theory has any such axiom, an explanation is surely called for.

Maddy's Identification of Physical Objects with their Singletons

Consider, now, Maddy's identification of the singleton of a physical object with the physical object itself. Recall that Maddy put forward this thesis in response to my discussion of the apple and its singleton. We seemed to be committed, by Maddy's original theory of mixed sets, to *a pair of distinct objects (the apple and its singleton) that are perceptually indistinguishable.* She attempted to blunt the force of this reasoning, by identifying the singleton with its member. But does this revision really help much?

Consider the following situation (which I shall call 'situation T'). I have on my desk two pieces of fruit, a banana and an apple. For purposes of reference, let us suppose that the banana is named 'Tom' and the apple 'Sue'. Let us name the pair set, whose elements are Tom and Sue, 'John'; and the triple, whose elements are Tom, Sue, and John, 'Mary'. Now John is a pair and Mary is a triple; yet, it is a consequence of Maddy's doctrine of set location that both objects are located in exactly the same place, namely where Tom and Sue are located. In fact, we can infer that John and Mary look exactly alike, feel the same, taste the same, and smell the same. In other words, we again seem to be committed to a pair of distinct objects that are perceptually indistinguishable. It would seem that the position in which we find ourselves, *vis-à-vis* John and Mary, is not significantly different from the one we were in, *vis-à-vis* the apple and its singleton.

Now it may be denied that John and Mary are perceptually distinguishable. After all, Maddy holds that the cell-assemblies functioning when one sees John differs from the cell-assemblies functioning when one sees Mary. Surely, it could be said, the visual experience of seeing John would differ from the visual experience of seeing Mary. Consider, then, the case of the person who looks at a piece of plywood cut out in the shape of Kohler's duck-rabbit[6] and first sees something in the shape of a duck and later sees that same object as having the shape of a rabbit. There is no doubt that the person's visual experience in seeing the plywood as a rabbit differs from his visual experience in seeing the object as a duck. But this is no reason to suppose that the object seen as a rabbit differs from the object seen as a duck. The former cannot be perceptually distinguishable from the latter; since the former is identical to (and the very same thing as) the latter.

How do we Obtain Knowledge of the Properties of Sets?

I turn now to Maddy's account of how we obtain knowledge of the properties of sets and of the relations they are in. Supposedly, in perceiving sets, we just see what cardinality properties they have and we just see that one is (or is not) a member of the other. The perception of the sets includes the perception of their properties.

All of this strikes me as pure fantasy. Reconsider situation T. When I look to see how many pieces of fruit are on my desk, I do not try to see a set or a number. I just look for the fruit. I don't have to see John to determine how many pieces of fruit are on my desk. It is enough just to see the fruit and, on this basis, determine that there are two pieces of fruit on my desk.

How do I know that the cardinality of the set of paperclips in my house is 173? Is this a matter of seeing the set of paperclips in my house and seeing its cardinality property? The answer is: I just go through the rooms in my house that contain paperclips and count the paperclips. I don't have to see *the set of paperclips*. Having such a visual experience plays no role in my arriving at the knowledge in question.

How do I arrive at the knowledge that John is an element of Mary and that Mary is not an element of John? Is this a case of visual knowledge? Do I see both sets and see (with my eyes) which is a member of which? Surely not. I know that John is an element of Mary and Mary is not an element of John *because it was specified that John is the set whose elements are Tom and Sue, whereas Mary is the set whose elements are Tom, Sue, and John.* That specification does all the work; looking at John and Mary does nothing for

[6] See Wittgenstein 1953: Pt. II for a discussion of the duck-rabbit.

me. I would be no worse off, in this respect, were John and Mary located on Mars.

Maddy's Revisionism

Consider again the revisions in the ontology of set theory that she has adopted: *are these revisions all that minor*? Let us grant that the pure-setless structure one obtains, if one adopts her revisions, is not significantly different from the standard one. It is easy to see how one can reproduce, within her favoured structure, the standard hierarchy of pure sets. So *from the Structuralist's point of view*, Maddy's revised version of set theory is only trivially different from the standard version. But Maddy is not concerned merely to justify the standard set theoretical structure. In her article, she had claimed that

> from the point of view of the set theoretical realist, the treatment of sets as forming a kind is much more likely to be correct than a more traditional theory according to which a set is anything which satisfies certain conditions. (Maddy 1980: 183)

Maddy's views, in this respect, are closely related to those expressed in Gödel, 1964*b*, in which Gödel responded to the suggestion that, were the Continuum Hypothesis proved to be independent of the standard axioms of set theory, the question of the truth or falsity of the hypothesis would lose its meaning, just as the question of the truth or falsity of the Fifth Postulate lost its meaning with the discovery of its independence from the other postulates of Euclidean Geometry.[7] Gödel objected that such an independence result in set theory would render the question of the truth or falsity of the Continuum Hypothesis meaningless, *only if set theory were regarded as a hypothetico-deductive system in which the meanings of the primitives of set theory are left undetermined* (Gödel, 1964*a*: 271). But, for Gödel, set theory is not that sort of system. It is clear that Gödel believed that the objects of set theory "exist independently of our constructions"; that we have "an intuition of them individually" (an intuition that is something like a "perception" of individual sets); and that the general mathematical concepts we employ in set theory are "sufficiently clear for us to be able to recognize their soundness and the truth of the axioms concerning" these objects (Gödel, 1964*b*: 262). Thus, he declares that "the set-theoretical concepts and theorems describe some well-determined reality, in which Cantor's conjecture must either be true or false" (Gödel, 1964*b*: 263–4).

[7] Cf. "Probably we shall have in the future essentially different intuitive notions of sets just as we have different notions of space, and will base our discussions of sets on axioms which correspond to the kind of sets we wish to study . . . everything in the recent work on foundations of set theory points toward the situation which I just described" (Mostowski 1967: 94).

On this matter, Maddy is in substantial agreement with Gödel. Maddy's set theoretical realist holds that

the world consists of physical objects, sets of these, sets of physical objects and sets, and so on, through the transfinite levels of the iterative hierarchy. There is a fact of the matter about the cardinality of the set of Dedekind cuts . . . and it is the set theorist's job to discover it. (Maddy 1990: 128)

And as we saw earlier, Maddy postulates a perceptual connection between us and individual sets similar to the one Gödel postulated. She emphasizes that it is a crucial part of her naturalistic world view that there are natural kinds. Indeed, she indicates that the existence of natural kinds is a key ingredient of her solution to the problem of how finite minds are able to grasp and refer to the infinite set theoretical hierarchy—a problem that arises from the thought that "I might just as well be related to any one of a host of another structures that agree with the standard hierarchy only on the minuscule finite portion I've managed to grasp" (Maddy 1990: 79). It is crucial for Maddy's naturalistic solution to the above problem, then, that sets form a natural kind and that the membership relation not be just any relationship that conforms to certain structural constraints.

Thus, one is forced to conclude that Maddy's book implies that practising set theorists are actually making a number of significant errors. First of all, the rejection of the existence of pure sets certainly implies that the assertions of practising set theorists that such sets exist are just false. If there is neither a null set nor any hereditarily finite sets built up from the null set, then practising set theorists are just wrong in thinking such pure sets exist. It follows that set theorists are promulgating falsehoods in asserting the existence of the von Neumann ordinals. Also, in view of Maddy's identification of a physical object x with its singleton $\{x\}$, she is, in effect, maintaining that *every physical object is a set which is an element of itself*—a conclusion that contradicts the belief of most set theorists that no set is an element of itself. Interestingly enough, she herself asserts (in a previous chapter): "no set can be self-membered" (Maddy 1990: 104).

There are many other widely accepted "truths" of set theory that would have to be rejected, were we to adopt Maddy's suggestions. For example, it is a consequence of Maddy's revisions that the Regularity Axiom of Zermelo-Fraenkel set theory does not hold. (Consider the set $\{x, y\}$, where x and y are physical objects.) One of a number of mathematical theorems that would have to be abandoned, were Maddy's revisions accepted, is the theorem that for any finite set a, if the cardinality of a is n, then the cardinality of the power set of a is 2^n. Another "truth" of standard set theory that would have to be given up is the intersection axiom that tells us, for any sets a and b, the intersection of a and b exists.[8]

[8] This was pointed out by someone in the audience at my talk at King's College, London.

It can be seen that, although Mathematical Platonism is generally thought to provide us with an account that justifies actual mathematical practice, Maddy's Set Theoretic Realism is a *revisionist theory.* According to Maddy's new account of sets, many of the standard axioms and theorems of set theory turn out to be simply false. In other words, Maddy is strongly suggesting, *on purely philosophical grounds*, that mathematical practice must be altered if the practice is to arrive at only truths. In effect, she has relinquished one of the striking advantages of Mathematical Realism. For, in her original article (Maddy 1980), she had staked out the following advantages for Set Theoretic Realism. In addition to allowing a straightforward Tarskian semantics for set theoretic discourse,

> it squares with the prephilosophical views of most working mathematicians; and . . . it allows set theoretic practice to remain as it is; it does not demand reform. (Maddy 1980: 163–4)

We have seen above that her version of Mathematical Realism does demand reform. Does it square with the *prephilosophical* views of most working mathematicians? How many working mathematicians believe that the number ten is located where my fingers are?

4. The Ontology of Mathematics

My own views about the ontology of mathematics stand in contrast to Maddy's Set Theoretic Realism. To provide a more detailed explanation of the position I developed in Chihara 1990, I shall take a quick dip into the history of mathematics.

Once upon a time, mathematicians talked as if they believed in the existence of all sorts of abstract entities; besides sets, functions, and ordered n-tuples, there were the various kinds of numbers: natural numbers, integers, rational numbers, real numbers, and even complex numbers. Complex numbers are numbers of the form $a + bi$, where a and b are real numbers and i is supposed to the be an "imaginary number" whose square is –1. The sixteenth-century mathematician Geronimo Cardano felt the need to speak of such numbers when he applied a method for solving certain sorts of cubic equations.[9] This method yielded the result that the equation '$x^3 - 15x - 4 = 0$' has the solution x = the cube root of $(2 + \sqrt{-121}$ + the cube root of $(2 - \sqrt{-121})$. The problem was: how could a number be the square root of a negative number? Cardano referred to square roots of negative numbers as "sophistic" and decided that a result such as the above is "as subtile as it is useless".[10] It wasn't long, however, before mathemati-

[9] Boyer 1985: 312. [10] Boyer 1985: 314.

cians were finding such results useful and important, even if somewhat mysterious. The practice arose of calling the square roots of negative numbers "imaginary numbers"; and Gottfried Leibniz is credited with remarking that imaginary numbers are like amphibians—halfway between existence and non-existence—resembling in this respect the Holy Ghost![11] By the early nineteenth century, mathematicians had gained a significant amount of confidence working with complex numbers, even treating them as visualizable. Thus, Carl Friedrich Gauss regarded the real and imaginary parts of the complex number $a + bi$ as the rectangular coordinates of a point in a plane.[12] William Rowan Hamilton then took the small step of showing how the complex numbers can be taken to be ordered pairs of real numbers, for which addition and multiplication are defined in such a way that all the required properties of the complex number system would result.[13] By this time, imaginary numbers had lost their mystery. One might say (although it is somewhat misleading to do so) that complex numbers had been ontologically reduced to ordered pairs of real numbers.

Later in that century, real numbers were similarly "reduced" to sets of rational numbers: real numbers were defined to be "Dedekind cuts" of rational numbers.[14] So the stage was set for "reducing" all the numbers to the natural numbers. The rational numbers were defined by some algebraists to be certain equivalence classes of ordered pairs of integers; and integers were defined to be certain equivalence classes of ordered pairs of natural numbers. Thus, the only numbers that were needed to get all the numbers were the natural numbers.[15] Gottlob Frege then took this process of ontological reduction one step further by defining natural numbers to be certain extensions of concepts, the extension of a concept *C* being, basically, the set of things that satisfy the concept *C*. In Frege's system, the only abstract objects one needed in mathematics were concepts and extensions of concepts. Since extensions of concepts can be regarded as classes or sets, Bertrand Russell carried out an even more parsimonious analysis of the ontology of mathematics with his "no-class" theory by eliminating even the need for classes or extensions of concepts. The only abstract entities

[11] Boyer 1985: 444.

[12] Boyer notes that "this simple step made mathematicians feel much more comfortable about imaginary numbers" (Boyer 1985: 548). Boyer also notes there that the graphical representation of complex numbers had been discovered by Caspar Wessel some thirty years before Gauss published his views about complex numbers.

[13] Boyer 1985: 634–5.

[14] Dedekind is generally credited with the main idea behind this "reduction". See, for example, Boyer 1985: 607–8. However, the contemporary notion of Dedekind cut is due in part to Bertrand Russell (Boyer 1985: 608).

[15] Cf. Kronecker's statement that "God made the integers, and all the rest is man's work" (Boyer 1985: 616).

needed in Russell's system were his "propositional functions"—roughly, the concepts of Frege's system.

I regard my own work as carrying on this process of ontological elimination. Consider again the ontology of Frege's logical analysis of arithmetic. The simple cardinality statement 'There are five apples on your table' was analysed by Frege in terms of:

(i) *ordinary objects* such as apples;
(ii) *linguistic objects* such as predicates and open-sentences;
(iii) abstract objects that predicates and open-sentences denote or refer to, i. e., *concepts*;
(iv) *extensions of concepts*.

Russell's "no-class" theory was aimed at showing how we can obtain essentially the same results without items of the fourth sort.[16] Going one step further, in Chihara 1990, I showed how to obtain a workable version of classical mathematics that does not require the postulation of items of the second or third kind. In other words, the mathematical system developed in the earlier work did not require either concepts or extensions. And although carrying out the details of this sort of "ontological reduction" may seem complicated, the basic ideas are simple: where Frege appealed to properties or concepts in his analysis of specific cardinality statements, I substitute open-sentences.

There are two apparent problems with substituting open-sentences for concepts: (1) If one asserts the existence of open-sentences whenever Frege asserted the existence of concepts, won't one be forced into the position of having to assert the existence of open-sentences for whose existence we have no good evidence? (2) There do not seem to be enough open-sentences to yield an acceptable mathematical theory. But these two apparent problems arise because, as I have described my version of classical mathematics, the assertions of the system are existential in nature: they are assertions of the existence of open-sentences. In fact, there are no such existential assertions in the system; what is asserted are the *constructibility* of open-sentences. Where Frege asserts the existence of a concept, my theory asserts the constructibility of an appropriate open-sentence.

Now, at first sight, it may be hard to see how a theorem asserting the constructibility of an open-sentence of a certain sort can tell us anything about the truth or falsity of a complicated statement of physics. But it is not at all hard to see how a theorem asserting the constructibility of a finite sequence of formulas of a certain sort can tell us something about the truth or falsity of a complicated statement of physics formulated in first-order logic (such

[16] For a discussion of the "no-class" theory, see Chihara 1973: ch.1, sect. 4.

a theorem can tell us that the complicated statement of physics is *derivable* from other sentences we already know to be true).

As one can find the details of these ideas worked out in Chihara 1990, the above overview should be sufficient to give the reader a general idea of how the constructibility theory avoids a commitment to mathematical entities.[17] Assertions of the existence of mathematical functions, in my discussions of possible worlds semantics, get transformed, by means of my constructibility theory, into assertions of the constructibility of open-sentences. In short, in model theory, as in mathematics in general, strictly speaking, there are no genuine assertions of the existence of mathematical entities. There are only assertions of the constructibility of open-sentences. True, I used the Platonic manner of speaking of mathematical entities in previous chapters, but this was done merely to simplify the logical and semantical developments: most readers would have found the discussion uncomfortably complex and unperspicuous had I carried out the reasoning using constructibility quantifiers and open-sentences.

5. Some Misunderstandings and Criticisms

A clearer understanding of my views on mathematics should result from a discussion of some of the misunderstandings of my writings that have arisen in the literature and from seeing how I respond to objections and criticisms of my view that have been brought to my attention.

The Constructivistic Interpretation

Despite several explicit warnings against taking the constructibility quantifiers of my system to be constructivistic—that is, to be those of mathematical constructivism—many readers still insist on giving these quantifiers just that constructivistic reading. (No doubt, the tendency to interpret my views in this way is due to the title of my book.) So again, I should emphasize that, in my system, to assert that it is possible to construct an open-sentence ϕ such that . . ., one need not have a method for constructing such a ϕ or even know that there is such a method. Perhaps, misunderstanding of my position can be best avoided by emphasizing that in developing my version of the quantifiers I did not intend to take sides in the debate between constructivists and non-constructivists *within mathematics*.[18]

[17] For an enlightening comparison between my negative approach to Mathematical Realism and those of other "nominalists", see Burgess and Rosen 1997: Pt II, sect. B and Pt. III, sect. B.2.

[18] This was suggested to me by Michael Levin.

Resnik's Anthropomorphic Interpretation

Michael Resnik has raised a number of objections to my views based upon his interpretation that my constructibility theory is concerned with "what it is logically possible for possible humans to do" (Resnik 1992: 651). Thus, he suggests that, in order to determine what axioms of the theory one should accept, one would have to turn to biology, anthropology, and linguistics.

I do not know how he arrived at his radical interpretation of my theory since he cites no textual evidence in support of his position. I certainly never said that my constructibility theory is concerned with what it is logically possible for possible humans to do. But to set the record straight, let me now state categorically that I never intended the constructibility theory to be understood in the anthropomorphic way Resnik has advocated. One has only to examine my proof of the Abstraction Axiom given in Chapter 4 to find evidence of the incorrectness of Resnik's interpretation: such a proof could not be stated as it is, if the constructibility quantifiers were given Resnik's reading. It should be clear from the proof that no constraints were being placed on the type of agent doing the constructing.[19] Fortunately, no other reviewer that I know of has made Resnik's mistake.

Mathematical Induction

Returning to my proof of the Abstraction Axiom, some readers have noted that I make use of a form of mathematical induction. It might be thought that mathematical induction concerns the natural numbers and hence that my proof presupposes the existence of mathematical objects.[20] To see that this is not so, let us consider in more detail the form of mathematical induction I employ. I start with a specification of a rule for constructing the arabic numerals, '1', '2', '3', . . . One first proves that it is not possible to construct an open-sentence ϕ of the level given by the numeral '1' which is such that ϕ does not satisfy condition *F*. One then proves that if it is not possible to construct an Arabic numeral *n* such that it is not possible to construct an open-sentence θ of the level given by *n* such that θ does not satisfy *F*, then it is not possible to construct an Arabic numeral *m* such that *m* immediately follows *n* and it is not possible to construct an open-sentence θ of the level given by *m* such that θ does not satisfy *F*. The principle of inference I employ, then, allows one to infer that it is not possible to construct an open-sentence ϕ of any level such that ϕ does not satisfy *F*. None of this presupposes the existence of the natural numbers. Thus, as I use mathematical induction, the principle is not a principle about certain mathematical objects (the natural numbers); rather, it is a modal principle.[21]

[19] Cf. p. 327 below.

[20] This was suggested to me by Lisa Hales.

[21] Cf. Chihara 1973: 178–81.

A Logical Reconstruction?

Another criticism of my views begins from the assumption that my constructibility theory is a "logical reconstruction" of mathematics. It is argued that such an analysis, far from telling us "what mathematics really is", distorts the true nature of mathematics and hides from us what genuine mathematics understanding is. The trouble is (it is claimed), most "logical reconstructions" of mathematics are directed at justifying mathematical theorems and not at understanding mathematical practice. As a result, we get a distortion of mathematics.

I do not deny that there are some philosophical analyses of mathematics to which the above criticisms justly apply. However, when directed at my views, these objections are simply misguided. For my constructibility theory was never intended to tell us "what mathematics really is". It was never my intention to be providing analyses of mathematical propositions. I never thought for a minute that mathematical theorems expressed the constructibility of open-sentences. My goals in developing my constructibility theory will become apparent from my reply to the following related objection.

Am I a Revisionist?

According to Maddy, the standard axioms of Zermelo-Fraenkel set theory assert the existence of entities of a certain sort, namely pure sets; and she holds that these existence assertions are false. Contrary to what the set theorist believes, she maintained, there is no such thing as the null set. There are no such entities as the von Neumann ordinals. The set theorist who believes that the Regularity Axiom is a true statement is just wrong. Now I am sceptical of the Mathematical Realist's belief that mathematical objects exist. So it may seem that I agree, to some extent at least, with some of the revisionist pronouncements of Maddy. Am I not claiming, in effect, that the set theorist's existential axioms are false? Not at all. It is not clear to me that the Mathematical Realist's (and Maddy's) referential analysis of mathematical language is correct. As I shall emphasize in the following sections, it is not clear to me that we should understand the semantics of the language of set theory in the way the Mathematical Realist advocates.[22] It still seems possible to me that there is a way of understanding mathematical languages and practices such that none of the mathematical theorems turns out to be a genuine assertion of the *existence* of mathematical objects. My own constructibility theory does not commit me to any sort of meaning-analysis of the assertions of contemporary mathematics. So I do not commit myself to holding that the axioms of Zermelo-Fraenkel set

[22] See, in this connection, Sections 6 and 7 below, where my position on this matter is developed in more detail.

theory are simply false or that set theorists make false claims about the existence of mathematical objects.

Still, is not this constructibility account severely revisionist? Am I not, in effect, advocating a serious "constructibility" *revision* of mathematics? No. I certainly do not think mathematicians should do their mathematical theorizing within my constructibility theory. Nor do I maintain that standard versions of mathematics are mere myths or even false. The point of my constructibility theory is to undermine the belief of those who have been convinced by various indispensability arguments that contemporary science requires a belief in mathematical objects. To those who use the indispensability argument to claim that reasonable people must believe in mathematical objects, I respond that, so far as I can see, contemporary science could be formulated and practised using the constructibility version of mathematics with no serious scientific loss. If so, then why must rational people believe in mathematical objects?

The following example may shed some light on that last point. Suppose that some philosophers argued that rational people must believe in the existence of certain abstract objects called derivations, because the Completeness Theorem of First-Order Logic tells us that, for every valid sentence θ, *there exists* a derivation of θ from the empty set of premisses. I would respond that the following version of the Completeness Theorem would allow one to draw essentially the same inferences as the Platonic version: for every valid sentence θ, *it is possible to construct* a derivation of θ from the empty set of premisses.

Is Classical Mathematics True?

This example is relevant to another objection critics have raised to my account of mathematics. Some have expressed wonder that I could allow that there is anything reasonable and correct in classical mathematics, if my constructibility account is true.[23] Resnik has such an objection in mind when he writes:

> Chihara's mapping of standard mathematics into constructibility theory shows only that the former is consistent relative to the latter. The mapping must preserve more for us to conclude that it establishes that mathematics is also (approximately) true. Chihara does not tell us what this is. (Resnik 1992: 650)

It should be pointed out that I never said that classical mathematics is "approximately" true. What I did allow was that there is much that is true, sound, or correct in classical mathematics. It seems to me that even if we

[23] This objection was originally raised to me by John Burgess. See Burgess 1992: 218, where it is touched on.

give the phrase 'there exists' the standard referential sense of Tarskian semantics, the modalist of the above example, who is sceptical of the existence of abstract objects called 'derivations', could still allow that there is much that is sound, correct, or true about the Completeness Theorem. In terms of *applications*, the modal version has essentially the same implications as the Platonistic version. For example, given some proof-procedure **P**, one can infer from either version of the Completeness Theorem that, for any valid sentence ϕ of the first-order language, if an agent (correctly) follows **P** long enough, she would eventually produce a derivation of ϕ.

Here is another way of seeing the same basic point. Take the case of Euclidean geometry. As I emphasized in my book, it was originally a modal theory: the axioms told us what geometric constructions it is possible to make. For well over a thousand years, geometry was developed and practised as a constructibility theory. At some point, however, geometry was transformed from a theory of constructibility into a theory of mathematical existence: instead of telling us what it is possible to construct, the axioms asserted the *existence* of such geometric objects as points, lines, and arcs. But from the point of view of one concerned only with applications of geometry, very little changed. The shift from constructibility to existence made hardly a ripple of difference when it came to applications. Thus, the advocates of the constructibility version did not regard the new existential version of geometry as all wrong; similarly, the old version of Euclidean geometry was not regarded as completely false by the practitioners of the existential version. When it came to applying the geometry, for the most part, it did not make any difference which version one used.

But there is another respect in which classical mathematics could be considered to be fundamentally sound, even within the framework of the constructibility theory. Consider Frege's analysis of arithmetic. Should I declare, from the perspective of my constructibility theory, that it is a completely wrong-headed analysis? Those who have studied my book will realize that the constructibility version of arithmetic found there is very close to Frege's. Of course, I do not accept the *ontological* presuppositions of Frege's account, but the way in which my constructibility version of arithmetic is applied is essentially the way in which Frege's arithmetic is applied. The point is: we can abstract the crucial ideas of Frege's analysis of arithmetic from its ontological framework and apply them in the constructivistic setting. Surely, it is not difficult to see why I would want to allow that there is much that is true in Frege's account of arithmetic.[24]

[24] This is the basic point being made in my discussion of Lebesgue's ideas on integration (see Chihara 1990: 257–8). I was not attempting to show that standard mathematics can be mapped into constructibility theory.

6. The Vagueness Argument

I now turn to a formidable objection raised by Geoffrey Hellman (in Hellman 1993). At bottom, this is a sort of complaint about the vagueness of my constructibility quantifiers. But Hellman gives this objection an especially pointed expression by comparing the idea of *set of natural numbers* and with my concept of *constructible open-sentence*. He imagines a classical mathematician reasoning as follows. Corresponding to the mathematician's notion of *set of natural numbers*, there is Chihara's notion of *constructible open-sentence satisfiable only by finite cardinality open-sentences*. Now I have a clear idea of what counts as a set of natural numbers; but I have only a vague idea of what counts as a constructible open-sentence satisfiable only by finite cardinality open-sentences. Hellman concludes with the words:

> Each reader must judge for herself or himself, but this one feels far more secure in his grasp of 'set of natural number' than in his grasp of the corresponding notion of constructible open sentence. (Hellman 1993: 87)

I must admit that Hellman's attack struck me as daunting indeed when I first read it. But as I began to analyse it, the persuasiveness of the reasoning began to dissipate. Note, first of all, that 'constructible open sentence' is not an expression that I use in developing my constructibility theory. Hellman brought in that expression to stand for a notion that I do use in developing my theory, namely that notion expressed by the phrase consisting of the constructibility quantifier combined with the noun 'open-sentence' ('it is possible to construct an open-sentence x such that'). But when one sticks to my terminology, it becomes evident that Hellman is not comparing the appropriate pair of concepts: what should be compared with 'it is possible to construct an open-sentence such that' is 'there exists a set such that'; and what should be compared with 'set of natural numbers' is 'open-sentence satisfiable only by finite cardinality open-sentences'. Now when the *appropriate* pairs of concepts are compared for clarity or vagueness, it is not so obvious to me that the Platonist's favoured concepts come out winners.

Let us start with the latter pair. Within the framework of, say Zermelo-Fraenkel set theory, the notion of *set of natural numbers* appears to be clear enough. Similarly, if one assumes some well-studied encompassing mathematical structure, then the notion will appear precise and definite. It is this presupposed theoretical framework or encompassing structure that gives rise to the "clear intuition" of the set of natural numbers; because, within the mathematician's theoretical framework or encompassing structure, the various embarrassing questions that philosophers are apt to ask simply do not arise. Ask a practising set theorist such questions as: "What sort of object is a set of natural numbers?" "Is the membership relation external or internal?" "If sets are not perceivable, how have mathematicians arrived at their supposed knowledge of these objects?" "How does one know that the axioms of ZF are true?" Chances are great that no satisfactory answer will be forthcoming. Thus, the apparent limpid clarity of the notion of set is largely illusory: if one removes the theoretical framework or encompassing structure which is needed to forestall the philosopher's questions, the vaunted sharpness of the notion dwindles. After all, who knows what this membership relation really is? We do know that every cat is in the membership relation to the set of all cats; but that is because we are taught that the set of all cats is that set to which a thing is in the membership relation if and only if that thing is a cat. It is not as if we had some independent way of determining whether a cat is in the membership relation to the set of all cats. In this respect, the membership relation is very different from the spatial relation of *being in*. In the case of the latter relation, we can determine by just looking that a particular cherry is in a jar. We have ways of picking out or referring to the jar that are independent of the things that are in it; and we have ways of determining what things are in the jar that are independent of how we refer to it. It is not surprising that some philosophers have found the nature of sets to be puzzling or mysterious. Lewis expressed his dismay at grasping the membership relation this way:

> [W]e have no alternative but to suppose that the relation of member to singleton holds in virtue of qualities or external relations of which we have no conception whatsoever. Do we really understand what it means for a singleton to have a member?
>
> Singletons, hence all classes, and worst of all the relation of membership, are profoundly mysterious. (Lewis 1993: 14)

(Recall the discussion of the concept of set found in Chapter 3, where Lewis—a Mathematical Realist—expressed genuine bewilderment about the nature of sets and about how we could possibly understand the concept of set.)[25] And it should not be thought that only philosophers are sceptical about the general concept of set. For example, the mathematical logician Solomon Feferman has expressed doubts about the notion of arbitrary set, suggesting that the Continuum Hypothesis is inherently vague and that the Platonic philosophy, according to which every sentence of Zermelo-Fraenkel set theory is either true or false, is unsatisfactory.[26] That the

[25] A noted modal logician once said to me that he didn't know if he was or was not a set. Such an expression of uncertainty indicates that this logician did not believe that the concept of set was without vagueness or unclarity.

[26] These ideas were expressed, in Berkeley, during his talk at the 12 March 1997 meeting of the Logic Club of the Group in Logic and the Methodology of Science.

concept of set is not all that clear is also supported by the concluding lines of Michael Hallet's study of Cantorian set theory:

The set concept was *always* unclear, at least after the 'Cantorian transition'; and it was axiomatized in large part because of that unclarity (the paradoxes made a bad situation worse). The mounting theoretical difficulties have simply forced us to go back and face that fact. (Hallett 1986: 305)

Furthermore, the opposing open-sentence concept has some features that give it definite conceptual advantages over its Platonic rival. For example, I can actually construct (write down) open-sentences that are only satisfied by finite cardinality open-sentences. I can point to such open-sentences. On the other hand, unless one resorts to a Maddy-type view of sets, I do not see how one can literally construct and point to any set of natural numbers. So in response to Hellman, I would say that I am much more confident of my grasp of the nature of open-sentences than I am of the nature of sets.

Consider, in this connection, the justification I gave (in Chihara 1990: 65–8) for the Abstraction Axiom of my constructibility theory. The main idea was this: if one replaces the expression '—*x*—' in the schema

$$(\mathbf{C}\alpha)(x)(\mathrm{S}x\alpha \leftrightarrow \text{—}x\text{—})$$

with an open-sentence of the formal language of the constructibility theory, which contains only free occurrences of '*x*', it can be seen that the resulting sentence will be true. This is because the open-sentence to the right of the double arrow will clearly be constructible. Indeed, if one actually writes down a specific example of such an instance of the schema, one will thereby construct the very open-sentence of the kind that is being asserted to be constructible. Such considerations point to the fact that our understanding of many principles of open-sentence constructibility is more definite and developed than it may first appear.

Contrast the situation in set theory where even the most basic and fundamental existence assumptions are under attack: we have seen, for example, that Maddy has denied the existence of the null set, and has even espoused a position that conflicts with the existence of some intersections; and I have already noted Lewis's questioning of the existence of unit sets. These views, in effect, contradict the Abstraction Axiom of Simple Type Theory.

Consider now the quantifier phrase 'there exists a set such that' which corresponds to my 'it is possible to construct an open-sentence such that'. Such quantifier phrases are to be found sprinkled throughout texts on set theory; and mathematicians are quick to make assertions using such phrases. But what does the phrase mean? The Intuitionist Arend Heyting thought that the existential assertions of classical mathematics were lacking in any clear sense.[27] In support of the view that such quantifier phrases are infected with a significant amount of vagueness or unclarity, one could cite the fact that eminent philosophers have given widely divergent analyses of such phrases. In the next section, I shall provide more detailed grounds for maintaining that mathematicians and philosophers have analysed these existential statements in radically different ways. For now, I shall content myself with noting that Wittgenstein, Carnap, and Quine thought long and hard about the existential statements of mathematics, but arrived at mutually incompatible accounts.[28]

Attempting to judge the relative clarity or vagueness of a pair of expressions or concepts involves so many subjective elements that it may be fruitless to continue arguing in the above manner. So I should like to return to a point that I made earlier. Remember, I have not been arguing that my constructibility theory is better than classical mathematics (construed as a theory about actually existing entities). Nor have I been advocating the replacement of classical mathematics with my constructibility theory. I was responding to Mathematical Realists who argued that contemporary science presupposes the existence of mathematical objects—that the very success of science shows that mathematical objects exist. Thus, I put forward my constructibility theory to show that, roughly, the "empirical content" of our scientific theories could be expressed without classical mathematics, by utilizing instead a theory of constructibility. Even if the Mathematical Realist's preferred terms are, in certain respects, clearer or less vague than those of the constructibility theory, it by no means follows that the added clarity of classical mathematics (Platonically construed) is sufficient to give us reasonable grounds for belief in mathematical objects. One might put it this way: the Mathematical Realist claims that we are forced to believe in mathematical objects on pain of being precluded from giving an adequate account of science. I respond: it is much too premature to draw any such conclusion. Look, we could do science this way (the constructibility way) without committing ourselves to mathematical objects. Subjectively based claims of superior clarity for the Realist's mathematical notions do not seem to me sufficiently weighty to justify taking on all the metaphysical baggage that we are asked to accept.

From the mathematical point of view, my constructibility theory looks very much like the set theory of Simple Type Theory (STT). Indeed, it is a simple matter to devise a translation scheme whereby each statement of STT is converted into a statement of the constructibility theory, in such a

[27] Heyting 1956: 2 and 11.

[28] See Wittgenstein 1956: Pt. 4, sects. 19–28 and Carnap 1958: Suppl. A; and Quine 1969: Essay 4. For more discussion of the different accounts philosophers have given of the existential statements of mathematics, see Section 7 below.

way that the axioms of the constructibility theory will end up corresponding to a theorem of STT. Thus, where STT asserts the existence of a set which has, as elements, sets of a certain kind K, my theory asserts the possibility of constructing an open-sentence which is true of the open-sentences that are both constructible and also the translational correlates of the sets of kind K. This has led some to suggest that my phrase 'it is possible to construct an open-sentence such that . . .' just means what the set theorist means by 'there exists a set such that . . .' After all, if the phrases in question have different meanings, then should not this difference have different mathematical consequences? The critic could then argue:

> If you have a genuinely new interpretation of the language of STT, that novelty has to show up somewhere. In the case of mathematical constructivists, it shows up in their rejection of various set-existence assertions. But you do not seem willing to reject any specific axiom or theorem of STT.[29]

Here, we need to distinguish what can be asserted in my constructibility theory from what I am willing to reject in mathematics. I do not claim that the translation of anything assertable in any mathematical theory, say STT with such axioms as Choice and the Continuum Hypothesis, can be asserted in my constructibility theory. On the other hand, if the translation of some theorem of such a set theory cannot be asserted in my constructibility theory, I would not consider that fact grounds for rejecting the theorem as false or meaningless. I again emphasize that I am not advocating a revisionist account of mathematics. I have no objections to anything purely mathematical that mathematicians may assert as mathematicians. In particular, I do not find the development of STT, even when augmented with a large array of axioms unobtainable in my constructibility theory, objectionable in the least. It is only when the mathematician begins to philosophize that my sceptical doubts begin to arise.

Now it may be objected that one cannot distinguish when a mathematician is making a purely mathematical assertion from when a mathematician is asserting something philosophical. However, I am not claiming that the distinction is a precise one, and I freely admit that one may not be able to determine whether some claim or other is purely mathematical. One very rough guide for judging that a claim is purely mathematical is this: all (or almost all) mathematicians who are knowledgeable about the area of mathematics in question should find the claim reasonably clear and understandable. Thus, if a claim is made which many experts are unable to make clear sense of, or if many conflicting analyses of what is being claimed are put forward, then there would be some reason to think that the claim is not

[29] This objection was suggested to me by John Steel.

a purely mathematical one. By this criterion, the set theorist who declares 'The Continuum Hypothesis and the Axiom of Constructibility ($V = L$), although both are independent of the axioms of ZF, are probably false' may possibly be judged to be making a kind of philosophical claim.[30] In any case, nothing I am claiming depends upon deciding whether such statements are or are not purely mathematical. However one might end up judging such assertions, there are clear cut cases of assertions which are clearly not purely mathematical. For example, such statements as 'There is as much reason to believe in the existence of mathematical objects as there is to believe in material objects', 'Our knowledge of sets is hardwired into our brains', and 'Humans have perceived sets since prehistoric times', are clearly not purely mathematical assertions, even if they were asserted by a mathematician.

Suppose that a set theorist says that, in addition to the cats that exist in this world, there exists a set of all cats. Such an assertion does not appear to be a purely mathematical assertion: it is not a theorem of any set theory that I know of. Rather, it appears to be an ontological claim. It is here that the difference in meaning between 'there exists a set such that . . .' and 'it is possible to construct an open-sentence such that . . .' clearly shows up. I have no doubt that it is possible to construct an open-sentence that is true of all the cats that exist in this world. Here is an example of such an open-sentence: '*x* is a cat'. And I had no problem constructing such an open-sentence. But the fact that such an open-sentence is constructible gives me no reason at all to think that, in addition to the cats that exist in this world, there exists another thing, namely, the set of all cats that exist in this world.

I now seem to be in the paradoxical position of not contesting the set theorist's purely mathematical assertion that there is a set of real numbers satisfying a certain mathematical condition, while at the same time, rejecting the set theorist's ontological claim that, in addition to cats, there is a set of all cats and hence there are sets. If I do not contest the former, am I not, in effect, allowing that there are sets after all? How I answer this question should be obvious from Section 5 above.

7. The Burgess–Rosen Evaluation

The previous section describes my position in a way that is highly relevant to a type of objection to "nominalism" that has been raised very recently by John Burgess and Gideon Rosen in their excellent book (Burgess and Rosen 1997). It should be mentioned straight off that these authors do not

[30] I have heard distinguished mathematical logicians say that such claims have no clear sense (or are just expressions of gut feelings about the axiom).

claim to be refuting my specific position: they are concerned, rather, to raise damaging questions about the scientific merits of the various "nominalistic reconstructions" of mathematics that they discuss in their book. Since they classify my position as "nominalist", and since they talk about my "reconstrual of classical mathematics", it is not unreasonable to suppose that at least some of their readers may come to the conclusion that they are aiming their pointed questions at my position, as well as at the others. So there are reasons for my responding to the Burgess–Rosen objection as if it were aimed at my views. In any case, the following discussion should shed further light on my philosophy of mathematics.

When evaluating the dispute between Mathematical Realists and nominalists, according to Burgess and Rosen, a crucial question that must be answered is: are any of the nominalistic reconstructions of classical mathematics so scientifically superior to the versions currently in actual use in science that they ought to be credited in preference to them? Before examining the Burgess–Rosen answer to this question, it will be useful to see what, if any, conclusions relevant to my anti-Realist position should be drawn, if it is agreed that none of the nominalist reconstructions of mathematics are scientifically superior to the mathematics currently in use by scientists. In that case, Burgess and Rosen maintain, the nominalist will have to fall back on arguments directly destructive of the Mathematical Realist's position—arguments that show on epistemological or semantical grounds that the Realist's position is untenable. As these authors are unconvinced by the nominalist's directly destructive arguments, the question of whether any of the various nominalistic reconstructions are scientifically superior to the mathematical systems currently in use in the sciences is, for them, crucial to the defence of nominalism.

A direct test of this crucial question—"the true test", they maintain—would consist in submitting a paper presenting the nominalist's reconstruction to a scientific journal (they suggest the *Physical Review*).[31] What would be the reaction? Burgess and Rosen are convinced that the nominalistic reconstruction would not be taken seriously to constitute a significant advance over the mathematical systems currently in use. They point out the madness of requiring the applied physicist, concerned with predicting the perturbations of Mars before a space shot, to carry out her mathematical reasoning within the framework of my constructibility theory. And they suggest that it would be equally insane to demand that the theoretical physicist's proposal to refine gravitational theory be couched in my modal language.[32]

A second line of argumentation is put forward to question the hypothe-

[31] Burgess and Rosen 1997: 210.
[32] Burgess and Rosen 1997: 211.

sis that some nominalistic reconstruction of mathematics is superior, on scientific grounds, to the mathematical systems actually in use in the empirical sciences. Burgess and Rosen list a number of criteria that are thought to be implicitly employed by scientists to determine whether some theory is superior to its rivals. Here are four:

(A) minimality or economy of assumptions;
(B) consistency or coherence with familiar, established theories;
(C) perspicuity of the basic notions and assumptions;
(D) fruitfulness, or capacity for being extended to answer new questions.

To judge some nominalistic reconstruction of mathematics to be scientifically superior to the mathematics currently in use, they argue, one would have to place great weight on (A) and discount the significance of (B), (C), and (D); and it is hard to see how the reconstructive nominalist could justify down-playing factors (B)–(C).[33]

Whether or not the above reasoning is destructive of any of the various nominalistic views these authors discuss in their book, I wish to argue here that it misses the mark if it is aimed at my position. Let me again emphasize that I have not been advocating the *replacement* of the mathematical systems, currently used by scientists, with my constructibility version of mathematics: I am not what Burgess and Rosen call a "revolutionary nominalist" (Burgess and Rosen 1997: Pt. III, sect. C.1). Nor have I put forward my constructibility theory as scientifically superior to these mathematical systems. I would not dream of submitting my constructibility work for publication in the *Physical Review*. My work is directed at *philosophers* who put forward ontological claims based upon the use of mathematics in the empirical sciences. It is not my aim to convince physicists to adopt my modal theories. If, as Mathematical Realists claim, mathematical objects must be postulated to account for physics, it is striking that physicists do not publish papers in which the existence of mathematical objects is postulated.

Here is how I view the situation. The empirical scientist, from the earliest stages of her education, learns a whole host of mathematical techniques, procedures, and skills. She learns to count and answer 'How many?' questions. She learns to add, subtract, multiply, and divide natural numbers; and she learns to use these techniques to answer all sorts of practical questions. She also learns similar techniques for dealing with what the teacher calls 'integers'. Then she learns to order, add, subtract, multiply, and divide fractions; later, she extends these techniques to the rational numbers; and later still, she learns analogous techniques for the case of

[33] Burgess and Rosen 1997: 209–11.

what are called 'irrational numbers'. Along the way, she begins to learn techniques that involve sets, relations, and functions of number. She learns how to find the sum of infinite series; and she masters the techniques of integration and differentiation. Of course, she is also trained in more theoretical aspects of mathematics: she learns to prove theorems, some of which begin with such phrases as 'There exists a number such that' and 'There are continuous functions which are such that'. Now Burgess and Rosen seem to believe that such theorems should be understood to be asserting, in a straightforwardly literal, existential quantifier sense, the existence of mathematical objects. In contrast to my position on the matter, Burgess and Rosen seem to believe that, unless the nominalist can produce a nominalistic version of mathematics that is superior, on purely scientific grounds, to the mathematical system currently used by empirical scientists, then in the absence of convincing reasons that are destructive of the realist's ontology, one is rationally required to believe in mathematical objects. As they put it, "any form of nominalism will have to be revisionary, and any revision demands motivation" (Burgess and Rosen 1997: 11).

My own view is that it is by no means obvious that the mathematical theorems in question should be understood in the straightforward manner given by the usual first-order versions of such theorems. It is not that I believe that the existential assertions of the practising mathematician should be construed to be constructibility assertions of the sort described in my book. I am no "hermeneutic nominalist" (to use the terminology of Burgess and Rosen).[34] I have no desire, at this time, to put forward any analysis of the mathematician's existential assertions. A study of the history of mathematics indicates the pitfalls of putting forward any such analysis.

Many mathematicians and philosophers have given quite different accounts of such assertions. The great mathematician-logician David Hilbert, for example, held that existence in mathematics is a matter of consistency:

> [I]f the arbitrarily given axioms do not contradict one another with all their consequences, then they are true and the things defined by the axioms exist. This for me is the criterion of truth and existence. The proposition 'Every equation has a root' is true, and the existence of a root is proven, as soon as the axiom 'Every equation has a root' can be added to the other arithmetical axioms, without raising the possibility of contradiction, no matter what conclusions are drawn.[35]

[34] See Burgess and Rosen 1997: Pt. III, sect. C.2.

[35] Frege 1980: 39–40. Given the above views, it is easy to see why Hilbert would also maintain that, as Mancosu puts it, "existence in the mathematical sense is always relative to a given system." See Mancosu 1998: 159.

Hilbert's views on mathematical existence are quite similar to those of the mathematician Henri Poincaré. In Poincaré 1952 he wrote:

[T]he word "existence" has not the same meaning when it refers to a mathematical entity as when it refers to a material object.

A mathematical entity exists provided there is no contradiction implied in its definition, either in itself, or with the proposition previously admitted. (p. 44)

A similar claim is made in Poincaré 1953:

In mathematics the word exist can only have one meaning; it signifies exemption from contradiction . . . In defining an object, we assert that the definition involves no contradiction. (pp. 151–2)

Here, we have the views of someone who was unquestionably one of the truly great mathematicians of all time and who was also a first-rate theoretical physicist. The fact that *he* did not understand the so-called "existential theorems" of mathematics in the way Burgess and Rosen do suggests that these theorems should not be simply *assumed* to have the straightforward existential quantifier sense: some serious analysis and investigation must be provided. (It can hardly be claimed that Hilbert and Poincaré were not sufficiently competent in mathematics to know what the existential assertions of mathematicians mean.)

Hilbert and Poincaré were by no means the first mathematicians to espouse a view of mathematical existence that deviates from the straightforward view presupposed by Burgess and Rosen. A study of the history of mathematics will uncover many views that deviate from the one Burgess and Rosen advocate. A detailed discussion of this point is not called for here, but a few more examples may be persuasive. In 1615, Giuseppe Biancani, professor of mathematics at the University of Parma, responded to the objection that mathematical entities do not exist in nature by arguing:

even though these [perfect mathematical figures] do not exist in the nature of things, since in the mind of the Author of Nature, as well as in the human mind, their ideas do exist as the exact archetypes of all things, indeed, as exact mathematical entities, the mathematician investigates their ideas, which are primarily intended per se, and which are [the] true entities. (Mancosu 1996: 180)

Thus, Biancani argued in his *De Mathematicarum Natura* that what the mathematician investigates are ideas in the human and Divine mind.

In 1683, the mathematician I. Barrows maintained that mathematicians may carry out demonstrations about things that never had (and never would) exist: "it is sufficient for a demonstration to assume a true hypothesis, i.e. such as to imply no inconsistence in themselves" (Mancosu 1996: 141). For Barrows, as for Poincaré, one can reason about entities in mathematics that do not exist, so long as they could exist: what is required is that the concepts of the entities in question be *consistent*. Obviously, mathematical existence assertions could not have been understood by this mathematician in the straightforward way advocated by Burgess and Rosen.

Why should there be any disputes about the meaning of existence assertions in mathematics? Is it not obvious what such assertions mean? Should not mathematicians and physicists know what they mean, especially given that they are continually dealing with such assertions in their theorizing? I noted in the previous section that Heyting could not attribute a clear sense to the existential assertions of classical mathematics. The brilliant mathematician Hermann Weyl seemed to have similar thoughts when he suggested that the existential assertions of mathematicians were not all that clear: "The concept of existence is overburdened by metaphysical enigmas" (Weyl 1987: 8).

There is something puzzling about mathematical existence statements. Suppose, for the sake of argument, that the phrases 'there are', 'there exists', etc. are given (in mathematical contexts) the straightforward meaning of the existential quantifier of predicate logic. Then from Euclid's theorem that there are infinitely many prime numbers, we can infer that there are numbers. But can we? Many have thought not. Paul Benacerraf, for example, claimed that "there are no such things as numbers; which is not to say that there are not at least two prime numbers between 15 and 20" (Benacerraf 1965: 73).

Of course, many philosophers who have investigated the nature of mathematical existence statements have arrived at positions that are at variance with what Burgess and Rosen advocate. Examples can be readily found in the writings of contemporary philosophers of mathematics—in particular Hellman 1989, Kitcher 1984, and Resnik 1981. Consider what Ludwig Wittgenstein said about such theorems after many years of research on topics in the philosophy of mathematics:

> A proof convinces you that there is a root of an equation (without giving you any idea *where*)—how do you know that you understand the proposition that there is a root? . . .
>
> Everything that I say really amounts to this, that one can know a proof thoroughly and follow it step by step, and yet at the same time not *understand* what it was that was proved.
>
> And this in turn is connected with the fact that one can form a mathematical proposition in a grammatically correct way without understanding its meaning.[36]

Michael Dummett's non-Fregean, non-classical analysis of the existential statements made by mathematicians also does not gybe with the

[36] Wittgenstein 1956: 146. As I indicated earlier, a related view was expressed by Arend Heyting, who suggested that the existential assertions of classical mathematics are "metaphysical" in nature and do not have a clear sense.

Burgess–Rosen view. In Dummett 1964 he holds that the sense of the existential quantifier is determined by the sort of fact that makes an existential statement true:

> and this means: the sort of fact which we have been taught to regard as justifying us in asserting an existential statement. . . . *We are thus justified in asserting that there is a number with a certain property only if we have a method for finding a particular number with that property*.[37]

Consider now the views of Quine, who more than any other philosopher is responsible for the currently widespread acceptance of Mathematical Realism. Quine did not think that we ought to believe in such objects just because scientists use classical mathematics in their theorizing: he certainly did not think that we can just assume that such English phrases as 'there exist' and 'there are', used in standard mathematics texts, are synonymous with the existential quantifier of classical mathematical logic. For Quine, the ontological commitments of our mathematical theories cannot be determined so simply: we need a *criterion of ontological commitment*.[38] The criterion Quine comes up with requires that the theory whose ontological commitments are to be assessed be formulated in a logical language with standard objectual quantifiers. To see what sorts of entities we ought to believe in, Quine suggests, we should *replace* our contemporary scientific theories with a theory that is couched in a canonical language that is both scientifically superior to natural languages and also more logically suited to the needs of the philosopher—a theory formulated in first-order logic, without modal or intentional operators. As we attempt to construct such a theory, we will see, argues Quine, that we will have to include mathematical objects within the range of the bound variables of the theory. In other words, the postulation of mathematical objects is held to be *indispensable* for the adequate formulation and use of scientific theories. Thus, it is argued, we have good reason for believing in mathematical entities—more specifically, we have good reasons for believing in the existence of sets, since all of the entities needed in classical mathematics can be defined in set theory.[39]

I find it ironic that the arch-Mathematical Realist, Quine, proposes just the sort of revolutionary reconstruction that Burgess and Rosen attribute to the nominalist. Yet they do not suggest that we test the scientific superiority of Quine's reconstruction by submitting Quine's proposals to the *Physical Review*. (Nor do they point out the madness of requiring the

[37] Dummett 1964: 109, italics mine.

[38] For a detailed discussion of Quine's criterion, see Chihara 1973: ch. 3, sect. 3.

[39] For the details of Quine's reasoning, see Chihara 1990: ch. 1, sect. 2. For a thorough discussion of indispensability argument of the above sort, see Hellman 1995.

applied physicist, concerned with predicting the perturbations of Mars before a space shot, to carry out her mathematical reasoning within the framework of Quine's set theoretical version of mathematics, formulated within his canonical language of science.) Nor do they complain that it is hard to see how the reconstructive Realists could justify their revolutionary proposals without down-playing (B), (C), and (D).

Maddy produces an argument similar to Quine's for believing in mathematical objects, writing:

> On the naturalized approach, we judge what entities there are by seeing what entities we need to produce the most effective theory of the world. (Maddy 1990: 29)

She agrees with Quine that the postulation of mathematical objects is good and useful in so far as it simplifies our account of physics (Maddy 1990: 29). But as we saw earlier, she feels that it is not enough to account for physics: we should also account for our mathematical practices. This is one reason she postulates mathematical objects that have some features distinctive of pure sets (such as having members), but which also have some features distinctive of physical objects (such as being located in space and being causally active). And as I noted above, the Set Theoretic Realism she ends up with is revisionist, in so far as it implies the falsity of significant portions of contemporary set theory.

The reasoning of these Mathematical Realists can be described along the following lines. We have the practice of physics, for which the Realist seeks an adequate formulation and explanation. As additional "data" to be explained, we can list the impressive successes we have achieved with physics. The Mathematical Realist postulates the existence of mathematical objects in order to obtain an acceptable formulation and account of this practice and its successes. (Maddy, as we noted above, feels the need to account for our mathematical practices, as well as physics.) The Realist then argues that belief in mathematical objects is justified or supported by its explanatory success—its success in explaining physics, our mathematical practices, and the like.

My response to this kind of reasoning proceeds along two fronts. On the one hand, I criticize the formulations and explanations of the mathematical theories advanced by the Realist, as I did above in the case of Maddy.[40] On the other hand, I provide an alternative way of seeing how physics (and the other empirical sciences) could be developed without presupposing the existence of mathematical objects. This is where the constructibility theory comes in: my constructibility theory has been devised not to replace current mathematical systems but to show how one could justify operating and calculating as we do in practical and theoretical contexts without pos-

[40] See, for other examples, Chihara 1982: Pt. 1 and Chihara 1990: ch.1.

tulating mathematical objects.[41] From my point of view, then, the grounds given by these philosophers for belief in mathematical objects are simply not sufficient. When it comes to belief in mathematical objects, I am (like the traditional nominalist) sceptical.

8. Reply to the Gillies–Wolenski Objection

The Objection

I would now like to return to the Gillies–Wolenski objection, discussed in the Introduction of this work, that my own constructibility response to the indispensability argument commits me to the existence of possible worlds. Evidently, my critics have based their objection on the heuristic and didactic explanations I gave of the constructibility quantifiers. I gave the formal semantics of these "quantifiers" in terms of S5-type structures; and I appealed to the idea of possible worlds as a way of understanding how these structures function. In other words, it could be argued, in giving the explanations that I did in my book, I committed myself to the reality of the entities I used in my explanation, thereby committing myself to the existence of possible worlds (recall the quote from Barwise and Perry in the Introduction). Evidently, this is why Gillies and Wolenski believe that my constructibility theory eliminates mathematical entities from our view of the world only at the expense of taking on the much more onerous ontological burden of possible worlds.

Consider the following situation. A physicist explains the phenomenon of tides by appeal to Newtonian physics. Now it might be argued that the physicist has committed herself to the truth of Newtonian physics and hence to the existence of all the entities that Newtonian physics postulates. We can imagine the physicist responding to such a charge in the following way:

> I use Newtonian physics to explain the tides in order to simplify the explanation. I realize, of course, that the laws of Newtonian physics are not, strictly speaking, true. But it is well known that these laws give good approximations when dealing with phenomena that are not too small (such as the sub-molecular) or not too large (such as the astronomical). The explanation I gave was not intended to be entirely accurate. Of course, my official explanation would be one using Relativity Theory. But for many purposes, the rough and ready explanation from Newtonian physics is to be preferred, especially since it is so much simpler and easier to grasp.

[41] Geoffrey Hellman's account of mathematics in Hellman 1989 can be regarded as having a similar motive.

My response to the charge that my explanations of the constructibility theory commit me to possible worlds is similar to the physicist's. My talk of possible worlds was given for didactic and heuristic purposes—these explanations were intended to supply a rough aid to one's intuitions in dealing with S5 structures and the semantical features of the quantifiers; they were not intended to be taken literally. Roughly speaking, the "official explanations" of the role of S5 structures, of the relationships that obtain between truth under S5 interpretations and truth (simpliciter), etc. are those given in Chapters 6, 7, and 8 of the present work.

To be more specific, let us consider the formal language in which the constructibility quantifier was introduced (in Chihara 1990: ch. 2, sect. 2). The interpretations or structures of the language were given set theoretically in such a way that each structure could be taken to be a pure set. Thus, the pure model theory of the language does not presuppose possible worlds. Of course, a connection must be made between the structures and applications of the model theory in virtue of which it is legitimate to draw conclusions about the truth or falsity of interpreted sentences of the language. How is that to be done? In Chihara 1990: 28 I said: "Intuitively, one can think of **W** as the set of all possible worlds, **a** as the actual world, and U(**w**) as the set of all things that exist in possible world **w**." In so saying, I wished to give readers familiar with standard versions of possible worlds semantics a handle on the semantics of the target language. I made it clear later that all this talk of possible worlds was not to be taken literally. Thus, I did not, and (in consistency) cannot, appeal to possible worlds in explaining how the model theory allows us to draw conclusions about the truth or falsity of interpreted sentences of the language. How then can I explain what is wanted? It should now be clear that my explanation would be along the lines given in Chapters 6, 7, and 8 of the present work.

To clarify this last point, let us examine my use of possible worlds semantics in Chihara 1990. An examination of the book makes it clear that possible worlds semantics is used primarily in explaining the logic of three formal languages: **L**, **L***, and **Lt**. For my purposes here, I can focus on **L***: Although it lacks the type-theoretical complications of the language **Lt**, in which the constructibility theory is formalized, it does have all the basic modal features of that language.

*The Language L**

The language **L*** was brought in to explain the logic of constructibility quantifiers. The syntax of **L*** differs, in the following respects, from that of **FL** (the language of first-order logic presented in Chapter 5): it uses two sorts of variables (starred and unstarred variables) and two kinds of quantifiers (standard and constructibility). One obvious question is whether the constructibility quantifier '(**C**x)' can be translated into '$\Diamond(\exists x)$'. To explore this question, let us consider the following example discussed in Chihara 1990: ch. 2, sect. 4. Any sentence consisting of the words 'There are no more than', followed by an Arabic numeral, which in turn is followed by the word 'star', is classified as an S-sentence. Now consider the sentence:

[1] $(\mathbf{C}x)(x$ is an S-sentence & x is true).

We wish to investigate whether we can translate [1] into:

[2] $\Diamond(\exists x)(x$ is an S-sentence & x is true).

Now [1] is true just in case there are finitely many stars, whereas [2] would be true even if there were infinitely many stars, since even if there were in fact infinitely many stars, the world could have been such that both there was only finitely many stars and also an S-sentence was constructed.

It can be seen that the trouble with treating '$(\mathbf{C}x)$' as '$\Diamond(\exists x)$' is that the diamond affects not merely the sentences that are constructible, but also the stars being talked about. Now an examination of the language **L*** will show that this problem would not arise in that language. **L*** is a two-sorted language in which the starred variables always range over things in **a*** (the set that is called, for heuristic purposes, "the actual world"). So in **L***, the constructibility quantifier '$(\mathbf{C}x)$' can indeed be treated as '$\Diamond(\exists x)$'—which makes it clear that the **L*** is basically an S5 language and that it would be a straightforward matter to prove a Fundamental Theorem for this language.

No Commitment to Worlds

The interpretations or structures of **L*** were given set theoretically in such a way that each structure could be taken to be a pure set. Thus, there is no temptation to suppose that the pure semantics of the language presupposes possible worlds. But as Plantinga argued, this pure semantics fails to assign any meaning to the modal operators or tell us what such sentences as

[3] $(\exists x)\Box x$ is a person

mean (Plantinga 1974: 127). For this reason, he argued, there is need of an "intended or associated applied semantics" according to which the set **W** of the structure is not just any set but is the set of all possible worlds, **a*** is not just any element of **W** but is the actual world, etc. In other words, it is thought that one of the possible worlds structures talked about in pure semantics must be the intended applied semantics. It is this intended applied semantics that gives meaning to the modal operators and allows us to use the logic to draw concrete modal inferences involving such sentences as [3] above. Needless to say, it is through the use of this intended applied semantics that the commitment to possible worlds is supposedly made.

What should be abundantly clear by now is that, except for didactic and heuristic purposes, I make no appeal to any such applied semantics. For me, the meaning of the modal operators is to be explained without reference either to pure semantics or to the sort of applied semantics Plantinga and Lewis had in mind (recall [T-d] and [T-b] of Chapter 6). It is by means of NL interpretations and the Fundamental Theorem that I am able to analyse such sentences as [3] and to justify drawing concrete inferences involving such sentences.

9. A Circularity?

One final objection should be investigated. In Chihara 1990 I used the notion of possible world in explaining the constructibility quantifiers. In the present work, I make use of constructibility quantifiers in my anti-Realist account of possible worlds semantics. It might be thought that my procedure involves a vicious circularity. Have I not, in effect, analysed constructibility quantifiers in terms of possible worlds, and then analysed possible worlds in terms of constructibility quantifiers?[42]

This circularity objection rests on a misunderstanding—one that is understandable, but that is a misunderstanding just the same. In my earlier work, I did not *define* my constructibility quantifiers in terms of possible worlds: in my system, constructibility quantifiers are *primitives*. The appeal to possible worlds was made in order to aid the reader in achieving a proper grasp of these quantifiers and the reasoning, involving these quantifiers, that is necessary to achieve the sort of "no-class" mathematical theory I was advocating. Indeed, in an even earlier work, in which I first set forth the basic features of these constructibility quantifiers, no mention was made of possible worlds at all.[43] As I emphasized above, talk of possible worlds in the explications of my mathematical system should be regarded as merely heuristic and didactic in nature, useful for aiding the uninitiated in getting the appropriate understanding of the key concepts and in learning to apply and to develop these concepts in a correct manner. I see no vicious circularity in applying possible world semantics in this way, and in then going on to explain apparent references to possible worlds in possible worlds semantical theory in terms of constructibility quantifiers, as I have done above.

[42] This objection was raised by Kazuyuki Nomoto (at my Keio University talk) and also, independently, by Ser-min Shei (at my National Chung-Cheng University talk).

[43] See Chihara 1984.

BIBLIOGRAPHY

ADAMS, R. M. (1979*a*). Primitive Thisness and Primitive Identity. *Journal of Philosophy*, **76**, 5–26.

—— (1979*b*). Theories of Actuality. In Michael J. Loux, ed., *The Possible and the Actual*, 190–209. Ithaca, N.Y.: Cornell University Press.

ARMSTRONG, D. M. (1989). *A Combinatorial Theory of Possibility*. Cambridge: Cambridge University Press.

BACH, C. (1995). Tarski and the Concept of Logical Consequence. Unpublished Ph.D. dissertation, University of California, Berkeley.

BALINSKY, B. I. (1981). *An Introduction to Embryology*, 5th edn. Philadelphia, Pa.: Saunders College.

BARWISE, J., and PERRY, J. (1985). Shifting Situations and Shaken Attitudes. *Linguistics and Philosophy*, **8**, 105–61.

BELNAP, N. (1962). Tonk, Plonk and Plink. *Analysis*, **22**, 130–4.

BENACERRAF, P. (1965). What Numbers Could Not Be. *Philosophical Review*, **74**, 47–73.

BERGMANN, M. (1996). A New Argument from Actualism to Serious Actualism. *Nous*, **30**, 356–9.

BIGELOW, J., and PARGETTER, R. (1987). Beyond the Blank Stare. *Theoria*, **53**, 97–119.

BOYER, C. B. (1985). *A History of Mathematics*. Princeton, N.J.: Princeton University Press.

BRADLEY, R., and SWARTZ, N. (1979). *Possible Worlds: An Introduction to Logic and its Philosophy*. Indianapolis, Ind.: Hackett.

BRICKER, P. (1996). Isolation and Unification: The Realist Analysis of Possible Worlds. *Philosophical Studies*, **84**, 225–38.

BRINGSJORD, S. (1985). Are There Set Theoretic Possible Worlds? *Analysis*, **45**, 64.

BROAD, C. D. (1933). *Examination of McTagggart's Philosophy*, vol. i. Cambridge: Cambridge University Press.

BROCK, S. (1993). Modal Fictionalism: a Reply to Rosen. *Mind*, **102**, 147–50.

BURGESS, J. (1992). Review of Chihara's *Constructibility and Mathematical Existence*. *Philosophical Review*, **101**, 916–18.

—— and ROSEN, G. (1997). *A Subject with no Object: Strategies for Nominalistic Interpretations of Mathematics*. Oxford: Oxford University Press.

CANFIELD, J. (1986), ed. *Criteria*, vol. vii. New York: Garland.

CARNAP, R. (1958). *Meaning and Necessity*. Chicago, Ill.: University of Chicago Press.

CARNAP, R. (1959). *The Logical Syntax of Language*. Paterson, N.J.: Littlefield, Adams.

CHELLIS, B. F. (1980). *Modal Logic: An Introduction*. Cambridge: Cambridge University Press.

CHIHARA, C. S. (1973). *Ontology and the Vicious-Circle Principle*. Ithaca, N.Y.: Cornell University Press.
—— (1979). The Semantic Paradoxes: A Diagnostic Investigation. *Philosophical Review*, **88**, 590–618.
—— (1980). Ramsey's Theory of Types: Suggestions for a Return to Fregean Sources. In D. H. Mellor, ed., *Prospects for Pragmatism: Essays in Memory of F. P. Ramsey*, 21–47. Cambridge: Cambridge University Press.
—— (1981). Quine and the Confirmational Paradoxes. In P. French, T. Uehling, and H. Wettstein, eds., *Midwest Studies in Philosophy*, vol. vi, *The Foundations of Analytic Philosophy*, 425–54. Minneapolis, Minn.: University of Minnesota Press.
—— (1982). A Gödelian Thesis Regarding Mathematical Objects: Do they Exist? And Can We Perceive Them? *Philosophical Review*, **91**, 211–27.
—— (1984). A Simple Type Theory without Platonic Domains. *Journal of Philosophical Logic*, **13**, 249–83.
—— (1990). *Constructibility and Mathematical Existence*. Oxford: Oxford University Press.
—— (1996). Poincaré and Logicism. In Jean-Louis Greffe, Gerhard Heinzmann, and Kuno Lorenz, eds., *Henri Poincaré: science et philosophie, congres international*, 435–46. Berlin: Akademie Verlag.
—— (1998). Tarksi's Thesis and the Ontology of Mathematics. In Matthias Schirn, ed., *Philosophy of Mathematics Today*. Oxford: Oxford University Press.
CHISHOLM, R. (1946). The Contrary-to-Fact Conditional. *Mind*, **55**, 289–307.
DEUTSCH, H. (1990). Contingency and Modal Logic. *Philosophical Studies*, **60**, 89–102.
DIVERS, J. (1995). Modal Fictionalism Cannot Deliver Possible Worlds Semantics. *Analysis*, **55**, 81–8.
DUMMETT, M. (1964). Truth. In George Pitcher, ed., *Truth*, 93–111. Englewood Cliffs, N.J.: Prentice-Hall.
ENDERTON, H. (1972). *A Mathematical Introduction to Logic*. New York: Academic Press.
—— (1977). *Elements of Set Theory*. New York: Academic Press.
ETCHEMENDY, J. (1990). *The Concept of Logical Consequence*. Cambridge, Mass.: Harvard University Press.
FINE, K. (1975). Some Connections Between Elementary and Modal Logic. In Stig Kanger, ed., *Proceedings of the Third Scandinavian Logic Symposium*, 15-31. Amsterdam: North-Holland.
—— (1978). Model Theory for Modal Logic: Part I—The De Re/De Dicto Distinction. *Journal of Philosophical Logic*, **7**, 125–56.
—— (1981). Model Theory for Modal Logic: Part III—Existence and Predication. *Journal of Philosophical Logic*, **10**, 293–307.
—— (1985). Plantinga on the Reduction of Possibilist Discourse. In James E. Tomberlin and Peter van Inwagen, eds., *Alvin Plantinga*, 145–86. Dordrecht: Reidel.
FORBES, G. (1980). Origin and Identity. *Philosophical Studies*, **37**, 352–62.
—— (1985). *The Metaphysics of Modality*. Oxford: Oxford University Press.

—— (1986). In Defense of Absolute Essentialism. In P. A. French, T. Uehling, and H. Wettstein, eds., *Midwest Studies in Philosophy*, vol. xi, *Studies in Essentialism*, 3–31. Minneapolis, Minn.: Minnesota University Press.

—— (1989). *Languages of Possibility*. Oxford: Blackwell.

—— (1994). A New Riddle of Existence. *Philosophical Perspectives*, **8**, 415–30.

FREGE, G. (1959). *The Foundations of Arithmetic*, trans. J. L. Austin, 2nd edn. Oxford: Blackwell.

—— (1979). Sources of Knowledge of Mathematics and the Mathematical Natural Sciences. In Hans Hermes, Friedrich Kambartel, and Friedrich Kaulbach, eds., *Gottlob Frege: Posthumous Writings*, 263–6. Chicago, Ill.: University of Chicago Press.

—— (1980). *Philosophical and Mathematical Correspondence*, trans. Hans Kaal. Chicago, Ill.: University of Chicago Press.

GARSON, J. (1991). Applications of Free Logic to Quantified Intensional Logic. In Karel Lambert, ed., *Philosophical Applications of Free Logic*, 111–42. New York: Oxford University Press.

GIBBARD, A. (1975). Contingent Identity. *Journal of Philosophical Logic*, **4**, 197–231.

GILLIES, D. A. (1992). Review of Chihara's *Constructibility and Mathematical Existence*. *British Journal for the Philosophy of Science*, **43**, 263–78.

GÖDEL, K. (1964*a*). Russell's Mathematical Logic. In P. Benacerraf and H. Putnam, eds., *Philosophy of Mathematics: Selected Readings*, 211–32. Englewood Cliffs, N.J.: Prentice-Hall.

—— (1964*b*). What is Cantor's Continuum Problem? In P. Benacerraf and H. Putnam, eds., *Philosophy of Mathematics: Selected Readings*, 258–73. Englewood Cliffs, N.J.: Prentice-Hall.

GOGUEN, J. A. (1969). The Logic of Inexact Concepts. *Synthese*, **19**, 325–78.

GOLDBLATT, R. (1993). *Mathematics of Modality*. Stanford, Calif.: CSLI Publications.

GOODMAN, N. (1973). *Fact, Fiction, and Forecast*, 3rd edn. Indianapolis, Ind.: Bobbs-Merrill.

GRIM, P. (1986). On Sets and Worlds: A Reply to Menzel. *Analysis*, **46**, 186–91.

GUPTA, A. (1980). *The Logic of Common Nouns*. New Haven, Conn.: Yale University Press.

HAJEK, A., and NED, H. (1994). The Hypothesis of the Conditional Construal of Conditional Probability. In Ellery Eells and Brian Skyrms, eds., *Probability and Conditionals*, 75–111. Cambridge: Cambridge University Press.

HALE, B. (1995). Modal Fictionalism—A Simple Dilemma. *Analysis*, **55**, 63–7.

HALLETT, M. (1986). *Cantorean Set Theory and Limitation of Size*. Oxford: Oxford University Press.

HARPER, W., STALNAKER, R., and PEARCE, G. (1981), eds. *Ifs: Conditionals, Beliefs, Decision, Chance, and Time*. Dordrecht: Reidel.

HAZEN, A. (1976). Expressive Incompleteness in Modal Logic. *Journal of Philosophical Logic*, **5**, 25–46.

—— (1979). Counterpart-Theoretic Semantics for Modal Logic. *Journal of Philosophy*, **76**, 319–38.

—— (1996). Actualism Again. *Philosophical Studies*, **84**, 155–81.

HELLMAN, G. (1989). *Mathematics Without Numbers*. Oxford: Oxford University Press.

—— (1993). Review of Chihara's *Constructibility and Mathematical Existence*. *Philosophia Mathematica*, **3**, 75–88.

—— (1995). Some Ins and Outs of Indispensability: A Modal-Structural Perspective. In A. Cantini, E. Casari, and P. Minari, eds., *Logic in Florence, 1995*. Dordrecht: Kluwer.

HEYTING, A. (1956). *Intuitionism: An Introduction*. Amsterdam: North-Holland.

HILBERT, D. (1983). On the Infinite. In Paul Benacerraf and Hilary Putnam, eds., *Philosophy of Mathematics: Selected Readings*, 2nd edn., 377–93. Cambridge: Cambridge University Press.

HINTIKKA, J. (1969). *Models for Modalities*. Dordrecht: Reidel.

—— and HINTIKKA, M. (1989). Towards a General Theory of Individuation and Identification. In J. Hintikka and M. Hintikka, eds., *The Logic of Epistemology and the Epistemology of Logic: Selected Essays*, 73–95. Dordrecht: Kluwer.

ISHIGURO, H. (1990). *Leibniz's Philosophy of Logic and Language*, 2nd edn. Cambridge: Cambridge University Press.

JAGER, T. (1982). An Actualist Semantics for Quantified Modal Logic. *Notre Dame Journal of Formal Logic*, **23**, 335–49.

JOURDAIN, P. (1912). The Development of the Theories of Mathematical Logic and the Principles of Mathematics. *Quarterly Journal of Pure and Applied Mathematics*, **43**, 219–314.

JUBIEN, M. (1988). Problems With Possible Worlds. In D. F. Austin, ed., *Philosophical Analysis*, 299–322. Dordrecht: Kluwer.

—— (1991). Could This Be Magic? *Philosophical Review*, **100**, 249–67.

—— (1996). Actualism and Iterated Modalities. *Philosophical Studies*, **84**, 109–25.

KAPLAN, D. (1979). Transworld Heir Lines. In Michael J. Loux, ed., *The Possible and the Actual*, 88–109. Ithaca, N.Y.: Cornell University Press.

KIM, J. (1986). Possible Worlds and Armstrong's Combinatorialism. *Canadian Journal of Philosophy*, **16**, 595–612.

KITCHER, P. (1984). *The Nature of Mathematical Knowledge*. New York: Oxford University Press.

KNEALE, W. and KNEALE, M. K. (1962). *The Development of Logic*. Oxford: Oxford University Press.

KONYNDYK, K. (1986). *Introduction to Modal Logic*. Notre Dame, Ind.: University of Notre Dame Press.

KRIPKE, S. (1971). Semantical Considerations on Modal Logic. In Leonard Linsky, ed., *Reference and Modality*, 63–72. Oxford: Oxford University Press.

—— (1980). *Naming and Necessity*, 2nd edn. Cambridge, Mass.: Harvard University Press.

KVART, I. (1986). *A Theory of Counterfactuals*. Indianapolis, Ind.: Hackett.

LAMBERT, K. (1991), ed. *Philosophical Applications of Free Logic*. New York: Oxford University Press.

—— and VAN FRAASSEN, B. (1972). *Derivation and Counterexample: An Introduction to Philosophical Logic*. Encino, Calif.: Dickenson.

LEAR, J. (1977). Sets and Semantics. *Journal of Philosophy*, **74**, 86–102.

LEONARD, H., and GOODMAN, N. (1940). The Calculus of Individuals and its Uses. *Journal of Symbolic Logic*, **5**, 45–55.

LEWIS, D. (1979*a*). Counterpart Theory and Quantified Modal Logic. In Michael J. Loux, ed., *The Possible and the Actual*, 110–28. Ithaca, N.Y.: Cornell University Press.

—— (1979*b*). Possible Worlds. In Michael J. Loux, ed., *The Possible and the Actual*, 182–9. Ithaca, N.Y.: Cornell University Press.

—— (1983). New Work for a Theory of Universals. *Australasian Journal of Philosophy*, **61**, 343–77.

—— (1986). *On the Plurality of Worlds*. Oxford: Blackwell.

—— (1991). *Parts of Classes*. Oxford: Blackwell.

—— (1993). Mathematics is Megathology. *Philosophia Mathematica*, **1**, 3–23.

LIGHTSTONE, A. H. (1965). *Symbolic Logic and the Real Number System*. New York: Harper & Row.

LOUX, M. J. (1979), ed. *The Possible and the Actual*. Ithaca, N.Y.: Cornell University Press.

LYCAN, W. and SHAPIRO, S. (1986). Actuality and Essence. In P. A. French, T. Uehling, and H. Wettstein, eds., *Midwest Studies in Philosophy*, vol. xi, *Studies in Essentialism*, 343–77. Minneapolis, Minn.: Minnesota University Press.

MACKIE, J. L. (1973). *Truth, Probability and Paradox*. Oxford: Oxford University Press.

MADDY, P. (1980). Perception and Mathematical Intuition. *Philosophical Review*, **89**, 163–96.

—— (1990). *Realism in Mathematics*. Oxford: Oxford University Press.

—— (1992). Indispensability and Practice. *Journal of Philosophy*, **89**, 275–89.

—— (1995). Naturalism and Ontology. *Philosophia Mathematica*, **3**, 248–70.

MANCOSU, P. (1996). *Philosophy of Mathematics and Mathematical Practice in the Seventeenth Century*. Oxford: Oxford University Press.

—— (1998). Hilbert and Bernays on Metamathematics. In P. Mancosu, ed., *From Brouwer to Hilbert: The Debate on the Foundations of Mathematics in the 1920s*, 149–88. Oxford: Oxford University Press.

MATES, B. (1972). *Elementary Logic*, 2nd edn. New York: Oxford University Press.

—— (1986). *The Philosophy of Leibniz: Metaphysics and Language*. New York: Oxford University Press.

MCGEE, V. (1992). Two Problems with Tarski's Theory of Consequence. *Proceedings of the Aristotelian Society*, **92**, 273–92.

MCKAY, T. J. (1986). Against Constitutional Sufficiency Principles. In P. French, T. Uehling, and H. Wettstein, eds., *Midwest Studies in Philosophy*, vol. xi, *Studies in Essentialism*, 295–304. Minneapolis, Minn.: University of Minnesota Press.

MCMICHAEL, A. (1983*a*). A New Actualist Modal Semantics. *Journal of Philosophical Logic*, **12**, 73–99.

—— (1983*b*). A Problem for Actualism About Possible Worlds. *Philosophical Review*, **92**, 49–66.

MELIA, J. (1992*a*). Against Modalism. *Philosophical Studies*, **68**, 35–56.

—— (1992*b*). A Note on Lewis's Ontology. *Analysis*, **52**, 191–2.

MENDELSON, E. (1987). *Introduction to Mathematical Logic*, 3rd edn. Monterey, Calif.: Wadsworth & Brooks/Cole Advanced Books.
MENZEL, C. (1986*a*). On Set Theoretic Possible Worlds. *Analysis*, **46**, 68–72.
—— (1986*b*). On the Iterative Explanation of the Paradoxes. *Philosophical Studies*, 49, 37–61.
—— (1988). On an Unsound Proof of the Existence of Possible Worlds. *Notre Dame Journal of Formal Logic*, **30**, 598–603.
—— (1990). Actualism, Ontological Commitment, and Possible Worlds Semantics. *Synthese*, **85**, 355–89.
—— (1991). The True Modal Logic. *Journal of Philosophical Logic*, **20**, 331–74.
MILLS, E. (1991). Forbes's Branching Conception of Possible Worlds. *Analysis*, **51**, 48–50.
MONTAGUE, R. (1963). Syntactical Treatments of Modality. *Acta Philosophica Fennica*, **16**, 153–67.
MOSTOWSKI, A. (1967). Recent Results in Set Theory. In I. Lakatos, ed., *Problems in the Philosophy of Mathematics*, 82–96. Amsterdam: North-Holland.
NEALE, S. (1990). *Descriptions*. Cambridge, Mass.: MIT Press.
NOONAN, H. W. (1994). In Defence of the Letter of Fictionalism. *Analysis*, **54**, 133–9.
NOZICK, R. (1981). *Philosophical Explanations*. Cambridge, Mass.: Harvard University Press.
PARSONS, C. (1983). What is the Iterative Conception of Set? In Paul Benacerraf and Hilary Putnam, eds., *Philosophy of Mathematics: Selected Readings*, 2nd edn., 503–29. Cambridge: Cambridge University Press.
PINTER, C. (1971). *Set Theory*. Reading, Mass.: Addison-Wesley.
PLANTINGA, A. (1974). *The Nature of Necessity*. Oxford: Oxford University Press.
—— (1979). Actualism and Possible Worlds. In Michael J. Loux, ed., *The Possible and the Actual*, 253–73. Ithaca, N.Y.: Cornell University Press.
—— (1985*a*). Replies to My Colleagues. In James Tomberlin and Peter van Inwagen, eds., *Alvin Plantinga*, 313–96. Dordrecht: Reidel.
—— (1985*b*). Self-Profile. In James Tomberlin and Peter van Inwagen, eds., *Alvin Plantinga*, 3–97. Dordrecht: Reidel.
—— and GRIM, PATRICK. (1993). Truth, Omniscience, and Cantorian Arguments: An Exchange. *Philosophical Studies*, **71**, 267–306.
POINCARÉ, H. (1952). *Science and Hypothesis*. New York: Dover.
—— (1953). *Science and Method*, trans. Francis Maitland. New York: Dover.
POLLOCK, J. (1985). Plantinga on Possible Worlds. In James Tomberlin and Peter van Inwagen, eds., *Alvin Plantinga*, 121–44. Dordrecht: Reidel.
PRIOR, A. N. (1960). The Runabout Inference Ticket. *Analysis*, **21**, 38–9.
PUTNAM, H. (1991). *Representation and Reality*. Cambridge, Mass.: MIT Press.
QUINE, W. (1959). *Methods of Logic*, revised edn. New York: Holt.
—— (1960). *Word and Object*. New York: Wiley.
—— (1969). *Ontological Relativity and Other Essays*. New York: Columbia University Press.
—— (1970). *Philosophy of Logic*. Englewood Cliffs, N.J.: Prentice-Hall.
—— (1976). Worlds Away. *Journal of Philosophy*, **73**, 859–63.

QUINN, P. (1982). Metaphysical Necessity and Modal Logic. *The Monist*, **65**, 444–55.
RAY, G. (1992). Modal Identities and De Re Necessities. Unpublished Ph.D. dissertation, University of California, Berkeley.
—— (1996*a*). Logical Consequence: A Defense of Tarski. *Journal of Philosophical Logic*, **25**, 617–77.
—— (1996*b*). Ontology-Free Modal Semantics. *Journal of Philosophical Logic*, **25**, 333–61.
RESNIK, M. (1980). *Frege and the Philosophy of Mathematics*. Ithaca, N.Y.: Cornell University Press.
—— (1981). Mathematics as a Science of Patterns: Ontology and Reference. *Nous*, **16**, 529–50.
—— (1992). Review of Chihara's *Constructibility and Mathematical Existence*. *Journal of Philosophy*, 648–51.
ROSEN, G. (1990). Modal Fictionalism. *Mind*, **99**, 327–54.
—— (1993). A Problem for Fictionalism About Possible Worlds. *Analysis*, **53**, 71–81.
—— (1995). Modal Fictionalism Fixed. *Analysis*, **55**, 67–73.
ROY, T. (1993). Worlds and Modality. *Philosophical Review*, **102**, 335–61.
SAINSBURY, R. M. (1988). *Paradoxes*. Cambridge: Cambridge University Press.
SALMON, N. (1981). *Reference and Essence*. Princeton, N.J.: Princeton University Press.
—— (1989). Logic of What Might Have Been. *Philosophical Review*, **98**, 3–34.
—— (1996). Trans-world Identification and Stipulation. *Philosophical Studies*, **84**, 203–23.
SHALKOWSKI, S. A. (1994). The Ontological Ground of the Alethic Modality. *Philosophical Review*, **103**, 669–88.
SHER, G. (1991). *The Bounds of Logic*. Cambridge, Mass.: MIT Press.
—— (1996). Did Tarski Commit "Tarski's Fallacy"? *Journal of Symbolic Logic*, **61**, 653–86.
SKYRMS, B. (1976). Possible Worlds, Physics and Metaphysics. *Philosophical Studies*, **30**, 232–332.
STALNAKER, R. (1984). *Inquiry*. Cambridge, Mass: MIT Press.
SUPPES, P. (1960). *Axiomatic Set Theory*. Princeton, N.J.: Van Nostrand.
TARSKI, A. (1956*a*). The Concept of Truth in Formalized Languages. *Logic, Semantics, Metamathematics*, 152–278. Oxford: Oxford University Press.
—— (1956*b*). On the Concept of Logical Consequence. *Logic, Semantics, Metamathematics*, 409–20. Oxford: Oxford University Press.
VAN BENTHEM, J. (1983). *Modal Logic and Classical Logic*. Naples: Bibliopolis.
VAN FRAASSEN, B. (1995). 'World' Is Not a Count Noun. *Nous*, **29**, 139–57.
VAN HEIJENOORT, J., ed. (1967). *From Frege to Gödel: A Source Book in Mathematical Logic, 1879–1931*. Cambridge, Mass.: Harvard University Press.
VAN INWAGEN, P. (1980). Indexicality and Actuality. *Philosophical Review*, **89**, 403–26.
—— (1985). Plantinga on Trans-World Identity. In James Tomberlin and Peter van Inwagen, eds., *Alvin Plantinga*, 101–20. Dordrecht: Reidel.

VAN INWAGEN, P. (1986). Two Concepts of Possible Worlds. In P. French, T. Uehling, and H. Wettstein, eds., *Midwest Studies in Philosophy*, vol. xi, *Studies in Essentialism*, 185–213. Minneapolis, Minn.: Minnesota University Press.

WALBOT, V., and HOLDER, N. (1987). *Developmental Biology*. New York: Random House.

WEYL, H. (1987). *The Continuum*, trans. Stephen Pollard and Thomas Bole. Kirksville, Mo.: Thomas Jefferson University Press.

WIGGINS, D. (1967). *Identity and Spatio-Temporal Continuity*. Oxford: Basil Blackwell.

—— (1980). *Sameness and Substance*. Oxford: Basil Blackwell.

WITTGENSTEIN, L. (1953). *Philosophical Investigations*. New York: Macmillan.

—— (1956). *Remarks on the Foundations of Mathematics*. New York: Macmillan.

WOLENSKI, J. (1992). Review of Chihara's *Constructibility and Mathematical Existence*. *History and Philosophy of Logic*, **13**, 233–4.

YABLO, S. (1988). Review of Forbes's *The Metaphysics of Modality*. *Journal of Philosophy*, 329–37.

ZERMELO, E. (1967). Investigations in the Foundations of Set Theory I. In Jean van Heijenoort, ed., *From Frege to Gödel: A Source-Book in Mathematical Logic, 1879–1931*, 199–215. Cambridge, Mass.: Harvard University Press.

INDEX